Antarctic Isolation as a Mars Habitat Analogue

STUDIES IN PHILOSOPHY, CULTURE
AND CONTEMPORARY SOCIETY

Edited by Bogusław Paź

VOLUME 34

Jan Felicjan Terelak

Antarctic Isolation as a Mars Habitat Analogue
A Psychological Perspective

Bibliographic Information published by the Deutsche Nationalbibliothek
The Deutsche Nationalbibliothek lists this publication in the Deutsche Nationalbibliografie; detailed bibliographic data is available online at http://dnb.d-nb.de.

Library of Congress Cataloging-in-Publication Data
A CIP catalog record for this book has been applied for at the Library of Congress.

Cover illustration: Courtesy of Benjamin Ben Chaim

This publication was financially supported by the Cardinal Stefan Wyszyński University in Warsaw.

ISSN 2196-0151
ISBN 978-3-631-85790-8 (Print)
E-ISBN 978-3-631-86058-8 (E-PDF)
E-ISBN 978-3-631-86107-3 (EPUB)
E-ISBN 978-3-631-86108-0 (MOBI)
DOI 10.3726/b18618

© Peter Lang GmbH
Internationaler Verlag der Wissenschaften
Berlin 2021
All rights reserved.

Peter Lang – Berlin · Bern · Bruxelles · New York ·
Oxford · Warszawa · Wien

All parts of this publication are protected by copyright. Any utilisation outside the strict limits of the copyright law, without the permission of the publisher, is forbidden and liable to prosecution. This applies in particular to reproductions, translations, microfilming, and storage and processing in electronic retrieval systems.

This publication has been peer reviewed.

www.peterlang.com

*I dedicate this book to
to the Members of the Antarctic Brotherhood,
who over the past few decades
wintered-over at the Polish Henryk Arctowski PAS station,
and especially my wintering-over companions from the Third Antarctic
Expedition of the Polish Academy of Sciences,
thanking them for taking part in "troublesome" psychological tests.*

Table of Contents

Preface I 11

Preface II 17

Preface III 19

Introduction 21

Part One: Theoretical Background 27

1. Man and the Environment 29
   1.1. Main statements of environmental psychology 29
   1.2. Situation as a basic concept of environmental psychology ... 34
   1.3. The environment as a stimulating system 35
      1.3.1. Neurocerebral concept of activation 35
      1.3.2. Psychological stimulation concept 38
      1.3.3. Individual differences in the need for stimulation 41

2. Data Sources Concerning The Effects of Sensory
   Deprivation and Social Isolation 51
   2.1. Taxonomic aspects of data sources 51
   2.2. Phenomenological aspect of data sources based on narrative ... 53
      2.2.1. Loneliness 55
      2.2.2. Imprisonment as punishment 58
      2.2.3. Incidental isolation 61
      2.2.4. Task isolation 67
   2.3. Experimental aspect of data sources 74

2.3.1. Laboratory experiments 74
2.3.2. Quasi laboratory experiment as an analogue of a space habitat 85
    2.3.2.1. Classical habitats 86
    2.3.2.2. Modern Habitats 89
2.3.3. Experiments under natural conditions 97
    2.3.3.1. Shelters 97
    2.3.3.2. Penthouses 98
    2.3.3.3. Underwater capsules 99
    2.3.3.4. Submarines 100
    2.3.3.5. Spacecraft and orbital stations 103
    2.3.3.6. Long-term Antarctic expeditions as an analogue of Martian habitat 115

Part Two: Own Research 133

3. Own Research into the Functioning of Humans in Antarctic Isolation 135

3.1. Theoretical basis for the research 135
3.2. Stress factors in the Antarctic environment 139
    3.2.1. Physical environment of the Antarctic 139
    3.2.2. Psychological and social environment 144
3.3. Assumptions of own research 154
    3.3.1. Aim of the research 154
    3.3.2. The issues of own research 155

4. Methods and Organisation of The Study 161

4.1. Characteristics of the study site 161
    4.1.1. Physical environment 162
    4.1.2. Logistical problems 163
    4.1.3. Psychological environment 166

4.2. Selection of individuals for the study and characteristics of the group 173
    4.2.1. Choice and selection process 173
    4.2.2. Characteristics of the persons studied 174
        4.2.2.1. Biodate 174
        4.2.2.2. Motivation 176
        4.2.2.3. Personality 177
    4.2.3. Tasks of the members of the wintering-over group 182
        4.2.3.1. Responsibilities of the technical group 182
        4.2.3.2. Responsibilities of the scientific group 184
4.3. Measurement techniques and organisation of surveys 185

5. Transactional Characteristics of the Winter-Over Syndrome 189

5.1. The stimulative aspect of Antarctic isolation 190
5.2. The functional aspect of the Antarctic isolation situation 196
    5.2.1. Clinical psychology perspective 197
    5.2.2. The personality perspective 205
        5.2.2.1. Dynamics of neuroticism and extraversion 205
        5.2.2.2. Dynamics of different types of aggression 211
        5.2.2.3. Dynamics of personality changes in selected factors of the 2nd degree Cattell's 16PF 216
        5.2.2.4. Dynamics of mental needs 219
    5.2.3. Emotional perspective 227
        5.2.3.1. Dynamics of moods 227
        5.2.3.2. Dynamic of anxiety 234
    5.2.4. Cognitive perspective 238
        5.2.4.2. Dynamics of attitude 241
        5.2.4.3. Dynamics of cognitive performance 243
5.3. The social aspect of Antarctic isolation 250
    5.3.1. Dynamics of interpersonal attraction 252
    5.3.2. Sociometric structure of group dynamics 261

    5.3.3. Dynamics of territoriality of behaviour 280
    5.3.4. Dynamics of perception of group 283
5.4. The chronobiological aspect of Antarctic isolation 285
    5.4.1. Photoecological aspects of Antarctic adaptation 286
    5.4.2. Dynamics of circadian rhythm of cognitive functioning 292
    5.4.2. Sleep disturbance 299
5.5. A medical perspective on the cost of adaptation to Antarctic
    isolation 306

Conclusion and Findings 311

Post Scriptum 331

References 343

Annex: Some Abbreviations Used in the Text 399

Index of Names 401

Preface I

Knowledge (almost) forgotten

Sometimes a thesis is formulated that "psychology develops through oblivion," i.e. we have forgotten our earlier research discoveries, and then we are proud to rediscover them. The problems of the importance of stimulation for human functioning and development were particularly intensely explored between the 1950s and 1980s. On the one hand, the research was inspired by the experience gathered by observing the effects of long-term isolation or inactivation among, for example, psychiatric patients, participants in polar expeditions, people serving prison sentences or soldiers imprisoned in POW camps. However, the main inspiration for collecting data from many sources was the then emerging space flight programme. It was then pointed out that there was a specific need for living organisms (not only humans, but also animals – because they were also affected by those findings), which was described as a 'need for stimulation.' It was pointed out that limiting the inflow of stimulation, in the form of isolation or even deprivation, treated in a multidimensional perspective, i.e. not only sensory, but also social in nature, leads to a reduction in the level of stimulation, which in turn results in a deterioration in the efficiency of the performance of tasks (deteriorating cognitive functions) and, paradoxically, to the excitation of negative emotions. Theoretical models were developed at that time, which indicated the existence of optimum stimulation (and excitation) and made it possible to explain and predict the consequences of sub-optimal stimulation – both the effects of short-term reduction in the inflow of stimuli or environmental monotony (e.g. for the work of pilots or power plant operators), as well as long-term isolation (because deprivation in the strict sense is rare, or at least it is difficult to study this phenomenon experimentally). To be precise, models have also been developed that indicate possible adaptation to various stimulation conditions, for example by "rescaling" the need for stimulation or changes occurring at the level of psychophysiological mechanisms of stimulation processing. These models were incorporated both by individual difference theories, which stressed the importance of temperament as the regulator of the level of stimulation (and determinant of the individual need for stimulation), and also by stress theories, which stressed the stressogenic nature of excessive, but also – as indicated earlier – too low stimulation, and therefore also sub-optimal stimulation. However, I am highlighting these problems solely for the purposes of this preface. I have no need to discuss them in detail, because Professor Jan Terelak's book presents these phenomena and theoretical models (together with his own original proposals) in a very learned and precise manner. In later years, the knowledge gathered in these studies and contained in the resulting scientific concepts ceased to arouse such great interest. It has not, of course, been completely forgotten. It is mentioned in psychology textbooks or in studies on the history of psychology. But, speaking the language of the modern Internet, it is no longer "trendy." Who has lately been interested in the phenomenon of stimulus isolation – in the era of an all-accessible Internet and the widespread ease of obtaining information? Or, more broadly, multidimensional isolation: sensory isolation, social isolation, activity restrictions, etc. Rather, the focus was on overload stress, known as informational stress – associated with the flood of information, the difficulty of selecting and assessing it. Research on these problems continued, however, although it concerned selected professional groups or specific phenomena, such as the consequences of isolation or the monotony of working with military and passenger aircraft pilots, submarine crews or habitat inhabitants. It would seem that this knowledge is no longer useful and has died away, but suddenly, life itself has again shown us the importance of this knowledge. One of the important premises has become the social experience of a pandemic and the need for epidemic isolation. The second – the planned space flights and space exploration programmes, abandoned after their successes in the 1960s and 1970s and now reactivated. And it turned out that the knowledge presented in Professor Jan Terelak's book and his unique research carried out during his one-year-long polar stay are extremely up-to-date and useful. It is particularly important to emphasise its application significance for planned space flights – the organisation of living and working conditions for crews and the prevention of possible negative mental phenomena associated with space limitations, the monotony of the environment, sensory and social isolation, etc., on the one hand, and – on the other – not only an environment that poses a challenge to physiological and psychological adaptation, but also hostile and threatening to health or life. Agreeing with the author's thesis that the "current experience with long-term space missions on space stations (e.g. MIR, ISS) has provided arguments that the longer a space flight takes, the more problems involve the so-called human factor" (p. 7), I personally believe that the future will rather be that of unmanned missions. Man in his biological as well as psychological structure may turn out to be the most unreliable element, and the problems involved in building a suitable artificial environment (essentially: simulating Earth conditions with a fairly narrow margin of freedom and tolerance for possible failures) may prove insurmountable. However, I have no doubt that such attempts will be made. And, at least in the initial phase, we have to take on the extremely difficult task of preparing them properly, so that the tragic failure at the outset does not discourage space exploration and does not destroy these ambitious plans. But can we not make an attempt at trying to explore the cosmos? Definitely not, because – as Stanisław Lem remarks in *Powrót z gwiazd* (Return from the Stars) – it is a part of our nature, just like discovering new lands, conquering mountain peaks or exploring uninhabited Arctic or unknown tropical areas... Because without it we would not be ourselves – we would lose an important part of human nature. It is therefore fortunate that we will not need to acquire this knowledge anew and that it has not been completely forgotten. In fact, thanks to Professor Jan Terelak's book, we can learn fascinating theories and research results, and use the knowledge contained in it to plan such expeditions.

However, what makes the scientific knowledge gathered in the book *Antarctic Isolation as an Analogue of Mars Habitat* so important? First of all, I think it is deeply theoretical in nature – the models presented allow explaining and anticipating the psychological consequences of long-term isolation. They are based on a very rich review of the results of scientific research, but also on colloquial observations of the effects of isolation and deprivation, in various groups of people, including clinical analyses, such as the consequences of developmental disorders in children with analytical disorders or children brought up outside the social environment. Secondly, the book presents the results of the author's unique own research, which was carried out during the Third Antarctic Expedition of the Polish Academy of Sciences at the Henryk Arctowski station on King George's Island in the years 1978–80, in a group of 21 winter-over participants, i.e. polar explorers who, together with the author of the monograph, stayed at the station for a year, including during the particularly difficult period of the polar night. This data is extremely important because polar conditions, due to their multidimensional isolation, but also threatening external conditions, can be an analogue of space expeditions. Professor Jan Terelak's research – unlike many other studies – was of a multi-faceted nature: it took into account many possible sources of stress and included various psychological functions, i.e. broadly defined emotional states, social reactions, changes in cognitive performance and efficiency, etc., examined both by use of objectivised methods and through the relation of subjective feelings. What is important, they were longitudinal studies, with multiple measurements, enabling analysis of the dynamics of mental phenomena over time and against the background of events taking place at the station. Thirdly, these studies have provided extremely valuable and important research results. They indicate – in general terms – that a polar expedition is a multifaceted stress situation in stimulative, social and chronobiological terms, with the winter period obviously being more stressful than the summer period (which is sometimes referred to as "winter-over syndrome"). The dynamics of various mental functions are different, which show periods indicating better and worse adaptation of polar explorers to different conditions at the station. This dynamics is characterised by quite significant intra- and inter-individual differences, i.e. the specificity of progress in individual polar explorers. Likewise, for social adaptation, stages of integration and polarisation have been identified which, if the group loses its task force nature, can lead to its disintegration. Finally, the changes that take place at the level of broader behavioural tendencies (e.g. personality traits) are of an adaptive nature and, as the author suggests, are reversible, i.e. they give way after returning to typical living conditions and do not lead to personality pathologisation. At the same time, these results do not fully confirm other observations of the particularly critical period of the "third quarter" – some indicators only reached their most unfavourable values paradoxically in the final period (by the way, are these aspects of the subjective perception of time and circumstances of the polar expeditions perhaps the ones that Roald Amundsen had in mind when he once said that "it is always darkest before the dawn")? Fourthly, the scientific part of the monograph is supplemented by the author's diary (*Antarctic Winter-Over Syndrome: Narrative Perspective*), in which personal notes made on an ongoing basis and almost overnight, and not ex post reminiscences – relations made from behind a desk in the warmth of the home) are presented. Thanks to them we can – although still in a limited way – see the life of a polar explorer, feel the atmosphere of the station, experience isolation and better understand the author's theses, presented in the scientific part. This is an extremely valuable addition to scientific considerations. Fifthly, it is also especially worth noting that this study has been elaborated by a scientist who not only has extensive knowledge of this area of research, but who, above all else, in the course of conducting Antarctic isolation research has been directly involved in the said Antarctic isolation (by the way, with good social attractiveness indicators obtained in final sociometric analyses!). This type of research, together with the so-called "participatory observation," refers to the classics of psychology and its best traditions. It is also a guarantee that the author knows and has in-depth understanding of the specificity of the phenomena studied, and the report itself is a credible analysis and a "first-hand" report. To conclude this preface, I would like to point out that I consider this book to be extremely valuable and very current. It is good that Professor Jan Terelak has saved this knowledge from oblivion and is now presenting it in such an erudite form. I consider the idea of publishing the book in English to be an excellent one, so that it will also be available to foreign readers. I wholeheartedly recommend reading it – not only to specialists, but also to a wide circle of readers. The content of this book may not be easy for them, but it will certainly prove worth the effort of reading it.

Warsaw, 5 December 2020
Prof. Bogdan Zawadzki
University of Warsaw

Preface II

Isolation of a human being causes certain psychological effects, gives time to think about behaviours, events, experiences, memories. One can be isolated by choice – just like cosmonauts, hermits, polar explorers, or by force – like prisoners, the sick or for quarantine purposes. Polar experiences confirm me in my belief in the psychological changes taking place as a result of a long-term stay in a small isolated group with limited external contacts. Extreme weather conditions, frost, hurricane winds, prolonged darkness of the polar night and continuous day can also be a stress factor. This is well illustrated by the head of the American Antarctic expeditions, Richard E. Byrd: "The uniformity of life and monotony nowhere in the world are as tiring as during a polar winter. When polar night passes, it's fortunate the man cannot see his inner self in the mirror – he would wipe the glass, thinking it is fogged over."

I personally experienced this state twice. Fortunately, it is almost reversible – as my relatives claim. Psychological research on isolation was carried out at the H. Arctowski Antarctic station as early as in the 1980s and 1990s and beyond, on spacecraft, by Jan Terelak Many years later, an American team commissioned by NASA and chaired by Professor L. Palinkas for three years conducted surveys at H. Arctowski's station, and they were also continued at American, Russian, Chinese and Indian stations. The conclusions were about the effects of long-term isolation of people of various cultures and different genders, which may be relevant for future long and distant space flights.

The book by Prof. on stress and social isolation written from the perspective of future interplanetary flights is worth special attention on the eve of the expedition to Mars via the Moon.

Prof. Stanisław Rakusa-Suszczewski
Polar Explorer

Preface III

During my eight-day space mission in 1978, I circled the Earth sixteen times a day. Time was scrupulously divided into life activities, technical activities and scientific work. I constantly had not enough time, and there were so many temptations to look at our blue and white planet for a moment. It was only night that gave me such a chance; then, at the expense of my sleep, I could enjoy the beauty of the Earth, I also felt moments of great joy, and when cosmic night came and my colleagues went to sleep, I felt loneliness and sadness. I often had thoughts of my loved ones. I watched every corner of our cradle, the joyful inhabited places full of various colours, the other mysterious and quite alien ones, single-coloured and without any trace of human activity – this is endless white Antarctica. It was there, in a small scientific station, in isolation from civilisation and in a harsh climate, that Polish scientists worked on their own for many months. Their situation was probably more difficult than that of ours. On the next overflight, I gave them greetings from orbit. How pleased I was to hear in response the familiar voice of Jan Terelak, a psychologist who was a member of the committee qualifying Polish candidates for the flight to space. Dr Jan Terelak, now a professor of psychology, has carried out tremendous work in the Antarctic station, the results of which he published in his work *Effects of Extremely Social Isolation: Perspective of space psychology*. The content of the study goes far beyond the "Antarctic theme." The conclusions from the study can be useful in psychological preparation of crews for long-term stay in a single team in a cosmic habitat or on another extraterrestrial planet. We are now entering a new dynamic stage of space exploration. We have long-term scientific missions to the Moon, and soon to Mars, ahead of us. That is why Professor Jan Terelak's study is timeless.

Mirosław Hermaszewski
Warsaw, 14.12.2020
Cosmonaut

Introduction

> Effective behaviour stems not from "good" people. It is called forth from "good" environments.
> – Stanley Milgram
>
> No one comes to Antarctica by chance.
> – Jean McNeil

At the beginning of 1953, *American Psychologist* published a summary on mental dysfunction in a situation of limited inflow of stimuli. The authors of the article "Cognitive Effects of a Decreased Variation in the Sensory Environment" were D. Hebb, W. Heron and W. Bexton. The paper presents a two-year experimental study and verifies some hypotheses on the relationship between stimulation and activation levels. It was, therefore, a continuation of the earlier reflections included by D. Hebb in his book *The Organization of Behavior* published in 1949. The author formulated, among other things, a thesis that both over-stimulation and under-stimulation, having an aversive character, are associated with discomfort and therefore prefer behaviours leading to maintaining an optimal level of stimulation. The Hebb's deprivatory experiment is considered a breakthrough in the literature. Many scientific centres, especially in Canada and the USA, were then interested in experimental research on the so-called effects of sensory deprivation. However, in spite of many studies carried out by serious scientific centres on the effects of sensory deprivation and social isolation, aside from the empirically established facts, a number of myths and scientific legends related primarily to the interpretation of the results obtained have emerged.

The bibliography on the issues of deprivation is very extensive, which means that the attempt to classify and analyse this work faces serious obstacles, mainly related to different terminology, heterogeneous experimental conditions and the lack of a coherent theoretical concept explaining the results obtained. The development of modern civilisation often forces man to act in the so-called artificial conditions, which are sometimes characterised by a limited supply of stimulation and information. We can already mention a fairly large group of astronauts, long-term residents of the international space station, who are isolated from the outside world, and whose numbers will systematically increase with the development of space technology and planned flights between planets. This also applies to many operators of ground and underwater technical equipment who usually work alone (e.g. crane operators) or in small task forces (e.g. submarine crews, soldiers in shelters, offshore drilling rig crews, etc.), performing extremely responsible tasks in isolation from the external environment. The problems of deprivation and social isolation are still of particular interest to the contemporary astronautics. The success of long-term space flights with human crew, especially interplanetary flights, depends largely on learning about the mechanisms of human functioning in an artificial habitat of an orbital space station, or that on the Moon or Mars. No wonder that the effects of social deprivation and isolation have been and will be analysed in the future, from the empirical and theoretical point of view, by specialists from various disciplines: psychologists, psychiatrists, sociologists, physiologists, etc. Psychologists in this case are mainly interested in the role of external control through information from the environment, which is important in the process of behaviour regulation. Contemporary psychology, especially environmental psychology, emphasizes that a huge role in the regulation of relations between man and the environment is played by man's expectations, ways of assessing and forecasting external phenomena, shaped on the basis of the whole life experience to date. Meanwhile, in situations of isolation from the outside world, situations, which are new for people, the network of the existing cognitive structures becomes disorganized, putting into question their usefulness and value in regulating behaviour. From this point of view, among other things, situations of social deprivation and isolation should be classified as stressful situations. A lot of pre-scientific data (e.g. logs of survivors and lonely sailors, clinical data, narratives, etc.) as well as scientific data has already been collected, but detailed knowledge of the role of external stimulation in behavioural regulation is still an open issue. Although there have also been attempts to systematise the theoretical facts collected, the results, usually collected within a single narrow field (e.g. psychology), are still unordered and often ambiguous. It must be stated, however, that these facts are consistent with one of the basic statements of environmental psychology: that man functions properly provided that they are properly stimulated from the outside under the conditions of earthly gravity.

The issues of deprivation and isolation were not properly reflected in the Polish literature on the subject until the 1970s. The first research in Polish space psychology was published in my book entitle: *Człowiek w sytuacjach ekstremalnych: izolacja antarktyczna* [Man in extreme situations: Antarctic isolation] (1982),[1] which presents the results of a natural experiment carried out during a year-long scientific query at the Henryk Arctowski Antarctic Station of

---

1   The first edition, at NASA's request through the U.S. Embassy in Warsaw, was sent to the Library of the U.S. Congress immediately after it was released.

Introduction 23

the Polish Academy of Sciences in Warsaw on King George Island (South Shetland Islands). I decided to present the second edition of the book in English, supplemented by the latest literature on the subject and new analyses of group dynamics graphs, which in recent years has become the subject of experimental research in Earth's space habits as analogues of missions on the planet Mars (e.g. MARS-ONE, MARS-500, MARS-520).

I was encouraged to do so a few years ago by the outstanding Antarctic psychologist, Professor A.J.W Taylor of the Victoria University of Wellington in New Zealand, who "wintered" at Scott Base N.Z., located in the southern part of Antarctica and conducted as a clinical psychologist research on the effects of social isolation from the perspective of interests NASA.[2] Professor came to Poland in 1984 to invite me to join the elite "Club of Psychologists of Antarctic Winter-Over." I gave him my book as a goodbye. After a few weeks, I received a letter in which Professor Taylor assured that he had read it and suggested that it be published soon in English, because it contains the results of longitudinal research, unique in the psychological literature, carried out over a period of more than a year in a two-week cycle with the use of many tools and only psychometric (e.g. situational experiments, narrative method[3]). For objective reasons, it was not possible to fulfil the pledge made for the second edition. For many years, participating in the work of the Space and Satellite Research Committee, I became convinced that the interest in the development of research in the field of Polish medicine and space psychology has been dominated by space technology.[4] Therefore, it is worth recalling the results of psychological research, documenting empirically the psychological barriers, concerning the effects of long-term perceptual deprivation and social isolation from the perspective of plans to "colonize" the planets closest to Earth in the Solar System. Moreover, not without significance is the reflection on the dynamic development of contemporary tools of interpersonal communication based on satellite communications at the turn of the 20th and 21st century, which will disallow the use of natural experimentation at present to study information deprivation and social isolation in the conditions of Antarctic expeditions. Although the results collected from

---

2 The results of these studies are presented in the book A.J.W. Taylor (1987). Antarctic Psychology. Wellington: Science Information Publishing Center.
3 Descriptions of these studies from a narrative perspective are included in the book Antarctic Winter-Over Syndrome: Narrative perspective, (Berlin 2021, Peter Lang).
4 K. Kwarecki, J. Terelak: *Medycyna i Psychologia kosmiczna* [Medicine and Space Psychology], (Warsaw, 1980, Wiedza Powszechna).

the 1980s and 1990s under the conditions of a natural Antarctic laboratory are experiencing their "renaissance," the dynamic development of space flight has unexpectedly quickly led to a paradigm shift, both in terms of spacecraft design and in terms of research concerning not only the individual health and mental fitness standards of cosmonauts/astronauts required for long-term space flight, like before, but also in terms of the interpersonal competences of the crew members, who are composed of people of various sexes, races, national cultures and professional status and role in the team. It is necessary to agree with H. Wichman (2011) from Aerospace Psychology Laboratory at Claremont McKenna College and Claremont Graduate University that completely new tasks are faced by space simulators, which under terrestrial conditions can create physical and psychosocial analogies of future space flights, especially long-range and interplanetary ones, which is well illustrated by experiments conducted in Martian habits (e.g. Dutch MARS One, Russian Mars-500, 520d, or American HI-SEAS I-IV, etc.). Current experience with long-term space missions on space stations (e.g. MIR, ISS) has provided arguments that the longer a space flight takes, the more problems involve the so-called human factor. This suggests that knowledge about man is always important to explain the factors adaptive to extreme situations, which include prolonging sensory deprivation and social isolation. Monitoring psychological factors begins with the selection and selection at the stage of preparation for space flight, but especially during the flight.[5] After the trip to the moon, new psychological problems over the effects of sensory deprivation and social isolation are related to the planned trip to Mars project.[6] This will largely depend on the development of research into protection against deadly cosmic rays and long-term weightlessness, as well as future technological developments such as "planet terraforming," nuclear propulsion or human hibernation, etc.

In the course of preparation of the second edition of this book I realised that a natural experiment designed years ago on human behaviour in the conditions of long-term sensory deprivation and social isolation in a small research station in Antarctica, will be repeated on a global scale, after the WHO declared the "Coronavirus" (Covid-19) pandemic, while at the same time verifying the previously empirically collected results of research by many polar psychologists

---

5   Bernd Johannes, Berna van Baarsen (2020). Psychological Monitoring. In: Alexander Choukèr Ed. *Stress Challenges and Immunity in Space* (pp. 421–432). Second Edition. Springer. Cham.
6   Heppener M. (2020) Moon, Mars and Beyond. In: Choukèr A. (eds) *Stress Challenges and Immunity in Space*. (pp. 709–733). Cite as Springer, Cham.

on a scale unprecedented in social sciences. Experience from years of polar data on the functioning of individuals and social groups in situations that can be briefly described as: Isolated, Confined and Extreme Environments (ICE), can be treated as an analogue of long term space flight with greater reliability than all previous results from laboratory experiments. This is evidenced, for example, by the early interest in these problems by the then young discipline of psychology, namely space psychology (Kubis, 1972); Petrov, Lomov & Samsonov, Eds. (1979), and earlier by polar psychology, whose research conducted so far in Antarctic stations by choice, proves that Antarctica is the only place on Earth that can be considered as an adequate analogue of a Martian habitat, for the reasons that it is "dead," i.e. deprived of life and its attributes such as sounds and smells, which significantly impairs the functioning of the brain. This constant "sensory hunger" leads to an imbalance between the rational and emotional structure of the brain and hinders the ability to adapt, known for example from the "Antarctic Winter-over Syndrome" also described in this book.

Part One: Theoretical Background

1. Man and the Environment

The claim about the complex relationship between man and his environment is accepted as a fundamental fact not only in environmental psychology. When considering the organism as a basic ecological unit, attention is drawn to the fact that the environment, both social, physical, chemical and biological, is characterised by a whole variety of specific and non-specific factors (Bańka, 2002). Individual environmental factors interact with each other, influencing each other, as well as modifying the body's reactions. This almost infinite number of interactions has led some environmental researchers to seek correlations between a single factor and an individual organism's response, while others have taken the view that the environment is such a compact whole that individual factors should not be broken down. The first of these approaches led to many controversial results, while the second one led to completely unscientific considerations. Therefore, contemporary environmental psychology, by creating its own research models, draws on knowledge describing the environment from other perspectives.

1.1. Main statements of environmental psychology

In environmental psychology, the researcher's situation is even more complicated if one considers both the multifaceted characteristics of the environment and one's own deliberate activity. It is known that when a process taking place in the body is influenced by several environmental factors, the reaction of the body can be very varied, often can be opposite to the expected one, depending on the action of mediation factors. From this point of view, environmental psychology should capture the interactions between environmental factors and various levels of psychological organisation. This demand is not easy to fulfil, especially at the methodological level.

Generally speaking, following the Polish psychologist Tadeusz Tomaszewski (1978), three main ways of considering the relationship between man and his environment can be distinguished. The first way accentuates, above all things, the importance of personal dispositions and is preferred by the so-called "dispositioners" or dispositional orientation. The second approach means focusing on the importance of the situation in human life, which is done by the so-called "situationalists." An excellent illustration of the search for the rationale to both extremes is the classic question: "heredity or environment," which has occupied an important place in the history of both philosophy and psychology. Contemporary psychology, taking advantage of the achievements of such disciplines as: physiology, genetics, pedagogy and others, believes that the opposition of both factors is unjustified. Human relationships are much more complex than what the proponents of accentuating interactions see, suggesting that situations are as much a function of the person as their behaviour is a function of the situation.

A transactional approach seems to be more appropriate, which poses systemic questions about the mechanisms of interaction between the two discussed factors: personal and environmental, explaining a new entity separate in terms of quality.

We already know a lot about the internal structure of man and his organisation of behaviour, although the knowledge about the relations that connect man with the surrounding world is still in its initial stage. No wonder that the issue of environmental psychology continues to arouse the interest of psychologists, as evidenced by the relatively rapid development of experimental social psychology on the one hand and personality psychology on the other. The main problem of contemporary psychology is not, therefore, to see man "against or outside the world," but in the context of events, because man's functioning is based not only on the which means that the primary consequence of man's maintaining a relationship with his environment is to treat these relationships in transactional terms: active environment – active man (Lazarus, 2006). Human activity resulting from specific transactions that take place between man and the environment is detailed in "The three-part world concept" by T. Tomaszewski (1980), referring to the "Kurt Lewin Equations,"[1] known in social psychology, the schematic presentation of which is presented in Fig. 1.

---

1 Kurt Lewin's Equation: $B = f(P, E)$, states that behaviour (B) is a function (f) of a person (P) in their environment (E).

Main statements of environmental psychology 31

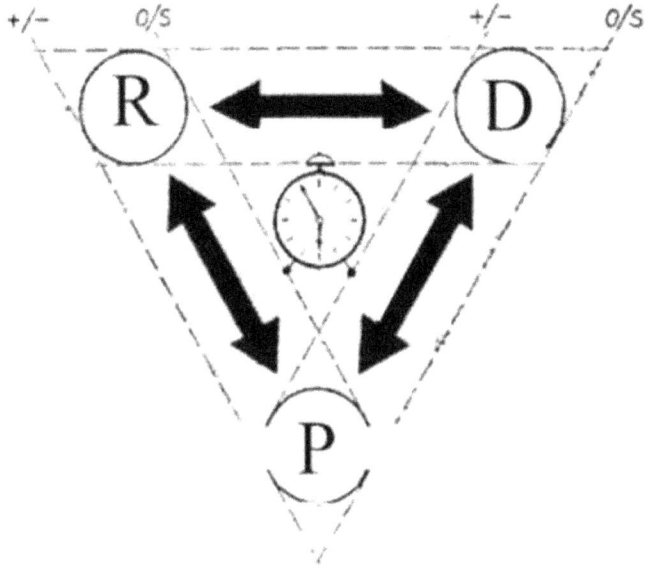

Fig. 1. The three-part world concept
Source: own elaboration based on Tomaszewski, 1980.
Legend: R – Reality ("as it is"); D – Duties ("as it should be"); P – Possibilities ("as it may be"); +/- emotional variables (positive vs. negative); O / S – variables objective vs. subjective; clock – biorhythms.

This concept assumes that the perception of the environment (i.e. the "as it is" statement) is a synthesis of the relationship that occurs between the assessment of requirements or expectations ("as it should be") and the assessment of one's capabilities ("as it may be"). However, for reality to become a representation of the world, a transaction between the subjective and objective states of affairs is needed, not just individual interactions between them: D – P, R – P, R – D, which as such do not provide a sufficient basis for the full representation of the surrounding environment, situation, event.

However, it would be too simplistic if we did not take into account the many variables that modify the perception of the surrounding reality. This is the third important claim of environmental psychology. Thus, both requirements (D) and possibilities (P) can be considered as a more or less stable arrangement between what is objective (O) and what is subjective (S). Data on e.g. self-esteem (level of aspiration, level of achievement) prove that these variables can effectively modify human activity, changing both the subjective assessment of one's abilities (overestimating or not believing in oneself), or overstating or lowering the level of self-imposed requirements (Hryniewicz, 2013).[2] The rich literature on the functioning of humans in stressful situations involving excessive demands characterises this part of the statement we are discussing (Lazarus, 2006; Terelak, 2019).

Finally, the "plus" and "minus" signs presented in Fig. 1.1 are to symbolise the modifying influence of positive and negative emotions. The literature on the experimental psychology of emotions provides many facts explaining the role of emotions in perceiving reality and regulating the level of activity (Gasiul, 2007; Strus, Cieciuch, 2017).

Any relationship between man and his environment must also take into account the spatial and temporal aspect (this is symbolised by the watch in the figure). Both physical distance (e.g. specific geographical latitude, distance from the perceived object) and time (e.g. around-the-clock or seasonal rhythms), affecting the psychophysical state of the body or external states of things, can effectively change both the layout of requirements and possibilities, and thus influence the current perception of the environment and modify the active adaptation to reality (Goldstein, 2004; Torbjörn, Göran, & Mats, 2007).

Summarising the above statements of environmental psychology, it should be said that the narrative of human relations with the environment (interactions) is not sufficient to explain these interactions without the knowledge of human relations with the environment in a psychological sense, taking into account the awareness of the significance of these relations for human life and development. Thus, the basis of all relations between man and his environment, which we tried to indicate, is the appropriate perception of the world, other people, ourselves by ourselves, by others, etc. From this point of view, specific research consequences stem – and not only for environmental psychology. However, this requires certain terminological arrangements.

The basic term used in environmental psychology is the concept of the "environment." However, it has a very broad scope and is therefore too ambiguous to be used without specifying more precisely what an environment is involved. The humanistic approach to ecology (LeMenager, Foote, 2012)[3] differentiates

---

2 This is the case, for example, according to clinical psychologists, in the case of a neurotic personality who does not adequately perceive his or her abilities and often places too low demands on himself or herself. Meanwhile, a mentally healthy personality develops according to their own activity, mainly by "creating" opportunities (cf. Horney, 1976.)
3 When in the 1960s, as a student of psychology at the Catholic University of Lublin, I was listening to a lecture on the "philosophy of nature," I did not realise that it was a

between three types of environments: (a) *natural*, covering those elements without which the human body cannot exist (e.g. atmosphere, sun, water, etc.); (b) *social*, covering all people with whom the human being has a specific or stable relationship (e.g. national, ethnic groups, family, social, professional, etc.); (c) *artificial*, as an analogy to the natural environment, enabling the human body to exist, for a limited time and to a limited extent, in a situation of extreme threat to life, such as during a prolonged space flight or on another planet (Terelak, 2004).

Thus, in general, the environment is defined as a relatively permanent arrangement of environmental elements, important for human life and behaviour. Such a definition stems from the fact that man remains in close relationship with the surrounding world and this fact is accepted by all the human sciences. However, psychology is interested in specific elements of the environment, i.e. both states of affairs and people in various relationships with each other and with other elements of the environment. These relationships are twofold: impact (cause-effect) and relationship (functional). Thus, based on the division of the types of environmental elements in human ecology, the following specific types of environment are distinguished: physical (e.g. ambient temperature), biological (e.g. forest, desert, mountain environment, etc.), social (e.g. urban, rural, family, school, etc.), cultural (e.g. artistic). However, when one treats the environment as a more or less compact organisation of its elements and man as an element functioning within a more general system, the above statement does not bring us any closer to understanding the mutual relations that take place between the wider system (environment) and its subsystem (man). It is therefore proposed to use the term "environment" not only to describe the whole of the relations between man and the world, but to tighten them up to meaningful systems, i.e. those that are relevant to the life and development of the human individual and to social activity. This arrangement as a relative by nature often leads to a terminological ambiguity which can be reduced by introducing a narrower concept: situation and position, suitable for the operationalisation of specific relations between a person and states of affairs that affect him/her (e.g. the situation of an employee, pupil, parent, etc.), and influence on basic activity.

Considering the situation in terms of basic activity is typical of modern psychology in general, and especially of Polish psychologists from the "Psychological

---

precursor of "human ecology." In the literature, *human ecology* is also often referred to as *ecological humanities, environmental humanities, sozology* or *sustainable humanities*, understood as a field actively involved in future-oriented sustainable development (Dołęga, 2004).

Warsaw School"[4] of the 1970s and 1980s, who perceive man as a complex regulatory system, and his actions – as acts of regulating relations between the subject and the environment.

1.2. Situation as a basic concept of environmental psychology

Consideration of human functioning in a specific situation must take into account two important findings. First of all, in line with D. Hebb you have to distinguish two basic aspects of behaviour: reactive and task-based (intentional). The first aspect is of interest to psychological concepts called "activation-based" (Hebb, 1965), while the second one is "stimulus-based" (Leuba,1965). These concepts and the results of the research that support them are discussed in detail, among others, by D.W. Fiske and S. Maddi (1967), drawing attention to their very important aspects of defining a stimulus from the perspective of its stimulative value (neuro-cerebral aspect, including strength or intensity) and/or from the perspective of content value (psychological aspect, including the meaning of information, resulting from the situational context). This makes it possible to distinguish three classes of situations: normal, optimal and difficult. These situations result from the relationships in which humans are sometimes included as part of the environment. These can be the systems most favorable for development and activity and then we are talking about optimal situations, they can be the most common, the most typical, to which a person has gotten used to, and then we call them normal and unfavorable and then we call them stressful (these can be difficult environmental situations, and also personal items).

When searching for the origin of environmental and situational difficulties, attention should be paid to three of their most important sources: (1) incompatibility between the characteristics or state of reality and the characteristics or state of man (e.g. incompatibility between the intensity of stimulation and the type of reactivity, etc.); (2) incompatibility between an individual's system of values and abilities and rules, social norms; (3) incompatibility between the state of reality and its subjective perception. The genesis of situational difficulties underlines the diversity of relations between man and his environment and

---

4   They created this school: Tadeusz Tomaszewski – "Master" and his "disciples:" Janusz Reykowski, Jan Strelau, Ida Kurcz, Stanisław Mika, Adam Frączek, Maria Materska, Xymena Gliszczyńska, Józef Kozielecki, Mariusz Maruszewski, Zbigniew Pietrasiński (Kurcz, Reykowski, ed., 1975).

allows ordering the arrangement of elements of the environment according to the following aspects: (1) The situation as a stimulus system; (2) The situation as a functional system, emphasising the values that give life and activity of a person a certain direction; (3) The situation as a social system, emphasising norms and rules of action; (4) The situation as a spatial-temporal system, emphasising the relativity of previous systems. The latter social aspect is often overlooked in psychological research, which leads to many so-called apparent effects of social isolation or imprisonment.

The aspects of the structure of the environment identified above concern both normal, optimal and stressful situations. Our interest is focused on stressful situations, i.e. situations in which, according to the theory adopted by T. Tomaszewski (1978), they are defined as situations that interfere with the probability of the task being carried out at the current (normal) level. These situations include e.g.: deprivation, overload, difficulties, conflicts, threats, etc.

We are most interested in the class of deprivation situations, i.e. situations of lack, deprivation, limitation of stimulation, i.e. sensory deprivation, perception and social isolation.

1.3. The environment as a stimulating system

When discussing the stimulating aspect, it should be remembered that in reality, the amount of stimulation is determined by two factors: the external situation and the reactivity of the subject and their own activity. The reactivity of a human being is primarily focused on psychophysiological concepts, while the regulatory role of the subject's activity is focused on psychological concepts.

1.3.1. Neurocerebral concept of activation

Internal stimulation is particularly emphasised by the activation concept (e.g. Hebb, 1965), which draws attention to the role of the reticular formation of the brain stem in the overall activation of the human body, which allowed to distinguish between specific and non-specific characteristics of a stimulating situation. A specific stimulative effect is that the situation (or only some of its elements) causes differently-sized reactions of the body. The mechanisms of these reactions (innate or learned) are extremely complicated when we are interested not so much in reactive behaviour as in task-based behaviour. It turns out that even under experimental conditions, when controlling the elements of the situation, various results are obtained. Data on this subject is provided e.g.

by electroencephalographic examinations in the so-called resting situation (i.e. with lack of external stimulation), which is illustrated by Fig. 2.

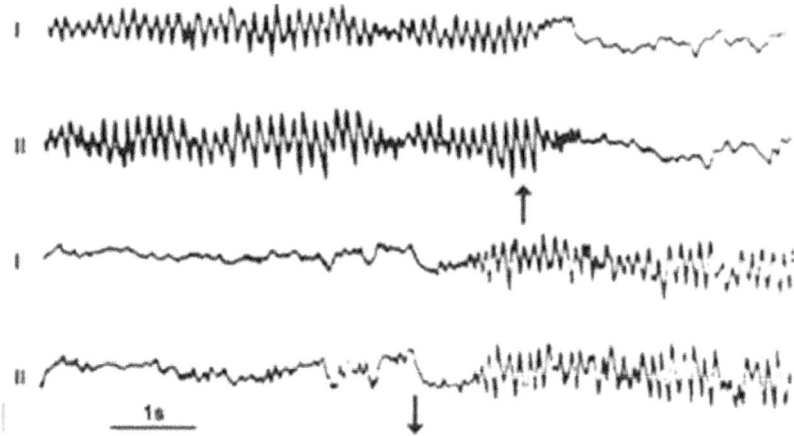

Fig. 2. Electroencephalogram changes from the occipital cortex area showing the transition from alpha to beta rhythm when the eyes open (up arrow) and the formation of the alpha rhythm when closing the eyes (down arrow)
Source: Archives of the Military Institute of Aviation Medicine in Warsaw.

As shown in Figure 2, it has been found, among other things, that such a sensitive indicator of response to stimulation as the EEG record, even in the absence of external stimulation, records the subject's own activity, which, after all, is not outside the experimental situation, but is one of its essential elements. For example, imaginative activities are sufficient internal stimulation to "block" the basic alpha rhythm, which is characteristic of a situation poor in stimuli. The characteristic electroencephalographic image of cortex area activation is based, among other things, on the appearance of a stimulus in the cerebral cortex to which the brain responds with the appearance of low amplitude fast waves (so-called beta waves), i.e. the desynchronisation of basic bioelectrical activity (Sadowski, 2000). These changes indicate a general mobilisation of energy resources of the whole organism. A number of studies suggest that alpha spectrum signal features causing rapid oscillations centred around the frequency of 10 Hz, are correlated with persistent temperamental differences (e.g. demand for stimulation) and also show differential modulation correlated with affective and cognitive states. Alpha rhythms occur in excess during relaxed wakefulness with eyes closed and show a reduction when the eyes are open which are correlated with temperamental traits in terms of overall activation level (activation, emotional reactivity, need for stimulation, etc.) (Terelak. 1976). Many electroencephalography tests show that alpha oscillations play a role in selective attention by "pulsed inhibition" of ongoing neural activity (higher alpha index is an indicator of selective attention, e.g. distractor suppression) (Mathewson et al., 2011; Ward, 2003). However, Basar (2012) cautions that alpha waves are not a single phenomenon but represent multiple processes that may show various effects in different areas of the cortex during different tasks, which is the subject of much detailed research in the field of electroencephalography.

However, during a state of relaxation, at rest, we observe the opposite phenomenon, the so-called alpha rhythm synchronisation, characterised by the appearance of high amplitude and low frequency waves (so-called alpha waves). The second aspect of activation refers to changes in the emotional part of the brain (the limbic system), characterised by, among other things, an increase in the level of excitation, which in the peripheral nervous system manifests itself between different physiological reactions that can be registered, such as: skin conduction (GSR), increase in total muscle tonus (EMG), heart activity (ECG), respiratory rate, etc. (Szczepańska-Sadowska, Łoń, Sadowski, 2000). These changes indicate a general mobilisation of energy resources of the whole organism.

The occurrence of various body reactions at the EEG level, even in such simple situations as the recording of resting EEG, drew the attention of many researchers to inter- and intra-individual differences in the overall level of activation (activation, stimulation, toning) (Terelak, 1976).[5] Furthermore, research on the stimulative structure of the situation has shown that sensory stimuli that do not have a signalling stimulating value have a lower stimulating value than stimuli that carry a specific information value. Also stimuli causing emotional states have a higher stimulative value than emotionally indifferent ones (Sadowski, 2000). Experimental research, mainly on human functioning in the situation of sensory deprivation and

---

5   The presence of various cortical rhythms, recorded by means of EEG, is therefore not only a manifestation of some hypothetical "starter" mechanism or a manifestation of spontaneity, but also an expression of interaction of external environment influences, acting through the receptors on the central nervous system and reaching the cortex of internal environment influences. The latter act on the cortex either through interoceptors or directly on nerve cells through automatic stimulation of centres (e.g. feedback afference).

stimulation overload, has allowed the formulation of some general regularities in terms of non-specific responses to these situations (Terelak, 1974). These problems are dealt with by psychological concepts, looking for a balance between the reactivity of the subject and the level of stimulation from outside the organism, called in the literature on the subject, similar to psychophysiological concepts of "optimum activation," "optimum stimulation" concepts.

1.3.2. Psychological stimulation concept

The concepts of "optimum activation" and "optimum stimulation" are mutually correlated and complementary in nature. It was found, among other things, that depending on the level of stimulation, some hypothetical zones (bands) of optimal functioning efficiency, suboptimal zones, as well as zones of disturbances and denials of activity, which overlap with the activation zones, can be distinguished. The following Fig. 3. illustrates such a theoretical curve.

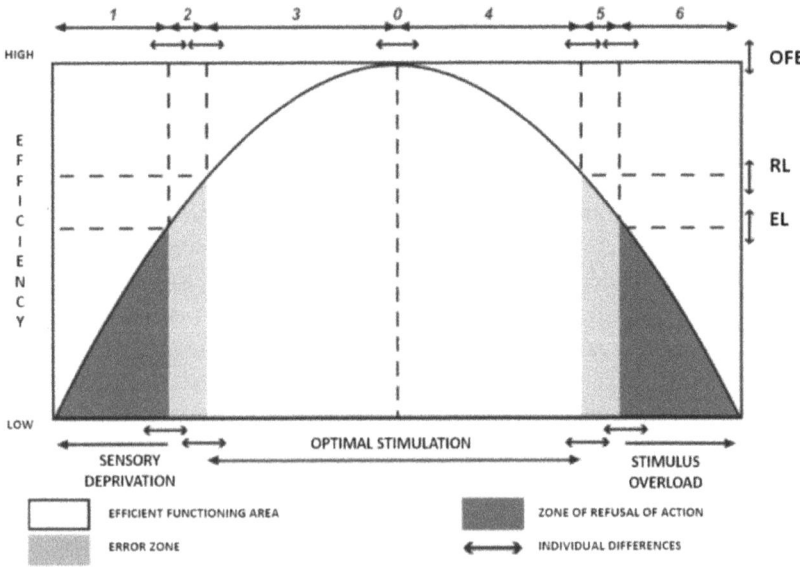

Fig. 3. Model of human functioning depending on the stimulation level
Source: own elaboration.
Legend: 1 – acute sensory deprivation range; 2 – stimulation limitation interval; 3–4 – the range of optimal conditions; 5 – overstimulation interval; 6 – stimulation overload interval; 0 – optimal level of stimulation; OFE – the optimum of functioning efficiency; RL – reliability limit; EL – efficiency limit

As shown in Figure 3, both in conditions of stimulation overload and in a situation of sensory deprivation (the so-called sensory hunger) one should reckon with deterioration of performance and mental discomfort as reactions to stress. This is confirmed by numerous empirical data provided from various sources.

Let us recall that the concept of "optimum activation" assumes that for the smooth course of cortical processes, which determine human activity, a certain level of stimulation is necessary, for which the reticular formation of the brain stem is primarily responsible. As Fig. 3 suggests, cortical processes only run smoothly if the level of stimulation (activation) is maintained around the optimal value. Below and above the optimum stimulation/activation, changes in well-being, emotional tension and various mental states are observed, which have been partially confirmed by physiological and biochemical studies. For instance, the studies by A.C. Ribeiro and D.W. Pfaff (2007) at the level of CNS activation aversiveness of sensory deprivation states (monotony) and overstimulation (information overload) were confirmed, while M. Zuckerman et al., 1970) pointed out the aversiveness of these states at the level of vegetative nervous system (Heart rate, Skin galvanic-reaction, Respiration rate/min). One can therefore expect to learn behaviours that lead to the search for or reduction of stimulation in situations of lack or excess of stimulation. However, this concept has some ambiguity when it is used to translate the so-called effects of sensory deprivation. If one assumes that the lack of stimuli leads to a strong decrease in the level of activation, then one should expect that the level of emotion and motivation will also be very low, which should be expressed, for example, in complete inactivity and sleep. Meanwhile, the results of deprivation experiments indicate rather the opposite phenomenon, i.e. the subjects show strong, unpleasant emotions and a strong need to be active (e.g. the need to look for a way out of this situation) (Sutherland, 2007), which means that too little consideration is given to the psychological significance of stimuli, which are directly related to emotions and motivation.

Looking for similarities and differences between the two concepts quoted, it should be noted that the optimum activation is defined by various authors not by indicating the physiological parameters of this state, but only by indicating its psychological effects, concerning both the well-being and effectiveness of the subject in the performance of tasks (Sutherland, 2007). Moreover, assuming that the optimum activation is the standard of regulation that a subject strives to achieve and maintain for the sake of effectiveness and well-being accompanying this state, it must be assumed that both the optimum activation and stimulation should be treated as "bands" and not "points" on the activation and stimulation continuum (Elijah, 1985). The location of optimal "bands" on this continuum is also a matter of discussion, as empirical data shows that the optimal level of activation and stimulation is the result of both personal experience, specific living conditions and relatively constant personality traits. Attention to this fact was drawn by D.W. Fiske and S. Maddi (1967, pp. 11–56), distinguishing between "normal" and "optimal" levels of activation and stimulation: The "normal" level occurs when there are no specific tasks, e.g. when it is related to the changing cycle of the body's daily rhythm (sleep-wake cycle), while the "optimal" level is determined situationally by a specific task. This allows one to look at the relationship between the "normal" and "optimal" levels of activation and stimulation as a system of mutual regulation. There are also researchers such as J. Matysiak (1993), who present an opposite point of view, defining optimum activation as a process (point on the scale) rather than as a regulation mechanism, colloquially calling this process a "stimulative hunger." There are many attempts to explain the mechanism responsible for individual differences "in the readiness to respond to stimuli." Among them, the Neopavlovian concept of temperament seems particularly interesting, in which reactivity, as a parameter of the temporal characteristics of behavior, next to the energy level, is the main component of temperament.[6] The optimal level of stimulation does not mean any one specific amount, and individual differences in the optimal level of stimulation can be reduced to differences in the level of reactivity of individuals.[7] According to this concept, highly reactive units, equipped with a "stimulation amplification mechanism," need less of it to achieve optimum performance compared to low-reactive units, equipped with a "stimulation suppression mechanism" (Strelau, Angleitner, Newberry, 1999).

If we assume that the statement that the behavior shows a tendency to maintain the level of stimulation in the optimal range is true, then this tendency should be manifested mainly in the so-called demand for stimulation, which may be manifested in the preferences[8] of situations or behaviors rich or poor in

---

6 This thesis was verified empirically in the doctoral dissertation of the author of the book, in which 24 features of temperament, derived from various concepts of temperament, were analyzed: J.P. Guilford and W.S., Zimmerman, L.L. Thurstone, J.A. Taylor, H.J. Eysenck and J. Strelau. On the basis of factor analysis using the Varimax Kaizer method with orthogonal rotation, two main components of temperament were distinguished: "energetics" and "emotional responsiveness," explaining about 60% of the variance (cf. Terelak, 1974).
7 Reactivity is understood as a relatively constant and individual-specific sensitivity to weak stimuli and tolerance to strong stimuli.
8 Preference may mean here the probability of choosing a certain behaviour from a set of behaviours in the individual's repertoire.

stimuli depending on the level of reactivity of individuals and / or differences in the level of dynamics of behavior (activity) or in the level of efficiency. Thus, reactivity as a temperamental factor can be considered as a moderator of achieving optimum activation through individual stimulation needs. This rather illustrative approach to the mechanisms responsible for "reinfocing" or "suppressing" levels of stimulation has been criticised by some psychologists, such as C. Leuby (1965, p. 174), who pointed out that in a situation of insufficient stimulation an increase in sensory sensitivity should be considered, and in a situation of excessive stimulation – a decrease, which assumes that the notion of optimum should be understood in terms of individual differences.

1.3.3. Individual differences in the need for stimulation

It is worthwhile to consider the learning mechanism, e.g. in the form of habituation, in the discussion on individual differences in terms of the need for stimulation. In the course of individual life experience, we have a repertoire of ways to avoid such situations and to deal with them when we cannot avoid them. This is possible by adjusting the preferred stimulation levels. The learned mechanism of maintaining the balance between reactivity to stimuli and the source of external stimulation, consisting of the the need for stimulation is illustrated by the proprietary model[9] shown in Fig. 4.

---

[9] The foundations of this model were created during my wintering at the Arctowski's Station in 1980 and on the basis of inspiring long discussions during the polar night with Andrew Ruta, MSc, co-creator of the mathematical formulas, whom I would like to thank.

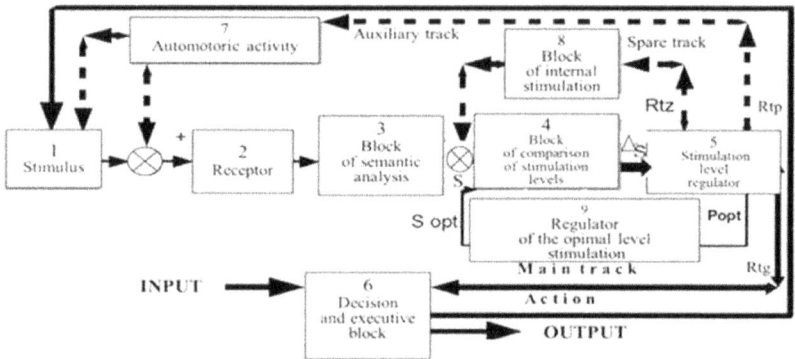

Fig. 4. Model of psychological mechanism of mutual regulation of stimulation levels of J.F. Terelak

Source: own elaboration.
Legend: S – current stimulation level signal; Sopt – optimal stimulation level signal; ΔS – error signal; Sp – stimulation level signal; Rtz – regulation of the signal by a back-up route; Rtp – regulation of the signal by an auxiliary route; Rtg – regulation of the signal by the main route.

The above scheme assumes that physical stimuli reaching from the environment (1) to individual receptors (2) are converted into signals, modified by the stimulus meaning value analysis block (3). The amplification or weakening of the signal passing through this block is influenced by the emotional responsiveness, experience, etc. The signal coming out of the block (3) together with the signal coming out of the internal stimulation block (8) form the signal of the current stimulation level S, which in turn enters the stimulation level comparison block (4). In this block, the stimulation level signal S is averaged and stored, and the transformed Sp signal is compared to the optimal stimulation level signal Sopt. The averaging time is relatively short (e.g. individual auditory stimuli after averaging constitute music with a certain stimulus value), while the time of stimulus extension is proportional to its strength (e.g. blinking of the control lamp signalling engine failure is a stimulus of high stimulus force, i.e. long prolongation time for the driver). Operation of the stimulation level comparison block is shown in Fig. 5.

The environment as a stimulating system 43

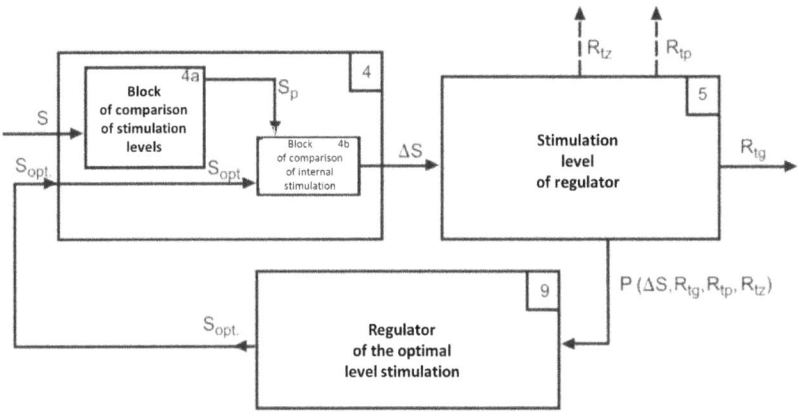

Fig. 5. Operation of the stimulation level comparison block 4
Source: own elaboration.

As indicated by Fig. 5, the error signal, i.e. the difference between the transformed stimulation level signal Sp and the optimal stimulation level signal Sopt, equal to ΔS (Δ S = Sp − Sopt), is transmitted to the stimulation level regulator block (5). There are three feedback loops in the system which enable to change the adjustable S signal. The main loop is used to both reduce and increase the S stimulus level (depending on the value of ΔS). The other two routes, marked in the figure with dotted lines, are auxiliary routes, used only with a low level of stimulation. The signal from the regulator (5) is transmitted via the main route to the decision-execution block (6). The block's (6) affects external stimulators (1) which causes a decrease or increase of the level of external stimulation (e.g. switching the tape recorder on or off causes a change in the level of sound stimulation). The magnitude of the signal flowing from the regulator (5) to the decision-execution block (6) simultaneously influences functioning of a person.

If it is not possible to increase the stimulation level via the main route, the auxiliary route is engaged. The signal then runs from the controller (5) to the automotor activity block (7), then it runs either through external stimulators (1) to receptors (2) (e.g. whistling, loud talking to each other, etc.), or directly from the block (7) to receptors (2) (e.g. scratching the head, tapping fingers on the table, etc.). In the case of the first one, not only the regulated subject but also the people in the immediate vicinity are stimulated. If the auxiliary route is not sufficient to achieve the optimum level of stimulation, or if the use of this route is not possible (as in the experiment in the laboratory), a back-up route for the internal stimulation block (8) is activated. The signals emerging of the internal stimulation block are summed up with the signals received by the receptors and transmitted to the block (4). The internal stimulus block is responsible for phenomena such as hallucinations and, more specifically, for all kinds of hallucinatory-like phenomena. The Sopt signal, led to the block (4), is generated by the optimum stimulation level regulator (9). It is an integral-inertial regulator. The regulator averages (integrates) the input signal P for quite a long time period and simultaneously delays the signal, and the delay time (inertia) is also relatively long. For example, due to the inertial and integrative nature of the regulator, after several days of perceptual deprivation, the individual optimal stimulation level of the subject should not change, as evidenced, i.a., by data on the so-called deprivation effects available in the literature. However, changes in the optimal stimulation level can be expected after a relatively long period of time, which requires empirical confirmation (Eliasz, 1985). The limitations of the magnitude of the signal of the optimal level of Sopt stimulation are: minimum and maximum stimulation levels, characteristic and relatively constant for a given individual (endogenous quantities) (Strelau, 2006). The activity of the stimulation level regulator block is shown in Fig. 6.

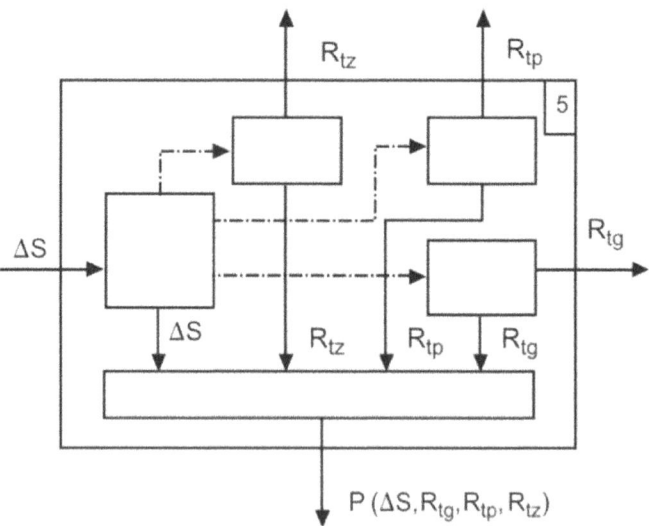

Fig. 6. Operation of the stimulation level regulator block 5
Source: own elaboration.
Legend: ΔS – stimulation level signal; Rtz – signal adjustment via a backup path; Rtp – signal regulation by an auxiliary path; Rtg – signal regulation via the main path.

The P signal, fed to the regulator (9) from the block (4), is a result of processing of the ΔS error signal and signals transmitted through three routes: to the block (6) by the Rtg main route, to the block (7) by the Rtp auxiliary route, to the block (8) by the Rtz back-up route (see Fig. 5). It is therefore possible to change the signal of the optimum Sopt level even if, as a result of the operation of the auxiliary and backup route, the ΔS error signal is equal to zero.

Summarising the considerations concerning the system of mutual regulation of stimulation levels, it should be stated that it consists in adapting the environment (external stimulators) to human needs (main and auxiliary routes), adapting to the state of the environment (the loop connecting blocks 4, 5, 9) and adapting the human being to the needs resulting from the previous state of the environment (the back-up route and part of the auxiliary route, avoiding the environment).

Although the analysis of human functioning presented in this study concerns mainly the situation characterised by limited stimulation, it seems that based on the proposed own theoretical model, the phrase "mutual regulation of stimulation levels" can be used, and the stimulation mechanism be treated as one of the important regulatory mechanisms of human adaptive behaviour to conditions deviating from the optimum stimulation range.

A review of empirical studies on the so-called psychological effects of sensory and perceptual deprivation, which illustrate the legitimacy of the proposed model, is presented in the second part of the book.

The previous considerations concerning the activation and stimulative aspect of the situation point to individual differences in the need for stimulation, which may determine the adaptability or preferences possibilities of certain behaviours, which should be taken into account especially in situations characterised by significant reduction of stimulation, i.e. situations of sensory, perceptual and Antarctic deprivation.

At this stage of research, it is not yet possible to formulate a general answer to the questions: What is the role of environmental information in human behaviour? Is external control more important than internal control by means of cognitive structures as relatively fixed disposals? However, research in this area can contribute a great deal, since the explanations of human behaviour so far in terms of habit as a result of learning or drive, as an internal, biologically determined force that directs human aspirations have proved insufficient. Especially in human cognitive concepts (as opposed to behavioural and psychoanalytical ones), the environment is an important, and sometimes the main determinant of behaviour. It is noted that human activities are governed by information from the environment, which provides knowledge not only about what can happen (probability assessment), but also important data on the assessment of the value of these events (gratification, subjective value). It is only in such an approach that the information coming from the environment has the right motivational power and the nature of the mechanism regulating human activity.

Similar views on the need for information conditioning human adaptability are characteristic of Polish psychology in the 1980s mainly in terms of general statements. The authors of these views emphasise that a constant stream of information flows between man and his environment, which is processed and stored, creating a "cognitive network" which is the basis for maintaining the necessary balance between man and environment (cf. e.g. Tomaszewski, 1978; Reykowski, 1975). According to the authors, cognitive mechanisms, although they are often treated as "instruments of orientation," play an important motivational role, which is the result of the functioning of cognitive networks.[10] This view is consistent with J. W. Atkinson's achievement motivating theory, who argues, among other things, that attributing information only to a function of assessing the probability of achieving results, without evaluating the subjective value of the gratification of the goal, would lead to the creation of a psychological portrait of a human being who is familiar with the environment but who is not able to change it (Atkinson, Feather, ed., 1996).

The "self" structure (as a subsystem of the cognitive network) plays a special role in the regulation of human relations with the environment, being an instrument for orientation and differentiation between the subject and the environment. Many data indicate that as a result of "informative" contacts with the environment, a person learns to anticipate the inflow of standard information by analogy, as well as learns new forms of behaviour that can provide an influx of the necessary information, i.e. learns adaptive behaviours, including the need for stimulation/information (Biela, 1981).

The notion of adaptation is taken from the biological sciences and means optimisation of mutual relations between the organism and its environment, consisting in the regulation of mutual relations between man and his bi-directional

---

10 We can imagine the cognitive network as a system of interconnected cognitive structures representing particular objects and events distributed within a multidimensional space. Thus, it can be said that the cognitive network reflects the multiplicity of relationships in which objects and states of the outside world remain with each other, i.e., it reflects the so-called "multidimensional psychological space" in which events within cognitive structures can be at different "psychological distances" from each other (cf. Reykowski, 1975).

environment: (a) human adaptation to the environment (homeostasis); (b) adaptation, taking into account human impact on the environment (allostasis) (Goldstein, 2004). Behaviour is assessed as adaptive if it meets the requirements of the external situation and the behavioural variability available to the body (plasticity). Thus, the analysis of adaptation phenomena should cover two categories:

(1) *Phylogenetic adaptation*, i.e. behaviours common to the entire population, transmitted by genetic code and constituting a relatively permanent means of organising the behaviour of an individual in the environment.
(2) *Ontogenetic adaptation*, which includes new adaptive responses and new types of responses that are based on phylogenetic adaptation as a result of individual learning during an individual's life. These behaviours are not common to the population as a whole, nor are they hereditary, although through social messages they are a cultural heritage, a subject of learning by members of a cultural group in the past.

The above classification seems to be too general for our purposes, as it only allows us to conclude that the phenomena, we are interested in fall primarily within the category of learned behaviours. However, this distinction is of particular importance in the study of adaptability to extreme conditions (situations), as the physiological mechanism underlying the adaptive behaviour is shaped by the combined influence of hereditary and environmental factors. In the literature it is assumed that the most general indicator of adaptation is an individual's level of functioning in given conditions (e.g. the Aborigines' resistance to cold or resistance to high-altitude hypoxia of Peruvians and Gurkhas from Nepal).[11]

As stated earlier, both development and proper functioning of the organism are closely dependent on the inflow of external stimuli. Relatively early on, the determining role of stimulation in the development of the nervous system was realised, with particular emphasis on the role of the reticular formation of the brain stem), as evidenced, among others, by the results of numerous empirical works confirming the thesis that the development of nervous structures depends on functional requirements, whose carrier (in both formal and content sense) are external stimuli. And so, the underdevelopment of afferent structures is associated with disorders of higher integration activities, i.e. the ability to process and synthesise sensory information. A dramatic example of this are the developmental disorders of deaf and blind children from birth (Jezierska, 1963). These

---

11 Due to their exceptional resistance to fatigue, the Gurkhas were readily recruited into the British Army.

narratives show that the deprivation of two basic information channels, hearing and sight, has led to such deficits in cognitive and emotional development that specialist therapy generally takes several years to reproduce a relatively adequate mental "picture of the world." No less drastic disruption of normal functioning and potential adaptability is caused by social isolation, especially for children and the elderly, but also in the artificial environment of the spacecraft. This is evidenced by numerous sources of both narrative and experimental data, which support the thesis that the need for stimulation/information and interpersonal relations is one of the basic human needs (Cacioppo, & Patrick, 2009), which is increasingly often referred to in the so-called positive psychology (Suedfeld, 2001).

Summing up the above considerations on the need for stimulation, we can safely say that it is one of our basic atavistic needs, because the supply of stimuli is a precondition for the normal functioning of our brain, and the state of sensory deprivation is also often referred to as a "hunger of the senses" (Terelak, 1978) or "hunger for stimulation" (Matysiak, 1993). We are convinced of this by views, apart from biological mechanisms, also psychological, explaining the adaptation to long-term sensory deprivation and social isolation. Among the various concepts explaining the mechanisms of human functioning in a situation of reduced stimulation, two groups of views can be mentioned, which are correlated with each other, namely that: (a) stimulation may come from the external or internal environment (i.e., the subject's own activity may also be a source of stimulation; (b) stimulation is needed for normal human functioning. However, the question about specific mechanisms of human functioning in a situation of limited stimulation differentiates individual researchers in emphasizing different aspects. Thus, the previously discussed neurophysiological concepts indicate the role of the reticular formation maintaining, inter alia, a relative balance between the external environment and the cerebral cortex. Although they are most useful for learning about the mechanisms of human functioning in deprivation situations, as D. Hebb claims (1965), they are not always adequate to explain complex behaviours. On the other hand, however, they form a good basis for psychological theories explaining mechanisms for regulating human behaviour not only in deprivation situations.

Let us pay attention only to those psychological concepts which refer to certain physiological mechanisms in the interpretation of empirical results. These include, first of all, temperamental concepts, among which H.J. Eysenck's Theory of Extra/Introversion (1990), assuming, among other things, that the threshold of sensitivity of introverts is lower than that of extroverts. The above hypothesis was the reason why introversion was associated with tolerance to pain and sensory deprivation.[12] Thus, it was emphasised that the same amount of stimulation is experienced by introverts as more effective than by extroverts. In addition, the optimal level of stimulation in introverts is lower than in ambiverts, who in turn have lower levels than extroverts. However, E.J. Ludvig III, & D. Happ (1974), in the course of evaluating the usefulness of Eysenck's concept for explaining depressive effects, suggest that it is more useful in the study of stimulative (informative) overload.

A review of studies and theories similar to the general hypothesis of cortical stimulation/braking suggests that the internal arousal process may modify the need for external stimulation. Thus, it is indirectly assumed that the maintaining "sensoristasis" is associated, among other things, with the control of internal stimulation. Secondly, D.W. Fiske and S. Maddi (1967) emphasise that the nature of stimulation depends on both the arousal value itself and the objective strength of the stimulus. Thirdly, the studies of M. Zuckerman et al. (1968) suggest that it is possible that real sensory deprivation situations may have a high arousal value and thus should be considered as stressful situations. Among other things, this is how the authors explain the results of their research on better tolerance of sensory deprivation situations in extroverts. The authors also stated that the situation of sensory deprivation does not significantly differentiate between extroverts and introverts. However, such differences are clearly visible in the case of high levels of stimulation. The authors believe that one of the reasons for the lack of differences in the deprivation situation is, among other things, the fact that empirical research to date does not cover the "extremes" of the deprivation (stimulation) continuum and is therefore not complete. So the research problem in this area is still open.

Leaving aside the detailed discussion of temperamental concepts in the context of the need for stimulation, as they are the subject of many separate studies (cf. Strelau, 1995), we will only draw attention to the general conclusions that come out of them in terms of such properties as the need for stimulation. First

---

12 I recall that when I was as a student of psychology, together with members of the Student Scientific Study Group of Clinical Psychology, I participated in the contemplative research of the Camaldolese monks in Krakow, who lived in complete isolation from the outside world and were unable to hold conversations apart from saying greetings using the wording of *Memento mori*. This research was based on a thesis that introverts are more resistant to social isolation. In the meantime, we were surprised by the results obtained, as it turned out that, in most cases, introverts left the monastery permanently, who could not stand such an extreme situation of sensory deprivation imposed by religious rules.

of all, it is believed that adaptation (understood as the adaptation of the whole organism with particular emphasis on the central nervous system) and proper functioning in a situation of limited inflow of stimuli are conditioned, among other things, by the individual demand for stimulation, which has a genetic background. It is determined by hypothetical mechanisms of stimulus intensity processing, which may have properties that suppress or amplify the objective (measured in physical units) stimulus charge (cf. Elijah, 1985). Secondly, the stimulus intensity processing mechanism often referred to by the above concepts is, in general, a neuro-brain mechanism formed by the combined influence of hereditary and environmental factors. On this basis, among other things, a hypothesis has been put forward that the stimulation processing mechanism responsible for the amount of demand for stimulation manifests itself as a formal feature of temperament (cf. Strelau, Angleitner, Newberry, 1999). Thus, the psychophysiological concepts discussed earlier are at the same time psychological concepts since the formal characteristics of behaviour do not have a direct influence on the content of behaviour but can form the basis on which certain personality traits are formed. This suggests that the individuals' demand for stimulation is also shaped by the learning process.

To sum up the discussion of individual differences in sensory deprivation tolerance and social isolation, one should agree with the suggestions of M.T. Orne and K.E. Scheibe (1964), who argue, among other things, that the controversies cited at the empirical level are the result of a number of "non-deprivative" factors, controlled not very precisely or not at all, and the use of various explanatory concepts.

2. Data Sources Concerning The Effects of Sensory Deprivation and Social Isolation

Sources of data on the psychological effects of deprivation can be found in fiction, popular science and scientific literature. The psychological analysis of these sources must be methodologically differentiated.

2.1. Taxonomic aspects of data sources

Before we move on to discussing sources of data on the stress of deprivation, we shall characterize its essence. The concept of "sensory deprivation" was promoted by J. Vernon and J. Hoffman (1956). It concerns three categories of situations, mainly experimental. The first category can include perceptual deprivation, operationally called "reduction of stimulus structuralisation." A typical example of this category of deprivation situations is the first experimental research by W. Heron, W.H., Bexton, T.H., Scott, (1954) from McGill University in Canada. The attributes of this situation of perceptual deprivation were usually: a comfortable bed or armchair, "white noise," semi-permeable goggles and special cylinders on the hands. The second category of sensory deprivation situations concerns "reduction of the absolute stimulus level." An example of experimental application of such situations is immersion research (Shurley, 1960). The third category of sensory deprivation situations includes "imposed stimulus structuralisation." This type of sensory deprivation situation was introduced, among other things, by researchers from Harvard Medical School in Boston using a respirator, the so-called iron lung (Zukerman et al., 1962).

The above categorization of sensory deprivation situations, based on the principle of reduction of stimulus structuralisation, is methodologically imprecise. The taxonomic difficulties caused by the terminological ambiguity have led many researchers to abandon the general classification of sensory deprivation, operationalising it according to the assumed research objectives of the specific experiment. In the psychophysiological concepts, attention is drawn to the fundamental fact that each stimulation affects a person in two ways: non-specific (general activation) and specific (meaning of the stimulus) (Hebb, 1965). Thus, both the restriction of stimulus inflow, as well as a rapid change in the meaning of the stimulus are of a stressful nature, as they somehow disturb the usual structure of human operation.

On the basis of the analysis of the literature on the subject, it can be proposed to separate four situations: sensory and perceptual deprivation, social isolation and imprisonment, and among them such sources of data that are interesting from an astronautic perspective. The pioneer of such classification from the point of view of cosmonautical interests was S.B. Sells (1973), who, taking into account the percentage of problems that can be used in astronautics, listed the following ranking taxonomy of data sources: 1. Submarines (79); 2. Polar expeditions (68); 3. Deep-sea vessels (61); 4. Atomic shelters (60); 5. Radar bunkers (59); 6. POW camps (39); 7. Athletic teams abroad (37); 8. Psychiatric hospitals (23); 9. Prison groups (20); 10. Isolated professional groups (16); 11. Maritime and mining disasters (11). The common feature of sensory deprivation and social isolation is the relative limitation of external stimulation, covering both its formal (strength, intensity of stimuli) and content (structure of stimuli, information, habits, needs, etc.) aspects. Human psychophysiological reactions to situations of sensory deprivation or social isolation were the subject of many laboratory tests and observations conducted in natural conditions. It was found that all types of deprivation are stressful – from impulse deprivation, monotony to cultural and social isolation and the imprisonment: not only of individuals but of entire societies, as in the case of the COVID-19 Pandemic.

K. Yang (2019), drawing on extensive data sources including literary work, case studies and large-scale surveys covering a wide range of countries (within and outside Europe), demonstrates the social nature of research on social isolation and its importance for the development of scientific fields such as medicine, sociology and psychology. Following this path, it should be added that the issue of psychological effects of sensory deprivation and social isolation has acquired a new meaning in space psychology, which is interested in long-term manned space flights (orbital, interplanetary, and perhaps interstellar ones in the future). These problems are presented in detail in another book (Kwarecki and Terelak, 1980).

Due to the lack of an unambiguous criterion of the theoretical taxonomy of data sources on the effects of broadly understood sensory deprivation and social isolation, I propose a methodological criterion of division into: (1) phenomenological approach, based on the narrative method; (2) empirical approach, based on psychometrics.

2.2. Phenomenological aspect of data sources based on narrative

In the analysis of the sources of data on stress of deprivation-isolation, a phenomenological approach was used alongside the method of historical analysis of source texts, one of the first methods of scientific psychology, namely the introspective method. This method is well characterised by its last supporter in Polish scientific psychology, M. Kreutz (1949), Professor of the University of Warsaw, who, among the so-called methods specific to scientific psychology, listed the introspective (narrative) method in the first place as a method which, unlike other psychological methods, allows for the study of psychological phenomena directly, i.e. subjectively, as opposed to indirect, i.e. objective methods. The differences between direct and indirect methods can be schematically shown in Figure 7 below.

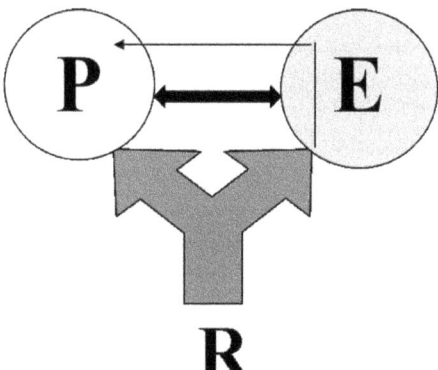

Fig. 7. Model of psychological research by direct methods (natural experiment, narrative)
Source: own elaboration based on Kreutz, 1949.
Legend: R – researcher; P – Psychological phenomena; E – Environmental phenomena.

As shown in Figure 7, according to Kreutz, "the introspective method is a basic, fundamental method, because if it were not, there would be no indirect methods, we would know nothing about mental phenomena at all, even if they existed as unconscious ones ... And then, to use indirect methods, we need to know the relationship between certain physical and mental phenomena"

(Kreutz, 1949, p. 26).[1] For example, Dan P. McAdams (2006) points out that past three decades over, narrative theories and methods have helped psychology of personality to better describe the meaning of human life, paying attention to the context of culture, gender, race, social and physical environment. Thus, there is a growing interest in narrative methods, which, together with the study of the characteristics and strategies of coping with one's own life, create a three-level conceptual framework that treat the personality as an integrated reconstruction of past experiences, perceived present and anticipated future.

Sources of data on the stress of deprivation can be found in fiction, journalism, popular science and scientific literature. An example taken from fiction can be the facts or legends described, such as the legend about the founders of Rome or about the so-called wolf children. A special case in fiction is the "science fiction" literature, in which one can also find many descriptions of human functioning in a situation of deprivation (isolation), which we are dealing with in the real world. An illustration may be the description of the "crazy bath of Pirx, the pilot," i.e. a typical experiment on human behaviour in immersion,[2] deprived of all sensory, visual and auditory stimuli, i.e. the extreme situation of sensory deprivation, which was described by S. Lem in his book entitled *Tales of the Pirx the Pilot*, (1978).

A contemporary version of journalism is K. Kąkolelewski's reportage (1976) entitled "The First Soldier of the Third World War," in which he describes the appearance and mental reactions of a man, hiding for 20 years after the Second World War in solitude in a shelter hidden in the thicket of the forest for political reasons: "I imagined Andrzej K. as a man different than everyone else, some extraordinary superhuman who spends days and nights with wolves. I can't tell the impression he made on me. He looked at me with feral eyes, he was wearing a dirty, torn shirt, he spoke very quickly, nervously, he had hardened hearing, sensitive to every rustling ... Those two people he used to see also noticed his slightly different manner of speaking, disorder, like a speech fade. He told them that he was talking to himself, but that he was talking less and less to himself and

---

1 The opinion of a well-known Viennese ethnologist Konrad Lorenz (1986) is worth mentioning here. Among other things, he claimed that one must speak out definitely against the epidemically-spreading misconception that only what can be counted, measured, deserves to be called real. It must be convincingly explained that the process of our subjective experiencing has the same degree of reality as everything that can be expressed in the terminology of natural sciences.
2 *De facto*, in literary language S. Lem describes the model of S.H. Shurley's scientific experiment (1960). Shurley (1960).

he rarely responded to himself: "Good morning, sir." – "Good morning, sir," ... "Enjoy" – "Enjoy," "Health" – "Health," "Good evening" –"Good evening," "What a life" – "What a life," "Good night, sir" – "Good night, sir" – that's the whole dialogue. He made the cards he played with himself. He complained that he often lost. He came closer to the animals, his eyesight was constantly improving, and his hearing was catching very distant murmurs and allowing him to distinguish them .... When he found out the exact date ... he was surprised that it had still been that old year, and he was impatient, another time he was amazed by the sudden progress of time" (Kąkolewski, 1976, pp. 128–139).

Examples of popular science and scientific narrative include a number of specific issues concerning human reactions in various life situations with limited stimulation.

2.2.1. Loneliness

Loneliness is the result of "separating" with a physical (e.g. wall), psychological (closing in) or sociological (language, group norms, religion, ideology, etc.) barrier from the external environment. Ballooners were the first to draw attention, as early as in the nineteenth century, to the psychological effects of loneliness and stimulus deprivation known as the *break-off phenomenon* (Ryan, 1995). Like D.G. Simons (1958), who spent 24 hours in the gondola of the "Manhigh II" balloon at an altitude of 30 000 m, stated that especially during the night hours he felt loneliness, a feeling of depersonalisation, a feeling of loss of contact with his own environment, a desire to quickly lower the altitude and return to the ground. Similar reactions of loneliness were described in the middle of the twentieth century by jet airmen, performing flights at the so-called altitude ceiling. They are subjectively perceived as a loss of contact with reality, detachment from the ground with a large component of anxiety, apathy and a desire to break out of this situation. For example, B. Clark and A. Graybiel (1957), when investigating American pilots of maritime aviation, found that in high altitude flights at the altitude of 10–12 thousand meters and more, especially at sea, where even in good visibility there is no reference point, the "break-off phenomenon" occurs in relation to 35% of pilots performing in single-seater planes.[3] A lot of data on monotony in *artificial satellites* is provided by industrial medicine and psychology, dealing with the impact of monotony (*industrial monotony*), which is one of the

---

3   No engine noise can be heard in a supersonic jet plane. In addition, a pilot does not need to control many instruments as much as in other types of flights and does not need to perform many motor manipulations when flying at an altitude ceiling.

attributes of sensory deprivation for the "productivity" of work.[4] For example, one of the motives for conducting research on the effects of sensory deprivation by a group from McGill University was, among other things, to explain the phenomenon of so-called apparent work, which was observed in soldiers who, locked in bunkers, operated radiolocation devices. Those workers, who were in small isolated rooms, did not detect all radar signals. The rate of decrease in their efficiency was higher than the classic fatigue curve would indicate (Bexton, Heron, Scott, 1954). I obtained similar results by examining soldiers serving in bunkers, protecting them from the effects of microwaves. The results obtained in a group of 50 healthy men aged 25–40 years, tested with the 16 PF Cattell Test and MMPI, allowed us to determine the "bunker syndrome," characterised by symptoms of neurosis, manifested by introverted processes, mood swings, apathy, aversion to mental effort and disturbed circadian rhythm, which we initially mistakenly called *microwave neurosis* (Maciejczyk & Terelak, 1978).

Many interesting data on "loneliness" is provided by observations and studies on the social isolation of ethnic groups during migration in metropolitan conditions (Hossen, 2012; De Jong Gierveld, Van der Pas, Keating, 2015; Wu, Penning, 2015). A lot of data on this subject is provided by American literature. New York City has been called by sociologists the "crucible of nations" of different socio-economic and cultural status. E.g. B. Malzberg and E. Lee (1956) gave statistical data for the years 1939–1941 illustrating the number of mental disorders recorded in New York, which are related to the reaction to the cultural isolation of foreigners in the first years of their adaptation to the "foreign" environment. These data show, among other things, that the basic symptoms are nostalgia and depression. Contemporary researchers of the effects of emigrant isolation prove

---

4   The situation of the influence of monotony on efficiency is particularly dangerous in transcontinental flights, during which after take-off the autopilot feature is switched on for several hours of flight. As an aeronautical psychologist, I've been analysing this phenomenon while in the cockpit on a flight from Sydney to London. Monotony appears after the first hour of flight, which takes place at 10 000 m above the Earth in complete silence (jet engines are located at the back of the aircraft fuselage) and in the absence of any visual stimuli in the field of vision. Pilots describe this phenomenon, like 19th-century ballooners, as "suspension in motion between Heaven and Earth" and "propensity to drowsiness," which obviously threatens flight safety. For this reason, aeronautical psychology and ergonomics has developed countermeasures in the form of cyclic commands to be sent out (e.g. current status of pilot and navigation devices) and their confirmation, sending information to the nearest "radio-beacon" on the flight route. This allows one to keep the pilot active along the entire flight path.

that in most cases these symptoms are reactive (so-called neurotic reactions or reactive psychoses) and decrease after an appropriate period of adaptation to the new environment (Jarekka, et al., 2014).

A separate problem, of great social importance, is the "loneliness" of the elderly, who are isolated from the family within their own family unit by assigning them tasks too narrow to their abilities or who are being completely isolated by placing them in special care facilities. This first problem has not yet been subjected to systematic psychological studies (Van Baarsen, et al., 2001; Aartsen, & Jylha, 2011; Cong, & Silverstein, 2015). The second problem is dealt with by clinical psychologists and geriatric doctors (Ong, Uchino, & Wethington, 2016).

The situation of children *separated from their normal educational environment* is even more drastic. Sources of data on developmental consequences of children isolated from their parents and staying in various educational institutions, such as orphanages, stress that isolating children from their parents, especially in the early stages of life, is a reason for reducing mental needs, which are important to life and development and leads to long-term mental and emotional deficits (Rotenberg, & Hymel, Eds., 1999). We are merely signalling this problem, since it has been found in the literature on the subject of a number of monographs and descriptions of many drastic cases, such as the 19th case of K. Hauser, who until the age of 17 was raised in a dark cellar in a situation of total deprivation. When he underwent a psychiatric examination in Nuremberg in 1828, an intellectual development corresponding to the level of a 3-year-old child was found (quoted after: Svab and Gross, 1964). An equally interesting case of developmental effects of long-term isolation is described by M.K. Mason (1942), which concerns an 8-year-old girl, Izabell, locked in a dark cell for 6.5 years. Her mental development, assessed on the basis of Sfanford-Binnet's non-verbal Intelligence Scale, corresponded to 19 months of development, i.e. practically zero scale level. After appropriate upbringing procedures, she reached only the developmental age of a 6-year-old child. It was only after subsequent six years that no significant deviation from the development standard was found.

A particular problem of isolation are children, who are simultaneously blind and deaf. As the A.V. Jarmolenko (1961) states, the case of the legendary deaf and blind Helena Keller clearly shows that in the event of damage or lack of adequate senses, prevention of social isolation plays a decisive role. Appropriate upbringing procedures can reduce the intellectual (but not emotional) developmental deficit of handicapped children after some time. In Polish psychological literature, a similar case of deafness and blindness og Krystyna Hryszkiewicz is described by S. Jezierska (1963), who, as a typhlopsychologist and educator at the same time, only after several years of working with her charge, led her intellectual development to a normal level, which allowed her to take up and complete her philosophical studies at the university. However, emotional deficits have persisted virtually throughout her adult life. An example of the negative impact of this type of deprivation stress is the so-called orphan disease described in the literature (Schaeffer, 2009).

2.2.2. Imprisonment as punishment

Imprisonment is a variant of forced social isolation. There is a lot of data on the functioning, not only of humans but also of animals, in a forced situation poor in stimuli, which is confirmed on the example of trapped animals and humans by A.W. Esser, (1971) in his overview monograph.

Prisons use isolation and deprivation as a moral punishment, which is indirect proof of their stressful nature. The preventive aspect based on the stress of deprivation raises many doubts, not only moral ones. This stress, especially in its chronic form (multi-year sentences), as is known from the experience of criminology, leads to personality deviation rather than improved behaviour (Whitehead, Steptoe, 2007). This is due, among other things, to the deprivation of many important sources of emotional gratification, and the monotony of life (day after day is planned the same way), which is reflected in aggressive behaviour, which is a good source of stimulation, accepted by the prison community, described by temperamental psychologists as "spontaneous aggression," compensating for the need for stimulation in conditions of social isolation and monotony (Reykowski, 1977).

B. Ł. Skowroński and E. Talik (2018) in empirical research on prisoners found that the stress of isolation significantly reduces all aspects of quality of life in all its dimensions: general, psychophysical, psychosocial, subjective and metaphysical ones, regardless of the differences between groups of prisoners and the strategies used to cope with the stress of isolation. Particularly interesting are data on the functioning of political prisoners and prisoners of war. For example, B. Bettelheim (1943), who was a concentration camp prisoner (from Hitler's rise to power until the outbreak of World War II), in his narratives he stated, among other things, that the first period of "adaptation" to the camp conditions was followed by a period of relative adaptation to forced isolation, in which the prisoners forgot about the life they led outside the camp. However, some people were soon experiencing the phenomenon of regression, i.e., according to the psychoanalytic concept, a retreat to a lower level of development. This conclusion is based on the change in the prisoners' behaviour. They manifested the so-called "childishness" characterised by a lack of distance, breaking friendships without a clear reason, excessive quarrelsomeness. The author believes that it is the phenomenon of regression that is the basic reaction to the very strong stress of forced camp isolation. This is confirmed in the observations of R.C. Spaulding and C. V. Ford, (1972), who describe the mental reactions of 82 American political prisoners who were in a political prison in North Korea for 11 months in 1967. The people investigated were the crew of the warship USS "Pueblo," which was taken prisoner by Korea. Psychiatric and psychological examinations carried out during hospitalisation after buying out several crew members after several months of captivity and arriving in the USA, showed in more than half of the respondents a high degree of anxiety or depression, symptoms of emotional imbalance, passive-dependence personality traits and obsessive-compulsive symptoms. The so-called repatriation syndrome initially included apathy, which turned into excessive irritation and hostility to the environment. The authors consider the above symptoms to be the effect of long-term cultural isolation (Far Eastern culture is completely alien to the average American) and poor nutrition, as well as the use of deliberate indoctrination and brainwashing. Similar results are quoted by other authors, who investigated American POWs returning from North Korea (e.g. Lifton, 1954) or from Japan (e.g. Nardini, 1952). For instance, E.G Schein (1957), based on the example of political prisoners, drew the attention to techniques that isolate from information from the outside world and are limited to acting with homogeneous information, calling them "psychological tortures," which sometimes lead to mental states that resemble the initial stage of schizophrenia. This is confirmed by H. Klonoff et al. (1976), who analysed the psychological and neurological reactions of Canadian soldiers returning after World War II, who were in Japanese prisons and German camps for about three years. Among other things, it was found that the group returning from Japan, where there was a longer and more acute cultural isolation, was characterised by greater mental disorders compared to repatriates from Germany. The authors concluded that imprisonment combined with complete cultural isolation, as well as the purposeful and consistent use of indoctrination, give first and foremost psychological effects, only on the basis of which a fairly rapid deterioration in health and mortality is pronounced. This is also confirmed by the observations and research carried out on prisoners of German concentration camps during World War II (Chodoff, 1970). The particularly drastic effects of imprisonment in concentration camps of children isolated from their mothers and systematically indoctrinated (the so-called Upbringing for the Great Reich) were described by B. Trossman (1968), who stressed that developmental disorders in such children (especially in the emotional sphere) are long-lasting and often irreversible. Finally, my colleague Kazimierz Godorowski (1985), a psychologist and prisoner of the Oświęcim Concentration Camp (listed in the camp census under number P 4460), characterises imprisonment in a concentration camp with its indefinite duration as being in conditions of extremely difficult stress situations and deprivation of many basic life needs, among them: detachment from the existing environment, being torn away from family and the existing cultural and social circle, the mother tongue, the entire set of cognitive patterns to date, breaking social ties and a sense of constant danger. These elements of isolation stress were intensified by a permanent existential fear of death threatening at any moment. Such acute stress leads to the relatively permanent disorder well-known in the literature called PTSD (Post-Traumatic Stress Disorder) (Yehuda, Wong, 2007).

From the data presented above it appears that even if such drastic psychological methods as "brainwashing" or torture are not applied to prisoners of war, the imprisonment situation itself is clearly of a stressful nature (Tennant, 2007), and as claimed by I. Genefke, H. Marcussen & O.V. Rasmussen (2007), the very nature of total sensory deprivation and social isolation is "inhumane torture." However, extreme information deprivation known in the literature as "menticide" was applied to many political prisoners (Spaulding and Ford, 1972), aimed at degrading subjectivity.

Contemporary psychological research on the reactions of people serving imprisonment is mainly focused on coping strategies. An example on individual differences in coping with imprisonment can be found in the research of J.F. Terelak & M. Steckiewicz (2007), which shows, among other things, that prisoners (males) with a sense of internal control in a situation of prison isolation stress to a greater extent prefer task-focused stress-coping styles than men with a sense of external control. In contrast, prisoners with a sense of external control have a greater preference for emotion-focused coping styles. Similar research from the perspective of individual differences was undertaken by B.Ł. Skowronski & E. Talik, E. (2018), which looked at coping with stress in relation to quality of life for inmates in correctional facilities. Among other things, the authors found that prisoners' ways of coping with stress are related not only to their overall sense of quality of life, but also to all of its dimensions: psychophysical, psychosocial, personal and existential. According to the adopted hypothesis, people with a high level of overall quality of life, more often than people with a low level of quality of life, prefer active coping strategies such as planning, seeking support (instrumental, emotional), positive reevaluation, personal development (physical, character). In another study by M. Mohino, T. Kirchner & M. Forns (2004), they related coping strategies to attributes of the prison environment such as time spent in prison, prior convictions (first-timers vs. recidivists), custodial status (remand vs. convicts), and cognitive level. Based on a questionnaire study of 107 prisoners aged between 18 and 25 years old, incarcerated at the Centre Penitenciari de Joves de Barcelona (Spain), the dominant coping strategies for prison stress were found to be cognitive strategies, modified mainly by the time spent in prison and partly by previous experiences (in recidivists). Of interest is the negative correlation of coping with intellectual level and lack of choice of emotion-oriented strategies. More detailed problems of human adaptation to prison conditions and ways of coping with the stress of social isolation are the subject of monographs by, among others, E. Zamble & F.J. Porporino (1988) and M. Ciosek (1993).

2.2.3. Incidental isolation

The incidental entrapment is random and concerns an unexpected situation of physical confinement in a limited space of a cave or a mine shaft (Angiboust and Saumande, 1965). Since 2008, ESA has been exploring the use of scientific expeditions into caves as a novel platform for astronaut training, using the natural analogies of the cave environment and associated technical operations with space missions. L. Bessone, J. De Waele & F. Sauro (2018) describe a training programme carried out between 2011 and 2016, using caves as an analogue of the space habitat under the codename CAVES (Cooperative Adventure for Valuing and Exercising human behaviour and performance Skills), in which 28 astronauts from partner space agencies participated: ESA, NASA, JAXA, ROSCOSMOS, CSA, CNSA. The programme has been recognised by all participating astronauts as a very realistic equivalent to space flight, providing unique multicultural training opportunities for operational teams, in one of the best analogue environments available on Earth, since in contrast to other analogues, communication inside the cave is difficult, and forces the development of very autonomous mission operations without rapid support from the MC, characteristic of interplanetary missions with delayed communication, requiring the crew to be autonomous in making quick decisions.

A spectacular example (publicised by all the world's television stations) was the entrapment of twelve football club players between the ages of 11 and 16 in the cave of Tham Luang Nang Non, located in northern Thailand. The narrations of the witnesses of this event show a high level of anxiety and apathy associated with helplessness, and the information coming from outside that a group of professional rescuers encountered serious obstacles to reach them due to the flooding of the escape route due to intense rainfall over several days. An interesting fact was a reduction of the symptoms of apathy and an increase in the motivation to survive this stressful situation, through a mobilisation attitude, of the trainer, who was slightly older than the group's average, and who did not allow panic, maintaining hope for an immediate release and reinforcing the will to fight for survival (Terelak, 2018, p. 13)

Mining catastrophes, in many cases, are comparable to the entrapment of speleologists in caves. In 2010, the media around the world reported for more than two months on the reactions of 33 Chilean miners who were trapped 688 metres below the surface. These narratives made people aware not only of the dramatic attempts to save miners' lives over the course of several dozen days but also of the psychological effects of long-term isolation experienced by healthy people and what makes it easier and more difficult to survive in such conditions. Miners' accounts show that in this particular traumatic situation of isolation they felt fear, uncertainty of tomorrow, boredom and deprivation of many basic physiological, psychological and social needs. Along with the duration of underground isolation, these symptoms intensified significantly and individual differences in dealing with this type of situation appeared, which manifested themselves primarily in the decreasing control of one's emotions (anxiety, irritability, aggressive behaviour, etc.), deterioration of interpersonal relations (e.g. blaming others for this state of affairs, appearance of selfish behaviours, defeatism, increase in conflicts, etc.) (Terelak, 2010, pp. 72–76). This narrative is confirmed by other researchers, suggesting that the unexpected cut-off of access to the outside world, apart from many physical threats, is generally burdensome in terms of social isolation. Although the reaction syndrome of a person trapped individually and in a group varies, many psychological studies, not only of miners trapped for a certain period of time, show that isolation from the outside world alone can cause adverse changes in behaviour. Behavioural disorders, ranging from claustrophobia to panic and group breakdown, can subjectively worsen an already objectively highly stressful situation (Beach, Lucas, 1960). This also applies to other types of catastrophes, such as passenger plane crashes in which passengers survived but were trapped, for example, in some area, which was difficult to reach (Li, 2007; Livingston, Livingston, 2007).

An additional factor increasing the isolation effects is the prevailing darkness, which also leads to disorders of circadian rhythms, important physiological and psychological functions. For example, in some miners, e.g., in a dozen or so days' imprisonment underground, the so-called drifting of daily biorhythms occurs in the direction of extending the physiological day in relation to the astronomical day (fluctuations between 25–26 hours) or shortening it to 22–23 hours (Kerkhof, 1987).

A specific variant of incidental entrapment are *alpine expeditions*, which are characterised by quite specific attributes of isolation. On the one hand, in the high parts of the mountains, people are completely isolated from the "rest of the world" in a spatial sense and left to themselves. On the other hand, the attributes of isolation are associated with the clash of camping, sometimes in a very limited space (e.g., rock shelves, etc.), or with hypoxic adaptation or a breakdown in the weather (Lieberman et al. 2005). As isolation progresses, symptoms of irritability appear, as well as negative character traits manifesting themselves in disobedience to group interests, conflicting, and sometimes irrational, risky decisions, etc., appear. (Kortko, Pietraszewski, 2016). The high mountain expedition in the Antarctic region is described in terms of psychological stress by A.J. Tylor & A.G. Frazier (1981). It was connected with the expedition for the corpses of passengers who died in November 1976 in the crash of a cruise plane from New Zealand, which in bad weather conditions crashed at the top of the Erebus volcano (3794 m) on the island of Ross in Antarctica (237 passengers and 20 crew members died). According to the authors, the stress described was the result of interaction between the territorial isolation and the extremely difficult Antarctic environmental conditions in this area of the world (low temperature and very high wind power), the difficulty of the task itself

There is a lot of empirical data on the narrative of adults who have been in a long period of forced isolation. These data come from two main sources: from studies and psychiatric observations of patients, of the so-called chroniclers or patients who have to stay in bed for some time for postoperative reasons and from observations of patients in quarantine for epidemic reasons. Such people experience mood disorders, reduced mental capacity, perception disorders and "live imagination," which often takes the form of hallucinations or other sensory illusions. As early as 1819 Dupuytren (quoted after Noyes, Kolb, 1969) described delirious type reactions with hallucinations and emotional disorders during the period of eye covering after cataract surgery. These symptoms were called "cataractic delirium." The observations of ophthalmologic patients confirmed the effects of sensory deprivation in hospital conditions. For example, E. Ziskind et al., (1960) observed 88 patients after cataract surgery and found that when the patients were isolated (eyes covered, bedside position, lack of contact with medical staff and other patients), all of them had typical depressive effects.[5] After changing the post-operative procedure and enabling communication with staff and family, these effects were significantly reduced.

---

5   Now that lasers have been used in ophthalmic surgery, this problem is practically non-existent.

Particularly fast and intensive disorders appear in patients immobilised in "iron lungs" i.e. ventilators. For example, we will quote the symptoms that J. Mendelson and J. Foley (1956) described in 9 patients immobilised in ventilators (the so-called iron lungs). They found, among other things, that after 24–48 hours primarily mental disorders with characteristic visual and auditory hallucinations[6] as well as intrusive thoughts appear. In their opinion, it was a typical reaction to the situation of sensory deprivation, which they characterised as follows: small area of the field of vision (a fragment of the ceiling), no possibility to change the position of the body, hearing stimuli reduced to a minimum (apart from the engine noise to which the patient is relatively quickly accustomed). The authors observed a number of interesting changes in the functioning of patients, such as: visual and auditory hallucinations occurring spontaneously at the beginning of deprivation; after 2–7 days of using the ventilator, hallucinations appear systematically; after 10–15 days symptoms of derealisation and depersonalisation appear (more often at night). J. Swartz (1960) significantly lowered the level of the above-mentioned disorders when he introduced on-call duty of medical personnel or someone from his family at the ventilator to maintain constant verbal communication with the patient (via microphone). Case study literature in this field until the 1989s is very extensive (Leiderman et al., 1958; Myklebust, 1969).[7]

This problem is presented in a slightly different manner in psychiatric hospitals, because the institutional isolation of psychiatric patients, which has an additional pathogenic effect (in addition to the so-called basic symptoms of axial mental illness) on psychiatric patients, has been known since ancient times and described by historians of medicine (Brzeziński, ed., 2014) and psychiatry (Shorter, 2005).

---

6   Most controversy in clinical observations and experimental studies concerns the existence of hallucinatory reactions (cf. Wheaton, 1959.) These are the most frequently observed hypnagogic hallucinations, which are a particular kind of hallucinations that occur during sleep when the body is calmed down and silenced before sleep. D. Hebb (1965) observed this type of hypnagogic hallucination in a situation of experimental sensory deprivation. Among other things, he stated that these hallucinations appear earlier in a situation of maximum separation from external stimuli, characteristic of immersion studies in comparison to perceptual deprivation, in which the structure of stimuli is limited.
7   The development of technical civilisation at the turn of the twentieth and twenty-first century, giving access to electronic communications and satellite television, practically solved this problem on a macro-social scale.

A brief historical outline of the situation of the mentally ill from the beginning to the present day is marked by isolation and imprisonment. For example, in the Middle Ages, "madmen" were not allowed to function in the society, and what is more – aggressive patients were locked in towers, prisons or cages located outside the city gates. In the fourteenth and fifteenth centuries in Western Europe they were thrown on the so-called "madmen's ships," which sailed nowhere, nobody wanted to let them into the harbours.[8] In the 18th century, special houses were established in Europe where socially excluded people were placed. For example, the General Hospital was opened in Paris in 1656, which was a chain of establishments more like prisons and work camps, and the "treatment" of patients consisted in their complete isolation and inhuman treatment. At the end of the 18th century and in the 19th century, the physical compulsion and immobilisation of patients (with shackles and chains) was completely abandoned and regular staff meetings with patients were initiated. In the 20th century, the humanist movement in psychiatry became the beginning of a prophylactic approach to the effects of long-term social isolation, as well as the beginning of systematic research. It was found that some of the symptoms, which were originally classified as the so-called axial symptoms, are rather a reaction to long-term hospital isolation.[9] So we will only draw attention to the study of K. Jankowski (1970), which

---

8   A brutal analogy to contemporary emigrants comes to mind, whose civilised communities in the 21st century are also isolated on "ghost ships," in transition camps or closed city quarters.

9   I have my own reminiscences on this subject when, as a young psychologist, right after my studies, I took up a job as a clinical psychologist at the State Neurosurgical Sanatorium in Komorów near Warsaw and participated in the Polish-American Research Program VRA-Pol-65 on the rehabilitation of chronic schizophrenics between 1966 and 1968, I had the opportunity to visit most of the psychiatric hospitals in Poland at that time. I found that the wards where schizophrenics were staying were isolated from the others, and the families did not usually visit them for years. On the basis of longitudinal psychometric tests (e.g. psychiatric symptom scales, MMPI, Rorschach, TAT, etc.) and the impact of occupational therapy (sports, bicycle trips, trips by suburban train to Warsaw for shopping, to the theatre or cinema, etc.), it was found that the reduction of isolation itself by moving the patient from a closed hospital ward to an open ward (no bars in the windows, normal doors with handles, open gate of the sanatorium fence, freedom of movement in the area, etc.).) and a sense of privacy (resumption of basic hygiene and cleaning habits, lack of pyjamas and care for "civil" clothing, possibility of shopping with pocket money, etc.) and self-esteem (visiting families in their place of residence), as well as a sense of social usefulness (participation in social work, participation in cultural life, attempts to return to normal work, etc.), have eliminated the symptoms characteristic of social isolation and imprisonment. Positive changes in

is very interesting in Polish psychiatric literature, concerning the psychological changes in chronic schizophrenics, in a situation of isolation, in the absence of normal physical activity. The author compared physiological indicators of 214 chronically ill patients with a control group of 100 healthy people. He found that a long-term stay in a psychiatric hospital results in a decrease in oxygen uptake, prolonged psychomotor reaction, increased skin electrical resistance (GSR), decreased thyroid and adrenal gland function. He did not find such symptoms in the control group of healthy people. These data clearly show that the influence of long-term deprivation (isolation) is not only limited to the deterioration of mental processes, but also includes physiological functions both at the level of the central nervous system as well as vegetative and hormonal functions. The description of these studies shows, among other things, that the knowledge of the effects and pathogenesis of the effects of long-term hospitalisation is an important social problem. Taking only the results of research on the effects of deprivation (isolation) as a starting point, it is worth reassessing the functions and tasks of psychiatric hospitals from the point of view of the need to stimulate the patient rather than the formal organisation of the place of isolation.[10]

The coronavirus pandemic in 2003 (SARS CoV-2) and 2020 (COVID-19) provided many new narratives on the isolation of both patients and medical staff under long-term quarantine. Analysis of medical and psychological literature (MEDLINE, PsycINFO, CINAHL and Cochrane Library), undertaken by, among others, W. Xing, et. al., (2010) and P. J. Gardner & P. Moallef (2015), drew attention to the following symptoms occurring in survivors of SARS and the medical staff caring for them during their illness: psychotic (Sheng et al., 2005), fear of survival (Mok, et al., 2005), fear of infecting others (Cheng, et al., 2004), a sense of stigmatisation (Lam, et al., 2009), reduced quality of life, lasting up to several years (Ngai, et al., 2010), emotional stress (Kwek, et al., 2006). Undoubtedly, the period of isolation in a life threatening situation is associated with numerous psychological and social stress problems related to the attempt to adapt to the conditions of hospital and post-hospital isolation. However, researchers

---

emotional and social behaviour were found in schizophrenics who discontinued taking medication during the experiment. Meanwhile, in a classical psychiatric textbook it was taught that "blunt feelings are an axial symptom of schizophrenia."

10 Criticism of the biological approach in psychiatry and the positive aspects of so-called humanistic psychiatry was once undertaken by K. Jankowski (1975), participant in the 1966–1968 Polish-American Programme for the Rehabilitation of Chronic Schizophrenics VRA-Pol-65.

such as, among others: S.B. Patten & J.V.A. Williams, 2007 and F.W. Lung et al., 2009, drew attention to the important fact that some of the symptoms of social isolation also affect the control group, which were healthy people in isolation outside the hospital. This problem has largely become an important challenge in the context of the COVID-19 coronavirus pandemic in 2020 (Subramanian, et al., 2020) and awaits new research, especially cross-cultural in nature.

2.2.4. Task isolation

Deep-sea fishing vessels are a good example of the stress of task isolation, as evidenced, inter alia, by the annual report of the Japanese Maritime Safety Agency (the so-called White Paper on Maritime Safety), which documents a significant number of murders or violence on board fishing vessels on long voyages of several months, which means that in this case the effects of long-term isolation have serious psychological and moral consequences. The literature on crew behaviour on fishing and merchant vessels is primarily of a narrative nature, describing a picture of "deep-sea fishermen's syndrome," characterised by such symptoms as: difficulties in focusing attention, aversion to mental effort, motor anxiety, irritation, psychomotor disorders, feeling of boredom, etc. These changes have a characteristic cycle. For example, the highest rates of anxiety and neuroticism are observed in seafarers between 80–120 days of sailing. This is confirmed by analyses of the logs kept by submarine commanders, which show, inter alia, that offences, neglect of duty, refusal to carry out orders, brawls and fights usually take place between 60–90 days of voyage, which has been called "the 3/4 of the way syndrome."[11]

Interesting observations were made by M. Ciosek (1977) on the change in the perception of the group and the situation under the influence of the duration of isolation associated with deep-sea voyage, on the basis of which he distinguished the following types of assessments in individual phases of the voyage: (a) initial phase of the voyage – the perception of both the group and the situation is definitely positive, which the author associates with the "novelty" of the situation and the attractiveness of the crew; (b) middle phase – is perceived in negative terms, as it is characterised by monotony; (c) final phase – is again positive,

---

11 Nowadays, thanks to psychological examinations, "health and safety at work" recommendations have been introduced even on deep sea voyages like submarines which accept 90 days of isolation as the limit of "comfort," obliging or requiring a crew change or submarine emergence for "recreational" purposes.

which is undoubtedly associated with the anticipation of once again taking on certain family, social, sexual roles, etc.

Another working in isolation in task isolation are workers from offshore oil rigs extracting gas or oil from the seabed, especially in the Arctic regions (Korneeva, Simonova, 2018).

An interesting example of task-based isolation is the isolation of polar stations documented for several decades. While research reports on the effects of this type of social isolation have been systematised (cf. e.g. Gunderson, 1973; Wood et al., 2005), one should not forget about the pioneering polar expeditions as sources of pre-scientific information on human adaptation to extreme conditions and health and psychological consequences. Let us recall that the first group of people wintering-over in Antarctica was the Belgian scientific expedition of "De Garlache" on the ship "Belgica." The ship's doctor, Frederick A. Cook (1909), made observations from the 13-month wintering-over on a ship trapped in ice. These observations show, among other things, that people suffered mainly from scurvy and from physical exhaustion. According to the doctor, however, the most burdensome were the signs of homesickness (nostalgia), boredom and mental disorders, which were described with the medical term of "polar anaemia" (Forster, 1959). This is one of the first data on the so-called winter-over syndrome, which in the following years became the subject of many detailed medical and psychological studies. And so, F. Jack & E. Jack (1988) remind the diaries of the Australian polar explorer, Sir Douglas Mawson – the greatest explorer of Antarctica.[12] Mawson's collected memoirs show a personal perspective on dealing with the stress of isolation and conflicts associated with the journey to terra incognita, achievements and failures, joys and tragedies. Interesting is the convergence of his observations with the results of research carried out in the Antarctic regions during the 20th century.

An interesting example of a narrative in the form of memories are the reflections of Leszek Moczydłowski – the head of the 4th Antarctic Expedition of the Polish Academy of Sciences to the Henryk Arctowski station – posted on the social media of the "Antarctic Brotherhood."[13]

---

12 Sir Douglas Mawson made four trips to Antarctica during his long and rich career. He travelled southwards as early as 1907 with Shackleton's British Antarctic expedition; in 1911 as head of the Australian expedition to Antarctica; and twice between 1929 and 1931 as leader of the British, Australian and New Zealand's research expedition to Antarctica.

13 I would like to thank Leszek Moczydłowski for consenting to the use of fragments of this narrative in this book, as they confirm the theses of the research that are verified

"Forty years have passed, like one day, since the end of the Fourth Expedition of the Polish Academy of Sciences to the Henryk Arctowski 1979–1981. ... I was 30 years old then and completely unreasonably accepted the proposal to lead the winter workers of the 4th Expedition. ... The main message of this memory is related to managing the team in such a mini-expedition to Mars. ... The measure of interpersonal relations of the 4th Expedition is the fact that for several years after the expedition we met almost every year to remember how Our resolved conflicts, cooperation and the joy of being in this team were an experience. Just don't think it wasn't rough. It was and sometimes very dangerous. ... Here, in a few sentences, I will describe the reasons for our common success in the conditions of Antarctic wintering. First of all, you need to be aware that in a separated group everyone knows after three months what their interlocutor will say in a moment and how annoying it is. ... Secondly – in unavoidable crisis situations, in an isolated community, no matter what the reasons are, it is forbidden to put, even those obviously guilty, against the wall. Regardless of the extent of the blame, it is imperative to give a man a chance to come out with face. ... Thirdly – the manager must be especially careful to use the manager's authority only in absolutely necessary situations. If the team is doing its job, there is no need, and it is even harmful when the manager tries to show others "who's in charge." People who do their job well feel great when the manager trusts them and will do much more than is expected of them. /... / It works like a Swiss watch based on the most natural psychic properties of human nature. Fourth – in a real crisis, it is the manager who has to cut the Gordian knot or ulcer and apply lotion. If he succeeds, they will love and respect him mainly because they are aware of what a difficult and delicate task it is. Simple? Quite trivial, but not in practice."

Of course, we will not analyse the diary materials that usually appeared after successive Antarctic expeditions in detail, but we will only draw attention to the important fact that the first data on human functioning in the situation of Antarctic isolation come from doctors and other participants of the expedition. Their scientific value is discussed in detail by R. Silverberg (1965), who claims, among other things, that data from observations in natural conditions, concerning general health problems and adaptation to extreme situations, are burdened with subjectivity, anecdotal and heroic threads (cf. also: Logan, ed., 2017).

---

in it. I saw Leszek for the last time at the Arctowski Station when, as a flag officer, I handed him the Polish banner, as a symbol of the end of the Third Expedition and the taking up of jurisdiction by the winter workers' manager from the 4th Expedition. We were thirty then.

The group of Antarctic authors using the narrative method during the annual wintering-over in Antarctica can include e.g. such polar explorers as: J.C. Behrendt (1988); L.A. Palinkas, M. Houseal & N.E. Rosenthal (1996); J.S.P. Mocellin & P. Suedfeld (1991); S. Rakusa-Suszczewski, 1973, 2012; T.Z. Qu, (2013); B. Ehmann, A. Altbäcker & L. Balázs (2018); R.J. Wróblewski (2017); Jean McNeil (2017). J.F. Terelak (2021), and others.

The group using the narrative method also includes Christian Ritter, who was one of the few women to describe her experiences in social isolation and sensory deprivation during a polar night on Spitsbergen (e.g. "dialogues" with the Moon, etc. are interestingly described), and Martha Martin, who spent the winter alone in Alaska and described visual and auditory hallucinations, characteristic of "kayak psychoses" (Kayak-Angst, Kayak-Phobia, Kayak-Dizziness), experienced by Inuit people who lost their spatial orientation on water – this is described, among others, by Z. Gussow (1963). They are similar to the so-called desert hallucinations, described, among others, by J. Baalsrud (cf. Svab & Gross, 1965), who stayed alone on a raft in the ocean for 27 days, or by A. Saint-Exupery, who in his autobiographical novel describes the desert experience of a pilot crash survivor, drawing attention to a hypnagogic hallucination. A. Bombard's diary (1953) provides a lot of interesting data. Among other things, he writes that the cause of death of lone survivors is surprisingly rarely hunger or thirst; what "kills" them is rather loneliness, inability to make contact with other people, lack of social environment.

In turn, John C. Behrendt (1999), as an assistant seismologist participating in the glaciology programme of the International Geophysical Year for eighteen months in Antarctica, at the emerging scientific Ellsworth Station on the Filchner Ice Shelf, kept a handwritten journal containing an overview of daily entries describing life and activities at the most isolated of the seven US Antarctic stations. The descriptions show that 9 civilians and 30 sailors lived in extreme conditions under the snow, and during the polar night in the winter of 1957 intense personal conflicts arose. In addition, the station was completely isolated from the outside world with no mail delivery and only occasional radio contact with families at home. The author describes the emotional stress associated with this difficult life situation. In turn, J.S.P. Mocellin & P. Suedfeld, P. (1991) presented a contextual analysis of the original diaries of 13 members of British Arctic and Antarctic expeditions from the heroic era of exploration of the polar regions at the turn of the 20th century, taking into account such areas as physical and social environments and their impact on positive and negative evaluations of experience and steadfastness of achievements. The authors conclude that the experiences described are burdened with variable social approval, affecting the over-generalisation and dramatisation of the narratives, which does not mean that they are not cognitively valuable.

Another aspect of the narrative, based on observations of the behaviour of 83 members of high altitude expeditions, was highlighted by N. Smith, et al. (2016). The authors highlight, above all things, the positive effects of stress, manifested by, among other things, an increase in well-being and resilience to life's hardships after the expedition, as well as agreeableness and openness in interpersonal relationships. This is consistent with Hans Sely's concept of eustress (Terelak, 2019). This is also confirmed by N. Smith, E.C., Barrett & G.M. Sandal (2018), who – using a self-report narrative method before and after the expedition of four participants crossing the Empty Quarter desert – found, among other things, that desert isolation in the narrative is perceived both in terms of distress (especially at the beginning) and also eustress (after the expedition).

P. Suedfeld (2017) undertook to assess the value of the scientific narrative method using the example of an analysis of the texts of the "Antarctic Diary" by Mark Wood, a British polar explorer who reached the geographic South Pole on skis alone and unsupported as part of the 2011/12 expedition. The analysis included cognitive, motivational and personality processes, as well as coping strategies for isolation and threat stress. The authors noted the importance of context in understanding a stressful situation, however, the study of one individual precludes any generalisations, if only because of the personality differences specific to solo adventurers as opposed to group participants. In his diary the Polish extreme polar explorer Marek Kaminski (1998), who was the first person in the world to reach both poles of the Earth on skis in the same year without any outside help (on 23 May 1995 the North Pole and on 27 December 1995 the South Pole), describes both aspects on stress and positive experiences. This multifaceted approach to narrative as a method of psychological research is emphasised by prof. Adam Biela, who sees the complementary utilitarian nature of narrative for scientific research when in the preface to my Antarctic Diary (Terelak, 2021, p. 11) he states: "It is precisely the narratives from such situations that, from the point of view of psychology, are an invaluable source of knowledge about complementary psychological laws in situations to which an experimental psychologist, having control over variables in laboratory or even quasi-natural conditions, has no access." Similar views are exposed by B. Ehmann, A. Altbäcker & L. Balázs (2018)

B. Ehmann, A. Altbäcker & L. Balázs (2018), in the course of analysing the narrative method used by polar explorers in their Antarctic diaries, points out their usefulness in documenting the dynamics of mental states accompanying adaptation to long-term social isolation. The authors refer to such outstanding authorities of Antarctic research as L.A. Palinkas & P. Suedfeld (2008), who stress that psychological adaptation to isolated, confined and extreme environments (ICE) is a multi-faceted field of research, covering four important aspects of social isolation: seasonal (summer vs. winter), situational (multiple risks), social (group dynamics) and salutogenetic (resistance to harmful agents).[14] Especially in long-term isolation, cross-sectional studies are of little use to assess emotional stability, including: mood, morale, anxiety, depression, aggression, hostility, life satisfaction, etc. (cf. Suedfeld et al., 2017).

Quantitative longitudinal studies and qualitative studies based on the narrative method are more useful in assessing the variability of these mental states in a situational context. Diaries can also be helpful for ex post interpretation of psychometric tests. This is indicated, among other things, by the studies of L.A. Palinkas, M. Houseal, N.E. Rosenthal (1996), noting the seasonal occurrence of the winter-over syndrome, characterised by overlapping: winter-over syndrome, fleece syndrome T3[15] and seasonal affective disorder (SAD). It was found that even clinically healthy people can experience SAD symptoms at polar stations located at high latitudes.

B. Ehmann, A., Altbäcker & L. Balázs, L. (2011) describe attempts to objectivise the narrative by using quantitative analysis of narrative content. They tested this method by monitoring the mental states of the 71-member Mars Desert Research Station (MDRS) space simulator crew. The analysis software was used to evaluate reactions to long-term social isolation at the level of: emotional states, team morale and subjective discomfort. Average results indicated characteristic patterns in individual subjects as well as in the whole group, as well as individual differences (Sandal, Leon & Palinkas (2006). The Hungarian researchers, while agreeing with the conclusion that there is still much to be done to integrate psychosocial and neurobiological adaptation rates, draw attention to their findings, indicating that not all patterns of adaptation to long-term Antarctic isolation that were observed in the Concordia and Halley VI experiments were confirmed. This is the case, for example, with Concordia and the complete suppression of emotions during isolation, which in the crew's diaries, gradually increased as the isolation increased. In addition, no significant changes were observed in any

---

14 P. Suedfeld, P. (2005) treats these four phases of long-term Antarctic isolation as the foundation of its cosmic analogue.
15 Polar T-3 syndrome, caused by the reduction of levels of thyroid T-3 hormone, characterised by, among others, such symptoms as: forgetfulness, cognitive disorders and mood disorders, caused, among others, by changes in the photoecological conditions.

category of emotions in the Halley VI experiment, although evidence was found for the existence of the "third quarter phenomenon,"[16] but it could be identified in anger emotions as other negative emotions showed different patterns. The creators of the content analysis believe that its novelty lies, among other things, in the fact that it makes it possible to distinguish seven different aspects of emotionality (affectivity, positive emotions, negative emotions, emotional value, anger, anxiety and calmness) in verbal descriptions. Differences in emotional patterns between different Antarctic stations (Concordia vs. Halley VI) can be attributed to many factors such as e.g. geographical location of the station, climatic conditions, cultural stereotype of expression or inhibition of emotions (Steel, 2001). This is largely confirmed by the studies by L.A. Palinkas et al. (2004) carried out over a period of 8 months (March to October) on 13 crews of winter-over participants from 5 countries serving research stations in Antarctica: United States (3 crews, n = 77), Poland (3 crews, n = 40), Russia (3 crews, n = 34), China (3 crews, n = 40) and India (1 crew, n = 26). The results of the research indicate inter-individual differences between the studied groups. For example, the Americans at South Pole station showed a significant increase in fatigue and anxiety and a significant decrease in strength during the winter, while Russia at Vostok station showed a significant decrease in depression, anxiety and entanglement, and the Indians at Maitri station showed a significant decrease in anger. In turn, the main decrease in social interactions with other crew members took place at South Pole, Vostok and Polish station Arctowski. Different patterns were also observed between the five stations in terms of mood and cultural orientation of the respondents (individualism vs. collectivism). Individualistic cultural orientation was correlated with low negative moods and low social support. This is also confirmed by experimental research, both laboratory and natural.

---

16 G.D. Steel (2001), in two studies at a remote base in Antarctica, confirmed that staff in isolated Antarctic environments are experiencing a decline in mood soon after the middle of the wintering-over period. This decline, called the "third quarter phenomenon," has been investigated in two studies in a remote base in Antarctica. In Study 1 and 2, 12 retrospective mood measurements were carried out, based on the Russell affect model. The results of both studies indicate moderate empirical support for the existence of the third quarter phenomenon, although some mood dimensions may be more or less prone to time effects.

2.3. Experimental aspect of data sources

The experimental approach in psychology, called the "royal way," concerning not only the description of human behaviour, but above all the explanation of its psychological mechanisms, includes two types of methods: laboratory experiments (tested variables with the intervention of the experimenter) and experiments conducted in natural conditions (without the experimenter's interference).

2.3.1. Laboratory experiments

Laboratory research on the effects of sensory deprivation is mainly connected with the names of Canadian psychologists from McGill University: D. Hebb, W. Heron and W. Bexton, who in 1953 published a report in the *American Psychologist* journal on mental dysfunction in situations of limited stimulus flow entitled "Cognitive Effects of a Decreased Variation in the Sensory Environment." In this paper, they reported two years of experimental research and verifications of hypotheses about aversiveness of both excess and lack of stimulation, and the body's desire to maintain an optimal level of stimulation. The deprivation experiment by Bexton, Heron and Scott (1954), considered a classic in the literature of the subject, consisted, among other things, in paying students $20 a day just for lying idly in comfortable beds. Their eyes were covered with a band which, although allowed light to pass through, made it impossible to see shapes. Their forearms were immobilised in special cylinders that prevented sensory impressions. Only the so-called white noise could be heard in their headphones. The conditions of the experiment provided only for meal breaks and going to the toilet. Despite this apparent comfort, only a few students have endured this situation for more than 2–3 days, with 6 days assumed and planned. The subjects reported that they had various sensory and visual hallucinations (scenes similar to cartoon flashes). In some people, the sense of identity was disintegrated and derealised. Some of these symptoms persisted for some time after the experiment.

Since then, hundreds of experiments have been carried out, some of which have confirmed the experiences described earlier, others have not. A number of legendary threads have accumulated around this problem, especially about the so-called hallucinatory-like effects. In general, it can be stated that in the laboratory conditions, various types of soundproof rooms, swimming pools, ventilators (so-called iron lungs), space simulators were used to model extreme deprivation situations (isolation). Despite many results, however, it is difficult to generalise them because of the variety of definitions of sensory deprivation and their experimental operationalisation.

Terminological problems result from imprecise definition of basic concepts: sensory deprivation, perceptual deprivation, social isolation or imprisonment. Let us recall that a classical study by a group from McGill University used a descriptive definition to describe a deprivation situation, emphasising its definiendum as a reduction in the range of volatility of sensory stimuli. However, another concept has already been used in subsequent work, namely: "perceptual isolation." Both concepts were given approximately the same empirical meaning. Other authors introduced a whole range of different concepts to define a situation characterised by a greater or lesser limitation of stimulus (information) flow. Chronologically speaking, it should be said that the term "sensory deprivation" was propagated by J. Vernon, and J., Hoffman (1956), among others. It concerned three categories of situations. The first category includes *the reduction of stimulus structuring*. A typical example of this category of situations of sensory deprivation are the first experimental works of a group from McGill University, in which the attributes of sensory deprivation were the earlier-described: comfortable bed or armchair, "white noise," semi-permeable goggles, and special cylinders on hands. The second category of deprivation situations concerned *the reduction of the absolute level of stimulus*. An example of experimental application of this type of situation was the immersion research by J.T. Shurley (1960) conducted at the Veterans Administration Hospital in Oklahoma City, whose attributes were: an eye band, an oxygen mask, a cylinder filled with water, the density of which allowed to keep the man without touching the bottom of the pool, and the temperature always corresponded to the body temperature. The experimenter controlled verbal (tape recorder) and motor activity (TV monitor). The described situation reduces to a minimum not only visual, auditory and tactile stimuli, but also proprioceptive stimuli (feelings of weight and spatial position of the body, etc.). The third category of deprivation situations includes the *imposed structuring of the stimulus*. This situation was introduced by M. Zuckeranan et al., (1962), researchers from *Harvard Medical School in Boston*, who used a respirator, the so-called *iron lungs*, for this purpose. The attributes in this experiment were: microphone, tape recorder, ventilator. After a cataract removal surgery, the patient was normally lying still in the ventilator, separated from the outside world by a special cabin. Since their head was immobile, their field of vision was limited to a small section of the white ceiling. In addition to the subjective experience, a number of physiological indicators were recorded.

76   Data Sources Concerning The Effects of Sensory Deprivation

The above classification of deprivation situations, based on the criterion of *reduction of stimulus structuring*, was discussed by C.A. Brownfield (1964), who thought, for example, that the first and third categories were identical and was rather inclined to accept the proposal for the classification of P.E. Kubzansky (1961), which included a division into: *sensory deprivation* and *perceptual isolation*. The first situation is clearly characterised by a reduction (limitation) of stimulus strength, and the second by a reduction (limitation) of modality[17] of stimuli (e.g. monotony, boredom, lack of novelty, invariability of the situation, etc.). For the reasons stated above, the concept of *isolation* is not appropriate in this context, as it is additionally burdened with traditional social implications (social isolation).

A different classification approach can be found in F.D. Fiske (1967), who introduced a new category of sensory deprivation, namely: *stimulative monotony*, whose empirical indicator is an adequate response to a recurring stimulus. However, this proposal poses new difficulties because it includes the way in which stimuli are given and their content significance in terms of novelty, and thus significantly broadens the concept of sensory deprivation situations instead of particularising it. Although, on the other hand, it indirectly draws attention to the fact that, alongside the formal approach, the content aspect of stimulation is important. In this situation, some researchers, such as J.T. Shurley (1962b), proposed not to use the concept of sensory deprivation at all, replacing it with a descriptive concept of "experimentally controlled range of changeability and constancy of a sensory stimulus." The introduction of ever-new concepts has contributed to the deepening of terminological and methodological difficulties. Let us quote the most common terms, which are in further or closer relation to the terms of "sensory deprivation" and "isolation:" (1) sensory deprivation; (2) perceptual deprivation; (3) social deprivation; 4) stimulus deprivation; (5) sensory isolation; (6) perceptual isolation; (7) social isolation; (8) stimulus isolation; (9) sensory limitation; (10) social limitation; (11) sensory reduction; (12) stimulus reduction; (13) environmental stimulus reduction; (14) decrease in sensory variability; (15) restricted stimulation; (16) controlled sensory input; (17) reduced sensory stimulation; (18) reduced sensory input; (19) sensory alternation; (20) Gannzfeld; (21) homogeneous stimulation; (22) solitude; (23) confinement; (24) isolation; (25) invariant input; (26) aloneness; (27) break-off phenomenon.

---

17 The term was introduced by the German physician and psychophysiologist Hermann von Helmholtz to describe the differences between impressions from various sensory organs.

Of course, these are not all possible concepts that various researchers use to describe a situation of reduced stimulation and information from the natural or experimental environment. In addition, from item 22, the terms refer to two broad categories of situations, which are also ambiguous in themselves. This was pointed out by, among others, S.B. Sells (1966), by analysing data on human functioning during imprisonment. According to the author, the very concept of "imprisonment" is also ambiguous. It is definitely different in scope when you talk about prisoners serving their sentences or "imprisonment" of a patient in a psychiatric hospital or in shelters and submarines, or, finally, when you talk about the "imprisonment" of cosmonauts for several hundred days in a spacecraft capsule. One has to agree with this view, as the differences considered from a psychological point of view concern not only the situational context, but also such variables as: intellectual level, level and type of motivation, degree of socialisation, identification with the situation, nature of stress, solidarity with the group (or loneliness), degree of experience. Similar reservations can be made about the concept of "isolation," which is also ambiguous, as we will say in detail later in our deliberations. In view of the ambiguity of the above-mentioned concepts, some researchers are operationalising them according to their research objectives. Still other researchers propose their own temporary operational concepts, such as J. Svab and J. Gross (1964), simply speaking of a "controlled field of stimulation," rejecting the theoretical concept of sensory deprivation in general.

Being aware of the ambiguity of the three basic slogans: deprivation, isolation and imprisonment, which are used in research on human functioning in a situation of limited inflow of stimuluss/information, we want to draw attention to our understanding of the above-mentioned concepts. This understanding is based on the premise that each stimulus plays a dual role: formal (energy) and content (informational, generally activating). Since the energy charge of stimulation (strength, intensity) is one of the most important features of the situation, it can be used for classification purposes. Thus, a continuum whose poles determine, on the one hand, a situation of sensory deprivation (understood as a total reduction of stimuli) and, on the other hand, a situation with an unusually high charge of stimulation, can be assumed. As numerous empirical studies have shown, both situations, being of aversive in nature, are therefore treated as sources of stress. Between these poles of the continuum, we can place the remaining types of stimulation as indicators that bring us closer to mental comfort (norm). Although the basis for the above-mentioned classification is very general, by referring to the Optimum Activation/Stimulation Theory, it can fulfil its role of ordering, especially in the case of laboratory tests, since, apart from it, the situation of total reduction of stimuli or their controlled intensification does not arise at all. As far as the content aspect of stimulation is concerned, it must be borne in mind that, on the one hand, it refers to the appropriate context created by the experimenter (instruction), which is largely controlled, and, on the other hand, to the context of the internal states of the subject, i.e. the psychological one, to which the experimenter does not always have access (cf. disturbances from the so-called variable social approval) (Ellenberger (1971). To sum up, the difficulties of the ambiguity of the terms concerning "barriers" separating us more or less from the outside world, in this work we adopt an understanding of three basic slogans: deprivation, isolation, imprisonment, similar to the views of N.M., Burns and D. Kimura (1963). However, we assume that the concepts of sensory deprivation and perceptual deprivation should only be used in the case of experimental research, where it is possible to guarantee the "measurability" of both the level of stimulation and the psychophysiological or psychological effects of deprivation. If the latter condition is not met, it is better to use the term "isolation" (as a separation from normal, customary situations), defining it through the content context of the situation itself or the mental state of the subject. This applies above all to research in natural experiments (such as in Antarctic isolation or on spacecraft), in which we are most often faced with a mixed situation – "isolation with perceptual deprivation."

Finally, it is worth mentioning that the use of the terms: deprivation, isolation, imprisonment in their typical dictionary sense is common in literature on the subject. These terms are usually assigned some words to describe what they refer to (e.g. need deprivation, sexual deprivation, informational isolation, cultural isolation, imprisonment in a mine, entrapment in caves etc.). We will skip the detailed discussion of laboratory studies on the effects of sensory deprivation, which created an "artificial mini world," carried out en masse until the 1970s and 1980s, as this approach is used in numerous monographs and reviews (cf. e.g. Zubek, 1969, Kubzanski, 1961, Fiske, 1967; Haythorn, 1973) and focus on the general conclusions of these experiments according to the following groups of symptoms: (1) hallucinatory-like sensations, (2) perception disorders, (3) mental capacity, (4) physiological changes.

(1) Hallucinatory-like experiences

Undoubtedly, the most spectacular achievement of the McGill University group's research was the description of various hallucinatory-like sensations previously known as symptoms during mescaline intoxication, but occurring in healthy, normal people who are in a situation of perceptual deprivation for several days (Heron, Doane & Scott, 1956). These sensations were mostly visual, taking forms ranging from simple geometric figures to complex, colourful "pictures." To this day date, the most controversial ones are the hallucinatory-like effects, which have not been confirmed in all experiments, except those conducted in immersion (Shurley, 1962a). The authors of the deprivation experiments only partially confirm the occurrence of the previously described hallucinatory-like symptoms. E.g. J. Vernon, T.R. Megill (1962) from the University of Princeton (New Jersey) describe the occurrence of hallucinations in a classical deprivation experiment in which the subjects were military personnel. In the first phase of perceptual deprivation (72 hours), the experimental situation was interrupted only three times per day (meals and toilet needs). During the second phase of the experiment (96 hours) meals and toilet needs took place in total darkness. In an interview conducted after the experiment, the subjects, although not all of them, described hallucinations, mainly visual, mainly in the second phase of the experiment. A controversial example are the studies by A.J. Silverman et al. (1962) – psychiatrists from Duke University Medical Center of Durham (North Carolina), who describe research on the effects of 2-hour perceptual deprivation conducted on pilots of both sexes. Hallucinatory-like experiences were found on the basis of an interview and a special questionnaire. However, there is reason to believe that these symptoms were provoked by researchers. They used a special reaction key, which when pressed, lighted up the phosphene, causing various visual illusions. Similarly, the noise of the fan provided many "ideas" for the auditory experience. Kinesthetic disorders were caused by the authors by enlarging or reducing the room (movable walls) and by changing the structure of their surface (e.g. smooth, soft, rough). This way, the authors argue that in a situation of perceptual deprivation, one can rather talk of illusions (visual, auditory, tactile). Completely extremely controversial views on the occurrence of hallucinatory symptoms are presented, among others. by A.C. Mundy-Castle (1958), who compares these symptoms directly with typical schizophrenic symptoms. Many researchers, however, did not find hallucinations in their deprivation experiments. E.g. G.E. Ruff, E.Z. Levy, V.H. Thaler (1961), working at the Laboratory of Aeronautical Medicine in Ohio, observed hallucinatory-like effects in only two cases per 60 subjects in various situations of sensory and perceptual deprivation. In turn, F.N. Arnhoff, H.V. Leon & C.A. Brownfield (1962) also found no hallucinations, illusions or other unusual "visual-image" states in their experiments with 48-hour perceptual deprivation. Moreover, J.P. Zubek, G., Welch, & M.G., Saunders (1963) did not observe hallucinations even in such a rare case of a 14-day perception deprivation.

Of the many reasons leading to controversial results in this regard, we will discuss some of the most important ones (in our opinion). Firstly, too little attention has been paid in research to the precise distinction between hallucinations, fantasies, hypnagogic states, etc.[18] There is simply talk of hallucinations in general – and this effect of deprivation, as the most sensational one, was exposed. Therefore, we should talk about hallucinatory-like effects in general rather than hallucinations. Secondly, the impact of important variables such as the suggestion and attitude[19] on results has not been taken into account in many studies. This is evidenced, among other things, by the research carried out by R.R. Short & S. Oskamp (1965), concerning a one-hour deprivation experiment involving two groups of men (12 each). In the experimental group, the content of the instructions suggested the occurrence of hallucinations, while in the control group no hallucinations were mentioned. In the "suggested" group, hallucinatory-like symptoms were present in all subjects, while in the control group in only 42% of participants. It was found on the basis of the alpha rhythm blockade in the EEG record and the recording of eyeball movements (electrooculograph). The authors provide an argument in favour of the previously exposed thesis that hallucinatory-like effects are largely related to the previous suggestion (direct or indirect). On the other hand, however, the above-mentioned data also support the thesis that in a situation of sensory deprivation, a person "switches" out of necessity to the sources of internal stimulation, activating own activity available at a given moment, and about individual differences in terms of vulnerability to deprivation situations, determined by the type of reactivity of the CNS. This is evidenced, among other things, by the results of experiments by E. Ziskind et al. (1960) who found that when people stayed only in the dark or in a chamber of silence for a period of 10 to 30 minutes (apart from that, the conditions of the experiment were normal), those people testified that they had various visual or auditory sensations, which they described as "amazing." These results indicate that hypnagogic reactions can now also be explained without the need to refer to sensory deprivation. However, it is important to explain why

---

18 A full review of the literature up to 1988 was carried out by M. Spitzer (1988), who clearly separated clinical (pathological) hallucinations from hypnagogic ones and other conditions characteristic of sensory deprivation.
19 The public perception of the sensational experience of normal people without the use of psychotropic drugs has extended the range of suggestions to potential participants in deprivation experiments. It is also not insignificant that, in some cases, the subjects were students of psychology who had previously known "hallucinations were to appear."

hypnagogic symptoms occur in some experiments, especially with the use of immersion. A model of such an explanatory mechanism has been presented earlier in Chapter 1.

As with the hallucinatory-like effects discussed above, there are a number of controversial results and views on the impact of deprivation on perceptual and psychomotor performance.

(2) Perceptual illusions

Another, no less spectacular aspect of research into the effects of sensory deprivation, concerned cases of perception disorders, which should be distinguished from hallucinatory-like effects, as they rather involve a change in the structure of sensory stimuli. For example, the subjects, who had been in perception isolation for several days, complained about the clear movement of the field of vision, changes (shape, sharpness and brightness of objects), mirroring of objects and deformations of the human face. These phenomena disappeared in 0.5 hours after completion of the experiment. However, it happened that after six days of isolation they sometimes appeared up to 24 hours after the end of the experiment (Heron, Doane, Scott, 1956). In these studies, the disorders described were found only on the basis of subjective relations. In other experiments in this field, such as S.J. Freedman, H.U. Grunebaum, M. Greenblatt (1961) from the Massachusetts Mental Health Center in Boston – they found, among other things, that an 8-hour period of sensory deprivation (30 normal people were examined) leads to impaired recognition of simple models of figures moving in space. In addition, a number of authors have found disturbances in the perception of time. For example, an Italian psychiatrist, M. Stroillo (1963), in a six-hour deprivation experiment (70 people aged 20–25) stated that the subjective sense of time in relation to real time is reduced by about 30 percent. Similar data are quoted by J.P. Zubek (1964), who stated in different sections of the isolation time (from 1 hour to several months) that the subjective evaluation of time oscillates around 60 percent of real time. This author suggests extreme caution in interpreting the data resulting from the narrative on perceptual illusions, as they only partially confirm the results of the McGill University group, and most studies using psychological tests have not confirmed their occurrence.

(3) Mental fitness

The research of the group from McGill University showed that during perceptual deprivation almost all subjects had: inability to concentrate, lack of clarity of thinking and difficulties in organising their thoughts (Bexton, Heron, Scott, 1954), which was also confirmed by T.H. Scott et al. (1959) in psychometric tests, indicating a significant reduction in mental performance. However, in addition to the general pattern that mental disorders are greater in situations of perceptual deprivation rather than those of sensory deprivation, the literature on the specific impact of these situations on mental performance is, as in previous cases, quite controversial. This is indicated by the results of the research carried out by J.P. Zubek, W. Sanson & A. Prysiazniuk (1960), in which they examined two groups: experimental (week-long perceptual deprivation) and control (without deprivation) with a battery of mental fitness tests, keeping the same time intervals for the tests. The results obtained show that there are no significant differences in the tests: verbal fluency, verbal reasoning, ease of counting, numerical reasoning, abstract reasoning, capturing spatial relationships and learning by heart. Among mental abilities, only direct memory (recollection and recognition) differentiated significantly on the first day after the deprivation experiment between people from the experimental group. These studies were repeated by J.P. Zubek et al. (1962) on a larger group of subjects (42 persons in week-long perceptual deprivations). 12 different mental and 8 perceptual skills were tested (before, during and after perceptual deprivation). The control group included 40 outpatient and 40 inpatient patients. This time the authors found a significant deficit in arithmetic tests, numerical reasoning, verbal fluidity, spatial orientation (with two and three dimensions) and abstract thinking. On the other hand, the tests for the scope of memory, memory learning and the recall of meaningless expressions have not deteriorated. However, as in the experiments of the Hebb's Group, there was no correlation between the degree of mental deficit and the time spent in a situation of perceptual deprivation. Clarification of this problem requires further detailed research. Partly this problem was raised by K. Fuerst & J.P. Zubek (1968), who used a battery of Guilford tests in their experiment, measuring three features of creative thinking: fluidity, flexibility and thoroughness. The study group consisted of 18 students of psychology (men with the average age of 19.2 years). The situation of perceptual deprivation lasted three days. The study did not show statistically significant differences between the experimental and control groups. The authors, citing the opinions contained in the literature, explain the lack of expected differences by the fact that tests on this type of thinking can only differentiate people in a situation of total sensory deprivation and that mental deficits have their dynamics, i.e. they are greatest on the first day of perceptual deprivation, followed by improved results. Meanwhile, in the studies described above, mental fitness was measured on the third day of deprivation.

To sum up the data on the mental effects of sensory (perceptual) deprivation to date, it can be said that the negative effects are not as serious as suggested by classical studies by Canadian psychologists. In addition, variables modifying the level of mental performance such as motivation and individual differences should be taken into account. One should accept the suggestion of C.A. Brownfield (1964), who spoke of two complementary, rather than opposing, hypotheses on the deterioration or improvement of mental performance in situations of sensory (perceptual) deprivation. The review of the literature on the subject does not allow us to draw general conclusions about better or worse intellectual functioning in conditions of short-term perceptual deprivation, but rather to empirically describe the type and extent of this functioning in specific deprivation situations.

(4) Neurocerebral correlates of responses to sensory deprivation

Physiological correlates of functioning in a deprivation situation have received less attention than other effects discussed so far. The reason for this was popular neurophysiological speculation, which tried to explain the mechanisms of emotional functioning in situations of deprivation based on narratives, without reference to empirical data. The earliest data concerns changes in the electroencephalographic recording. For instance, W. Heron (1961) research included a description of the alpha rhythm frequency in deprivation situations. The blockade of the alpha rhythm detected by Heron indicates an increase in cortical activity in a situation of deprivation, which "triggers" various types of fantasies, imaginations, as well as hallucinations, which in this situation play the role of internal stimuli, increasing adaptation to this situation. Interesting data on this subject is provided in particular by studies conducted by J.P. Zubek, G., Welch & M.G. Saunders (1963) on three people in a 14-day deprivation experiment, during which only an EEG record was recorded (this record was recorded before the experiment and on the seventh, 10th, 12th and 14th day of the experiment, and after three hours and on the first, second and seventh day after the end of the experiment). Sensory deprivation was of a partial nature, i.e. goggles (semi-permeable light), earphones on ears ("white" noise) and cylinders on hands (reduced tactile sensations) were used. Communication with the outside world has been kept to a minimum. EEG studies were carried out telemetrically and the behaviour was observed by means of a television system. It was found that the alpha rhythm was decreasing progressively during the subsequent 14 days of the experiment, and that the modified EEG persisted for a week after the end of the experiment. These results, according to the authors, show clearly that prolonged perceptual deprivation disorganises activation at the level of the central nervous system. This is confirmed by repeated studies carried out by Mr P. Zubek (1970), also in an experiment with 14-day perceptual deprivation (10 people participated in the study) with EEG registration and biochemical indicators that physiological changes in activation of the central nervous system (CNS), persist after 10 days from the end of the deprivation experiment.

All possible physiological correlates of the reaction to the deprivation situation have been used by E.A. Serafetinides et al. (1972) using a whole range of different physiological measurements: electroencephalographic (EEG), electromyographic (EMG), electroocographic (EOK), electrocardiographic (ECG), and skin and galvanic reaction (GSR). This way they studied changes at the CNS level as well as in the vegetative system. The study involved 16 men in four-hour perceptual deprivation. The results obtained show, among other things, physiological changes in all the examined indicators. The authors discuss the results obtained based on the changes in reactivity of the autonomic nervous system caused by the lack of external stimulation. The disadvantage of this experiment, however, is the lack of a control situation, although the data support the thesis that the deprivation situation should be regarded as stressful. This shortcoming was supplemented by H. Persky et al. (1966) in their earlier studies, who were interested in the psychoendocrine effects of perceptual deprivation compared to normal stimulation conditions. The authors state, among other things, that situations with limited stimulation can be considered as stressful, especially from the point of view of physiological correlates of emotional response (heart rate, skin galvanic reaction, respiration rate).

In turn, M. Zuckerman et al. (1970), based on the postulates of the theory of "optimal level of stimulation," demonstrated on the basis of psychological (anxiety, depression and hostility and a sense of unreality and positive contemplation) and physiological indicators (respiratory rate, ECG, GSR and metabolic factors: 17-ketosteroids, 17-hydrocorticoids, etc.), that the aversiveness of sensory deprivation states (eight hours of sensory deprivation) and excessive stimulation (eight hours of a stimulus situation), entitles to treat both situations in terms of stress. However, the physiological indicators of the functioning of the autonomic system were significantly higher in a stimulatory situation than during deprivation. In both situations, it can be seen that within 2–4 hours of the experiments, the activation level is adjusted and then the indicators increase in value.

To sum up the considerations to date on physiological correlates of functioning in a situation of sensory/perceptual deprivation, it should be generally stated that deprivation or reduction of stimulation is an argument at the level of physiological indicators that we are dealing with stressful situations. In addition, the rather considerable scattering of results indirectly demonstrates the existence of individual differences in adaptation behaviour to such situations.

2.3.2. Quasi laboratory experiment as an analogue of a space habitat

The first systematic attempts to investigate psychological factors of adaptation to entrapment in a simulated operating environment were made in the 1960s and 1970s, placing volunteers in several enclosed rooms, commissioning them to carry out repetitive studies to assess various aspects of reduced productivity. The tests were then carried out in specially constructed closed chambers,[20] in which small groups of 3–6 people were kept for weeks or months in order to study the biological adaptation and different mental responses of individuals and groups from the perspective of social isolation. No one expected that laboratory tests carried out for example at McDonnell Douglas Aerospace (now Boeing Space Systems) in Huntington Beach in 1996 in California could, on the one hand, boost knowledge of human adaptation mechanisms to long-term sensory deprivation and social isolation and, on the other hand, influence the rapid development of space technology and accelerate the next stage in the exploration of living conditions outside the planet Earth (Wichman, 1992).

Characteristic of this type of previously performed laboratory tests, described inter alia by the D.A. Rockwell et al. (1976), was the exploratory approach, involving the continuous control of many physiological and psychological variables throughout the entire period of isolation. The authors describe the psychological and psychophysiological reactions of nine subjects staying for 105 days in laboratory isolation. The study was conducted on three groups of healthy men (20–24 years). Two of these groups were placed in the experimental rooms and the third was a control group. Both experimental groups were controlled under conditions of complete social isolation. The possibility of indoor communication was left open, but these contacts were strictly controlled. The rooms contained records, tapes with recordings, books, video recorders and stationery. Local time and cycles: light and dark were regulated experimentally, which gave the opportunity to study the effects of interaction between isolation and circadian rhythms. The selection of people for research was multi-stage. Of the large number of candidates, healthy people with a high level of motivation were selected in the first place. The screening included: standardised psychiatric interview, MMPI and

---

20 Cf. for example, laboratories at Johns Hopkins University School of Medicine or simulators at the Marshall Space Fight Center in Huntsville, Alabama; McDonnell Douglas Corporation in Huntington Beach, California; or the Ames Research Center in Moffett Field, California, etc. (Zubek, 1969).

a number of personality questionnaires. The final selection of the three groups was based on the compatibility of preferences. During the experiment, great importance was attached to the assessment of the mood. This was done from time to time by a psychiatrist during a personal interview with a particular group of subjects. The number of complaints was recorded objectively. In addition, every person surveyed has noted the need for contact with the outside world every day. Of the 16 different questionnaires used (e.g. the list of Gough's adjectives, MMPI, mood scale, Taylor and Spielberger scale of anxiety, hostility scale and others) only four were used daily. The tests were solved remotely by means of monitors and microphones. Physiological indicators were recorded every 0.5 hours telemetrically. In terms of psychological parameters, there is a correlation between mood, depression, hostility, anxiety and light-dark cycles. For example, both experimental groups achieved a statistically significant increase in depression after a shift in the light-dark cycle. The peak of the increase in depression always occurred 48 hours after the time shift. The test subjects returned to normal after two to 12 days. Similarly, the level of aggressiveness in both experimental groups increased. However, in the control group (without isolation) the depressive symptoms and aggressiveness described were not found. The level of anxiety, as measured by the Taylor and Spielberger scales, has had a downward trend in all three groups. Unexpectedly, the highest level of anxiety emerged at the end of the experiment. However, despite certain tendencies, the scattering of results was quite large.

Since then, many ground-based laboratory experiments have been carried out using different types of habitats as analogues of real spacecraft or planetary (lunar and/or Martian) habitats.

2.3.2.1. Classical habitats

Classical ground or underwater habitats,[21] which are analogues of spacecraft, were generally intended for two or more participants. For example, D.E. Flinn et al. (1961) used a two-seater simulator to test the effects of six sessions with

---

21 By habitability we mean, in this study, any "habitable environments" in a situation of external threats and facilitating maintaining activities related to maintaining living conditions, both physical, biological, psychological and social, within a strict time frame (Cockell et al. 2016). Knowledge of biological and psychological mechanisms of adaptation to functioning in Earth's habitats allows to construct "artificial satellites" protecting against the dangers of life in space flight conditions or on other planets, named as spaceflight analogues (e.g. Moon, Mars, etc.) (Palinkas et al. 2000).

17-day isolation and one session with 14-day isolation. An example of the use of a hyperbaric chamber as an analogue of a space habitat is the research conducted by M. Tanaka, et al. (1998), on the psychological effects of five days of isolation for a group of men aged 22–26 who had never met before. The study was conducted three times: two days before entering the chamber, during isolation and after. Interpersonal communication was monitored in the chamber. It was found that on the third day of isolation, a three-person faction characterised by strong interrelationships emerged. The disadvantage of this type of study was that the isolation time was too short.

Leaving aside the detailed discussion of the results obtained, we will only draw attention to the general conclusions of these studies. Firstly, there were no auditory illusions or other perceptual aberrations that occurred during the isolation tests in the cabin of the single-person simulator. Secondly, there are no major emotional disturbances that would affect the level of performance of tasks or significantly disrupt social interaction. However, significant individual differences in social behaviour were found, depending, among other things, on the individual personality traits of individual pairs of respondents. For example, a person who is reticent was irritated when talking to a talkative member of the crew, while previously features such as "talkative" were not noticed. It turned out, for example, that during such long and forced contacts with another person, apparently harmless habits and manners lead to irritation and serious conflicts. During the experiments, each of the subjects could be designated by the experimenter to act as a leader. This also gave rise to hidden antagonisms. However, the fact that it was a task situation, a highly motivating one, meant that any "hostile" reactions were controlled to such an extent that they did not have a significant impact on the level of the tasks performed.

The research presented above, typical for two-seater simulators, has drawn attention to the social effects of cosmic isolation. An example of the use of a multi-person habitat was a complex programme, which was led by O.S. Adams & W.D. Chiies (1963) from Lockheed Laboratory (Georgia Division of Lockheed), on the following subject: (a) relationship between the level of performance of tasks and the duration of isolation; (b) type of motivation required for being in a situation of isolation; (c) group dynamics in different types of isolation; (d) psychological characteristics of people who tolerate situations of isolation well; (e) effects of isolation on emotional processes and mental performance; (f) adaptive behaviour to the isolation; (g) readaptive behaviour after long-term isolation; (h) optimal group size in a cosmic isolation situation; (i) cycle optimisation: work/rest in isolation situations; (j) role of experience and training, increasing individual resilience to long-term isolation situations. This programme was interdisciplinary in nature: (psychologists, physiologists and engineers). A five-person simulator created conditions for staying in social isolation for several to several dozen days. During the research, continuous observation was carried out by means of a closed-circuit TV system, as well as physiological and psychological telemetric examinations. These studies show, among other things, that with increasing isolation time, interpersonal behaviour tends to disintegrate. Moreover, some people become more susceptible to suggestions. Finally, in a situation of isolation, the subjective sense of time changes. This has been confirmed by other researchers, such as N.M. Burns, & E.C. Gifford (1961), representing another research centre – Naval Air Material Center in Philadelphia and K.R. Coburn (1967), representing another research centre in Philadelphia (U.S. Naval Engineering Center), conducting research in a 6- and 8-person space simulator. Thus, in a study carried out in an 8-person simulator, in which the crew was in isolation for 34 days, it was found, among other things, that the level of task motivation significantly decreased after 12 days of isolation and increased again after 18 days, to decrease significantly after about 30 days. In addition, it was found that overt anxiety was at its highest at the beginning of the isolation, while irritation in social situations increased progressively with the duration of the isolation.

G. Sandal, R. Vaernes & H. Ursin (1995) in turn described the long-term space flight simulation studies (30 and 60 days) they conducted for the European Space Agency (ESA). This group of scientists used decompression chambers in the naval base as space flight simulators. The analysis of Bales interactions showed, among other things, that negative interpersonal interactions reached their peak in the middle point and shortly before the end of isolation (Harrison, Clearwater, eds., 1991).

To sum up the examples of classic laboratory experiments to date, we have only pointed to the main research trends. A detailed review of the habitat types used in the study (for one to several dozen people), as well as the duration of isolation (from several hours to several dozen days) is the subject of many review studies (cf. Vernon, McGgill, 1962; Fraser, 1968; Vakoch, 2011). However, regardless of the differences in research approaches, it has been found conclusively that situations of sensory deprivation and social isolation are stressful and have a negative impact on certain mental functions (e.g. reduction in perceptual performance in many tests was in the range of 10–30%).

2.3.2.2. Modern Habitats

A.A. Harrison & E.R. Fiedler (2011) pointed out that astronauts living and working in real space flights have to withstand many stressors, for which, in addition to good space technology and medicine, a more dynamic development of psychology is required, taking into account the new conditions associated with space exploration plans, which require multicultural and gender-mixed crews able to cope with long-term interplanetary flights. The authors noted that although psychology played a significant role, especially in the initial phases of the space programme, the participation of American psychologists up to the 1990s was lower compared to technical programmes. A new openness to psychosocial problems appeared when astronauts and cosmonauts started working together on the Russian space station Mir and then on the International Space Station (ISS). A new era of natural habitats as analogues of long-term space missions has begun, taking into account the model of international crews, the role and place of women in the organisational structure of the crew, problems of group dynamics, psychological support, etc.

Before we make a brief review of research using contemporary Martian habits, we will start with their precursors from the late 1970s and early 1980s. A number of interesting laboratory experiments on the effects of long-term social isolation were conducted by Soviet psychologists. For example, N.A. Agajanian et al. (1965) describe the reactions of two people in isolation lasting 60 days. It turned out that irritation and emotional imbalance only appeared at the end of the experiment. Among others, 120-day insulation is described by O.O. Kozerenko & W.I. Miasnikow (1977). The authors emphasise that with the prolongation of the isolation time, the personality problems of individual crew members and the problems of the so-called polarisation of small task groups are becoming more and more visible. They also found that a personality trait such as self-esteem is largely responsible for effective or poor adaptation to the conditions of isolation, for the level of motivation, and for proper interpersonal communication.

Interesting experiments under the common name of "homeostat" are described by M.A. Nowikow, G.W. Izosimow, A.A. Gierasimowicz (1977). They aim to learn about the mechanisms for group decision-making that are risky in a situation of isolation. At the same time, these techniques can also serve as social training for future real-life space flights, as we wrote in detail at the time in another work (cf. Kwarecki & Terelak, 1980).

Admittedly, there are also opinions that thorough research on the effects of long-term isolation is unnecessary, because neuropharmacological mediators will be used in space flights to regulate sleep, anxiety and fatigue, and indirectly to stimulate "from within," they are isolated, and in research on deprivation (isolation) we are still looking for an answer to the question of what individual resistance to such situations depends on. For instance, W.M. Smith & M.N. Jones (1962) see individual differences in terms of personality traits such as autonomy vs. authoritarianism. They assume that the selection for the so-called male professions (astronauts, polar explorers and others) is made as if by itself. Using a specially constructed "Pensacola Z Scale" questionnaire, they examined such groups as: astronauts from the final selection stage of the "Mercury" programme (26 people), Antarctic researchers (57 people), cadets of the Maritime Aviation School (766 people) and prisoners from Portsmouth (407 people). The results obtained indicate differences between the individual groups, with the most pronounced autonomous personality characterised by two professional groups: astronauts and a scientific group of Antarctic researchers. The other two groups are closer to an authoritarian personality. It seems unreasonable for the authors to be optimistic that this scale can be a useful tool for forecasting adaptation to conditions of cosmic isolation. For example, the differences obtained may be determined not only by personality, but also, for example, by the level of intelligence, because the average IQ measured by the Wechsler test in cadets was 110, while in astronauts it was 133. The results may also be influenced by age and, consequently, social experience (e.g. groups of astronauts and polar explorers were much older than the other two). Finally, the level of motivation, which was the highest in astronauts, was not without significance here. Therefore, the so-called "social approval variable" could have an impact on the results of the questionnaire. This is confirmed by my own research, concerning Polish candidates for cosmonauts, whose high motivation influenced the falsification of their results in the direction of dissimulation (e.g. the high F-K Gough index in MMPI), i.e. in showing themselves in a better light (cf. Terelak & Błoszczyński, 1978).

In the 1970s, under the direction of V.I. Myasnikov & I.S. Zamaletdinov (1996), methods for the analysis of interactions of cosmonauts with ground control and between spacecraft crew members were developed to assess their mental state. The methodology of the analysis was based on quantitative and qualitative evaluation, which allowed the use of data concerning actual activity (regular radio talks of astronauts among themselves and with the Mission Control Centre). A number of laboratory experiments have been conducted using these methods. And so, N. Kanas and D.S. Weiss (1996) verified the hypothesis that interpersonal problems resulting from the tension of the spacecraft crew members are also transferred to relations with the ground-based command centre and are a source of crew inconsistencies and informal leadership roles. During a 135-day study of these problems under the conditions of the simulated Mir space station in Moscow, it was found, among other things, that, contrary to expectations, more cases of total mood and tension disorders occurred during the first nine weeks rather than during the subsequent 10 weeks of the simulation, with unchanged levels of crew cohesion. These problems only emerged with greater intensity in the last third of the isolation phase. There has also been a transfer of tension and dysphoria to the outside world, and that support from the ground control centre has maintained crew cohesion. The quantitative analysis of the effectiveness of communication between the space crew and the mission control (MC) was carried out, among others, by N. Kanas, V. Gushin and A. Yusupova (2008), who suggest that the features of an individual communication style reflect the psychological emotional status and individuality of the communicator, as well as their strategy for dealing with stress. V. Gushin et al. (1997), examining the reliability of the method of analysis of communication within and outside the spacecraft with a ground-based mission command centre, identified a phenomenon typical of the communicative behaviour of small groups in isolation, referred to in the sociological literature as "psychological closure and filtering of information," manifested by a gradual decrease in the amount of communication with the outside world, a decrease in the number of topics discussed, obvious preferences for selecting information flowing outside while maintaining "communicative autonomy." This is accompanied by attempts by the spacecraft crew to develop internal communication standards, which, among other things, are characterised by a reduction in discussions with the "outside world" and an increase in egocentricity and sensitivity to the "tone" of messages coming from outside. The subsequent experiment not only confirmed the previously identified phenomena, but also showed that they are characteristic of the communication of a small task force that normally adapts to extreme isolation stress conditions (Kozerenko, Gushin, Sled, 1999).

One of the first Martian experiments was the unsuccessful, essentially commercial, "Mars One" project led by the Dutch entrepreneur Bas Lansdorf, whose aim was to establish a permanent four-person human colony on Mars in 2027. The first stage of selection of participants was completed in 2013. The company went bankrupt in 2019. (Owens, Ho, Schreiner, 2016).

The subsequent laboratory experiments, known as Mars-500, started in 2009 and consisted of 500-day isolation (hence the name) of six volunteers, during which their functioning was comprehensively studied (Kuznetsova, et al. 2017). First, however, volunteers spent 105 days[22] in solitary confinement, and in June

---

22 During the 105-day experimental isolation, M. Nicolas et al. (2013) observed the

2010 the actual stage of the experiment simulating 520 days of flight to Mars began, which was completed in November 2011. The project, co-organised by Russia, ESA and China, was carried out at the Moscow Biological and Medical Institute.[23] As part of the experiment, the following simulations of specific flight elements were performed on the following dates: 2007 – 14-day preselection test; 2009 – 105-day test; 3 June 2010 – start of 500-day mission, December 2010 – power failure, 8 February 2011 – separation of capsule from ship, 12 February 2011 – landing on Mars, 14 February 2011 – first walk on Mars, 5 November 2011 – return to Earth. According to Peter Suedfeld of the University of British Columbia in Vancouver, one can have reservations about the reliability of the experiment because of the lack of a key psychological element, such as the inability to obtain emergency assistance (Suedfeld, Brcic, Legkaia 2009). C. Tafforin, et al. (2015) as part of this experiment, they conducted research on the correlation of questionnaire profiles (Personal Self-Perception and Attitudes; Luscher's sociometry and anxiety test) with video-behavioral profiles of group dynamics participating in the experiment with 520-day isolation and imprisonment (from June 3, 2010). until November 4, 2011) of an international crew (n = 6). All procedures were performed monthly. The results show that the observed level of non-verbal behavior is higher than the level of verbal behavior and that the behavioral dynamics profiles over the months differentiate the crew members. A negative correlation was also found between anxiety and interpersonal communication and popularity in the group and the level of anxiety, as well as a positive correlation between popularity in the group and interpersonal communication. The authors suggest the use of group behavior monitoring and discuss controversies related to the quantitative analysis of socio-psychological data.

M. Basner et al. (2013) on the example of MARS-520d, drew attention to the so far little-studied fact that the success of interplanetary space flight will depend on many factors moderating the functioning of astronauts, which include,

---

dynamics of coping with stress at the level of applied defense mechanisms, emotions and depression as key for adaptation to ICE environments. They found, inter alia, that during the 105-day isolation, positive emotions significantly decreased and the use of less mature defense mechanisms in coping strategies increased, which was also correlated with an increase in depression symptoms.

23 Composition of a 6-person Martian habitat crew: Alexei Sitiov (Russian, 38 years old, engineer, commander), Sukhrob Kamolov (Russian, 32 years old, cardiac surgeon), Alexandr Smolyevsky (Russian, 33 years old, doctor), Romain Charles (Frenchman, 31 years old, engineer), Diego Urbina (Italian/Colombian, 27 years old, engineer), Wang Yue (Chinese, 26 years old, doctor).

among others, Earth's biological circadian rhythms, responsible for the alternation of sleep and vigilance of crews exposed to the lack of Earth's photoecological synchronisers (24-hour astronomical day) and the lack of civilisational social synchronisers. In examining the behavioural effects in the Martian MARS-520d habitat, NASA researchers used a faithful simulation of the future mission to Mars, tracing the dynamics of sleep and vigilance in a six-person multinational crew. Measurements included continuous registrations of wrist actigraphy and light exposure (4.396 million minutes) and weekly computerised neurobehavioural assessments to identify changes in crew activity, sleep quantity and quality during the record 17 months of the experiment. Among other things, reduced movement rate and extended sleeping and rest periods have been found. With reduced exposure to light, sleep quality disorders and changes in the amplitude and phase of the circadian rhythm were also observed. The results indicate the need to maintain Earth's primary photoecological synchronisers (e.g., appropriately timed exposure to light) and civilisational synchronisers (e.g., food intake and work time) for astronauts to maintain optimal levels of activity during long-term space missions (Whitmire et al., 2009).

On the other hand, Iva Poláčková Šolcová et al. (2016), presented the Mars-500 simulation from start to finish as one narrative story. The authors of this experiment used a narrative method two weeks after the end of long-term isolation, the content of which was subject to horizontal thematic analysis, including quotations from crew diaries published on the Internet by the European Space Agency. Among other things, it was found that for the crew members the most enjoyable experiences were celebrations, video messages from important people and a simulated landing on Mars, and the most difficult – situations related to delay or loss of communication with the ground and monotony in the second half of the experiment.

The Americans conducted similar experiments in other habitats, the *Hawaii Space Exploration Analogue and Simulation* (HI-SEAS) project, which took place in an isolated habitat, on the large island of Hawaii, about 8200 feet above sea level, a site simulating a location on Mars on the side of Mauna Loa near a saddle. HI-SEAS is unique as an analogue of the Martian Mission because: (1) the crew is selected; (2) the analogue is deliberately designed to be similar to the conditions of the Martian exploratory mission; (3) the site allows for a year-long and longer isolation; (4) an environment similar to that of Martian as it allows for geological work in the field by humans and/or robots; (5) it is ideal for detailed observations of crew dynamics, roles and performance in space flight. By 2020, four missions were completed in HI-SEAS, lasting two to four months each, one eight months and the last one – a year (Binsted et al., 2016). Thus, during the

HI-SEAS 1 mission, the impact of support from the ground-based mission command centre on astronaut autonomy and interpersonal communication (internal and external) was studied. The trial included six volunteers, with the average age at the level of 39 years, staying in four-month Martian isolation. Studies on communication were carried out every week for eight weeks. The results of the analyses have shown a systematic differentiation between weeks – between autonomy linked to greater internalisation and acceptance of the instructions received from the Command Centre, and worsening cooperation between crew members and ground staff. It was also found that crew members who are satisfied with autonomy are more motivated to carry out their daily tasks and less frequently rebel against the task procedures in force.

In turn, the HI-SEAS 4 experiment, which lasted twelve months, aimed to investigate which of the basic mental needs of astronauts contribute to their well-being and the success of their mission, and how to support them in meeting them. The main focus was on examining, on a weekly basis, the changes in astronauts' autonomy and competence needs. Positive and negative motivations were also observed in relation to compliance with operational procedures, cooperation with mission members, efficiency of work and expected support. The results indicated a significant decrease in the need for autonomy, as well as a worsening of relations with crew members and loved ones at home along with the duration of the mission. In turn, the weekly increase in experience has been shown to correlate positively with autonomy and interpersonal relationships within the mission and productivity (Goemaere et al., 2019).

G.M. Sandal, H.H. Bye, F.J.R. Van De Vijver (2011) accepted a thesis that during the mission to Mars, the crew is exposed to the "Group Thinking Syndrome",[24] which is a consequence of the high degree of group cohesion, will tend to seek consensus at the expense of considering alternative modi operandi, which is a potential security risk for the mission. To this end, a 105-day isolation experiment, simulating a mission to Mars, used the "Crew Values Portrait" questionnaire in a multinational crew study, which was regularly provided to assess the attachment to personal values and their impact on crew cohesion. In addition, interviews were conducted before and after the mission. The analysis of multiple regressions did not show significant changes in the homogeneity of values over time, rather the opposite trend was indicated. Greater tension was attributed to differences in hedonism, kindness and tradition over the last 35

---

24 Detailed data on group thinking syndrome can be found in a classic monograph (Janis, 1972).

days, when the crew was allowed more autonomy. Three subgroups have been identified which are distinct in terms of personal values. Thus, the hypothesis of "group thinking" syndrome in small task forces has not been confirmed. Perhaps this is due to the lack of unequivocal informal and formal leadership among international space mission crews and respect for the expert competence of all members of Martian missions.

The problem of leadership is referred to in the studies by R. Wu et al. (2020), which confirm that, in an isolated spacecraft environment, leadership roles are important for the success of the mission and for good interpersonal relations. The research included 4 crew members (3 men and 1 woman aged 35 years each), who took part in a 180-day experiment called: "4 Subjects 180-Day Controlled Ecological Life Support System (CELSS) Integration Experiment," located in the city of Shenzhen, China, included a "Group Environmental Scale" filled in bi-weekly throughout the whole duration of the mission, concerning the perception of leader's support and leader's control by crew members. Among other things, it was found that the objective control of the crew by the leader increased significantly along with the duration of isolation, but the perception of the quality of the leader's control remained stable. The authors interpret this result *post factum* that the crew accepted the role of leadership, providing more support than control, because leadership had a positive impact on the group's climate and reduced aggressive behaviour. I think that this requires further experimental research, because the questionnaire survey in a small task force is burdened with the error of variable social approval, as emphasised as early as the 1960s by R.F. Bales (1951).

S. Bouchard, T. Martin, M. Perreault (2009) carried out a 110-day (July 1999 to April 2000) flight simulation of an international crew on a space station under the code name of SFINCS in the isolation chambers of the State Biomedical Institute in Moscow, with the aim of observing psychosocial adaptation and social interactions in long-term isolation. Data on interaction processes indicate some adaptation problems, which are indicated by stress in interpersonal relationships, identified as stress events. It has been noted that in a transcultural group quite significant differences in stress management strategies can be identified, which should be the subject of further research and selection and training in this area of group performance. N. Kanas et al. (2009) are of a similar opinion. In a review article of knowledge arising from exploration of ground-based space habits as well as from research on cultural, psychological, psychiatric, cognitive, interpersonal and organisational issues in orbital flight and space stations, they formulate a thesis that these issues should be central to research preparing astronauts for future, long-term human space missions, including transit and surface operations on the Moon or Mars: selection and training of astronauts, in-flight behaviour monitoring and support, and re-adaptation after the mission.

H. Wichman (2011) pointed out that since the time of the orbital stations allowing long-term space missions in the history of space flight, the way in which new habitats are designed and equipped had to be changed under the influence of experience gained. Changes have been made to allow the use of space stations as a simulator to train new habits of functioning in weightlessness conditions, as well as to study psychosocial conditions of functioning in long-term isolation in small task groups. According to the author, these historical changes in the paradigms of space ergonomics herald the imminent development of space tourism – and not only orbital but also interplanetary. However, the International Space Station should earlier be used as a simulator for behavioural research. The first such research was carried out for McDonnell Douglas Aerospace Corporation (now Boeing Space Systems) in Huntington Beach, California, in a space flight simulator on a group of civilian passengers in a 45-hour orbital flight. Social interactions of the subjects (verbal and non-verbal, such as attitudes, gestures and expressions) were observed on-line and analysed using the technique by R.F. Bales (1970). The author stated that a space flight simulator is an excellent tool, both for carrying out the necessary research into social psychology and for conducting interpersonal training. The observations obtained can be described and explained on the basis of the Attribution Theory, which suggests that people have the wrong tendency to attribute people's behaviour to their character and underestimate the influence of situational factors. This type of attribution error should be avoided when selecting candidates for space flights (cf. Meyers, 2006).

Training of social interactions before qualifying volunteers to take part e.g. in a simulated 520-day Mars mission is the beginning of the analysis of human behaviour in future long-term space flights and staying in habitats on other planets. The renaissance of space laboratory experiments is connected with the prospect of landing once again on the Moon and flying to Mars (Sipes & Fiedler, 2007). One should agree with the opinion of S.L. Bishop (2011) on the need to find suitable analogues or ground-based substitutes for future long-term space missions, which, to a greater extent than before, need to take into account, on the one hand, the shortcomings of space technology, such as in the case of the Challenger (1986) and Columbia (2003) disasters, and, on the other hand, the negligence of research teams operating in extreme space environments (e.g. psychosocial and cultural factors in international crews) as well as psychological predictors of the efficiency of astronauts (Rose et al, 1994; McFadden et al., 1994; McQuaid, 2007). These problems are perhaps less precisely described psychometrically, but more multidimensionally in experiments conducted in natural conditions.

2.3.3. Experiments under natural conditions

Natural experiments, as opposed to laboratory experiments, differ primarily in that the experimenter is a passive observer of a situation that is not manipulated by them in order to induce a certain behaviour. Below we will give examples of such studies, which for decades have been treated as partial analogues of space missions.

2.3.3.1. Shelters

Shelters – built in the years of the so-called "Cold War" in the 1960s and 1970s as an asylum in the case of nuclear warfare – were used as natural habitats for natural experiments on people's behaviour in the conditions of forced social isolation and threat to life. The leader of natural experiments in this field was the US Ministry of Defence and Navy, providing various types of anti-atomic shelters (Rasmussen, ed., 1963), which were used to observe the mental discomfort caused by prolonged staying in natural isolation. We will cite examples of studies from that period, which are described by E.A. Ramskill (1962), carried out in 1962 in winter and summer seasons, on a group of 100 military personnel. All were volunteers aged about 20. On the part of the organisers, the experiment was attended by the shelter's commander, an engineer, a doctor and an experimenter as well as two nurses. The conditions simulated an atomic hazard from outside and required survival in a situation of many limitations: only two meals a day (dryers and coffee in the morning, powder soup in the afternoon), rationed washing water, limited use of the toilet, communication with outside light completely stopped. The experiments lasted two weeks. The observations were aimed at learning about the mechanisms of functioning of a small group in a situation of limited stimulation and the action of an external nuclear threat. The list of the ranks of discomfortable feelings of the subjects, drawn up by the doctor, showed that the most important stress factors of isolation are: basic living conditions (e.g. hygienic treatments, bad food, etc.), aesthetic (e.g. tightness of rooms, etc.) and emotional (e.g. monotony, imprisonment, lack of contact with loved ones, etc.).[25] The dynamics of subjective discomfort are also described. Thus, already

---

25 This is in line with the results of similar studies carried out by A. Brand-Persson (1960) in Sweden two years prior.

in the first week, symptoms of apathy, decreased appetite, irritability, difficulty in concentrating on anything, etc. were observed. Experiments of this kind draw attention to the important fact that natural isolation should be considered more broadly than just in terms of reducing stimulation – rather as a complex stress situation. This is confirmed by other experiments which have taken into account the isolation effects in relation to the duration of isolation and the composition of the group. For example, I. Altman et al. (1960) studied mixed groups of 30 people (women vs. men, adults vs. children) placed in a shelter for 1–2 weeks. Apart from interpersonal conflicts and excessive irritability, no symptoms typical of homogeneous groups were observed. The authors conclude that the group's diversity in terms of gender and social roles is an important source of gratification for many psychosocial needs and thus a source of stimulation.[26]

It must be realised that the experiments described, for example, have many flaws. First of all, the research was carried out on volunteers, which creates a rather unrealistic situation, resulting from confidence in the experimenters, who take responsibility for their safety. Secondly, the selection of the groups was not random, as most of the research was carried out on students or soldiers.

2.3.3.2. Penthouses

Penthouses – that is annexes on the roofs of very tall buildings intended for temporary shelter (until evacuation) during a fire on the lower floors, were used as a habitat to study people's behaviour in a forced isolation and danger situation. And so T.A. Cowan & D.A. Strickland (1965) describe several subsequent natural experiments conducted in penthouses, covered by cryptonyms: "Penthouse II" and "Penthouse III," in which volunteers, well paid, aged 20–39 (6-person groups) participated. They took place on the roof of a university building in California and lasted 88 days. There were differences in the degree to which the stimulation was restricted. In the experiment P-II, no restrictions on socialising, using telephone communication, mail were present, whereas in the case of the experiment P-III no one was allowed to contact anyone from the outside, either by phone, mail or in person. Although both groups were voluntary and not task-based (no commissioned activity), the results showed worse functioning of the participants in experiment P-III, in which the inflow of stimulation from the outside was limited. The main symptoms were: excessive irritability,

---

26 Using such data, among other things, the Argentineans left 12 families for wintering-over at their Antarctic "Esperanza" station in 1979 (direct contact message) as part of a natural experiment.

confrontationism, hostility, depressive states, outbursts of anger and impulse, mood swings, etc. Meanwhile, the participants of the P-II experiment spent most of their time watching TV programmes passively. The above-mentioned experiments have drawn attention to the fact that limiting stimulation while limiting own activity makes social isolation more stressful.

2.3.3.3. Underwater capsules

Underwater capsules,[27] have been used to observe, in particular, the effects of social isolation on human behaviour in a very small social space and situations of threat from the external water environment. (Mears & Cleary, 1980). Two experiments, codenamed "Tektite I" (T-I) and "Tektite II," were the subject of many publications (Gilluly, 1970). The "Tektite-1" platform (constructed by *General Electric Company under the patronage of NASA and US Navy*) was placed in 1969 at a depth of 15 m and stayed there for 60 days. The crew consisted of two oceanographers, an ichthyologist and a marine geologist. Topics of psychological research conducted by E.H. Cliftona et al. (1970) were: recreational activities, a 17-hour cycle: work-recreation and the dynamics of crew behaviour. The following variables were observed: individual differences in adaptation to isolation, emotional balance, overt anxiety, level of motivation to perform tasks, morale, leader functions, types of aggressive behaviour. Among other things, the results of the research showed that only the constantly activated light was a source of permanent discomfort. In the absence of the so-called "photo-environmental and civilisation synchronisers," disturbances in circadian rhythms were found. Although the group was focused on a task, a number of conflicts were observed, which were the reason why in the subsequent "Tektite-II" experiment in 1970 the focus of psychological research was shifted to the problem of the structure of a small task force. It was found, among other things, that if a leader was chosen in the group, combining formal and informal values, then the vested interests of the members of the group were easily subordinated to the main objectives of the mission. However, conflicts occurred in every group. In the second half of the isolation period, almost every member of the group spent more time sleeping than working (the so-called drifting of circadian rhythms). It has also been found that longer isolation (20 days) gives more chance for relative adaptation than 14 days, although both men and women adapt to the situation of isolation

---

27 E.g. bathyscaphs – small, self-propelled underwater vessels for exploring the depths of the sea and bathyspheres – stationary cabins for underwater observation at depths of up to 330 m and for managing underwater work.

gradually. Preferred reactionary patterns took into account rather passive activities in terms of movement, but rich in stimulation, such as watching TV,[28] music, videophones, books.

Other underwater experiments were conducted by G. Weltman, T., Crooks and G.H. Egstrom (1969) under cryptonyms: Sealab 1 (S-1), Sealab Il (S-II) and Sealab III (S-III). Experiment S-1 concerned only 4 sailors who were underwater for 12 days. Experiments S-II and S-III included, among other things, a psychological programme, similar to that carried out in the "Tektite experiments. Behaviour of a group with a diversified personal composition (military vs. civilian and technical vs. scientists) was analysed. However, the too short duration of the experiments (for technical reasons) affected the results (experiment S-II lasted 15 days and experiment S-III only 12 days), which do not add anything to what we discussed earlier.

The thirty-day underwater isolation of a group of 6 people trapped in the "Ben Franklin" bathysphere is described among others by R.J. Del Vecchio et al. (1970). The psychological programme included a survey of the effects of isolation, motivation issues, psychosomatic symptoms and group morale.

To sum up the above research, it should be stated that it was a model study of physiological and psychological changes occurring in a small task force in a situation of short-term isolation and for this reason, among other things, they were treated in the 1970s as analogues of short-term space flights (Suedfeld & Steel, 2000). From the perspective of long-term space flights, on the other hand, submarine research was more interesting.

2.3.3.4. Submarines

According to many authors, submarines – staying in the submerged state for a few months – provide an opportunity to study humans functioning in long-term cosmic isolation (Sandal et al., 1999). As much of the source material from submarine research is unavailable due to military secrecy, the review work and materials from scientific conferences (Weybrew, 1991) are very useful. A typical review study on the functioning of submarine seamen in the field of functioning

---

28 The installation of a CCTV system in the "Tektite II" experiment was interesting, especially from the methodological point of view – it was used for conducting continuous observation of the behaviour of the subjects, which resulted in the awareness of continuous observation generating an increase in discomfort and reinforcing the effect of variable social approval, which could have a negative impact on the results of the research.

of an autonomous system (psychosomatics) and subjective evaluation of discomfort is the study by B.B. Weybrew (1963), describing psychological observations from a 30-day patrol voyage and 60-day combat submersion, in which he draws attention to cyclical changes in the level of cognitive tests, which, according to the author, is associated with fluctuations in motivation level. I think that the influence of photo-ecological factors and the lack of civilisation synchronisers should be added to this interpretation.

A more detailed description of the psychological effects resulting from the 60-day submersion of the vessel [29] was conducted by J.H. Earls (1969), which he calls the "Syndrome of adaptation to the isolation of a submarine," which has its own dynamics. Thus, the first stage of adaptation to the situation of social isolation on a submarine, which falls on a 4–5 week of submersion, is called "Half submersion syndrome," which is characterised by a significant reduction in mood, up to and including depression, a decrease in appetite, sleep disorders, withdrawal, etc. Week 8 is called the "End of voyage syndrome," which is characterised by individual differences in mood, i.e. an increase in some people's mood, and depressive tendencies associated with the anticipation of disliked social and sexual roles. The author interprets these symptoms in terms of psychological cost, which proves the actual stress of isolation. The most interesting thing from the perspective of long-term space flight is the dynamics of the symptoms observed by Earls, characterised by great diversity. The detailed dynamics of symptoms during a 60-day battle submersion is as follows: (1) Anticipation of submersion (28 days and one week before submersion) – slight mood elevation with a tendency to lower just before submersion; (2) Submersion – slight relaxation with signs of excitement; (3) "1/4 way syndrome – 2–3 weeks – depressive tendencies, symptoms of nausea, sleep disorders, friend-oriented behaviour, overt sexual humour; (4) "1/2 way syndrome" – 4–5 weeks – significant mood reduction, spreading depression, loud complaints, decreased appetite and stomach disorders, complaints of headaches and muscle pains, difficulty in concentrating attention, sleep disorders, withdrawal, escape from interpersonal contacts, intense sexual humour, disorders of circadian rhythms (in several cases episodes of depersonalisation); (5) "3/4 way syndrome" – 6–7 week – unexpected improvement in the group's mood, return of sexual humour, tendency to anticipate upcoming heterosexual activity; (6) "Eighth week syndrome" – most people have increased mood, and some people have experienced depressive tendencies, linked to the anticipation

---

29 This concerns research carried out during a 60-day submersion of a nuclear-powered submarine. USS "Nautilus" with a 140-person crew.

of disliked social and sexual roles; (7) "Harbour entry fever;" (8) "Docking" – 60th day – hypomaniacal states, which may manifest themselves in the form of work disturbances and morale reduction and the return of depressive states, fatigue, mood reduction linked to the crew "break-up." G.M. Sandal et al. (1999) conducted, on a sample of three submarine crews, classic studies into the relationship between personality profiles coping strategies with the stress of isolation and discomfort. They found, among other things, that there was no such relationship, implying that the personality profile may play more of a mediating role in adaptive interpersonal behaviour. The authors suggest that pro-social attitudes should be taken into account when selecting submarine crew members. In turn, T.L Thomas et al. (2003) presented an interesting study on crew health and medical support during a long-term mission of US Navy submarines. The study population consisted of the crews of 240 submarines over the period of 1997–2000 and included 1389 officers and 11,952 other crew members. It was found that the most common categories of medical events were upper respiratory tract infections, injuries, musculoskeletal conditions, infectious diseases and skin diseases. Some of these symptoms are characteristic of the artificial environment of space habitats.

B.B. Weybrew (1961), discussing the impact of isolation on submarine personnel, believes that of all the research issues, the selection techniques are the least developed and especially their predictive value is estimated at only about 10%. However, the author believes that, if only because of the real danger caused by the knowledge that submarine rescue is not effective, this type of an isolation situation can be successfully compared to a spacecraft cabin.

To summarise the common attributes of all terrestrial space analogues, S.L. Bishop (2011) believes that of the many space analogues, submarines have many common features: e.g. threat from the physical characteristics of the external environment, the possibility of survival by maintaining the parameters of the artificial environment (pressure, humidity, air composition, etc.), the catastrophic effects of loss of propulsion power, the concentration of the crew in a small space, social isolation, the constant need to revitalise the habitat environment (e.g. decontamination of the atmosphere, radiation effects), etc. From this point of view, the problems of psychiatric and mental health of seafarers also appeared in research (Sexner, 1968). Some authors stress that, from the perspective of space analogues, further research should be carried out in real conditions during the many months of submarine submersion (Weybrew, 1991), as well as in the analogues of Antarctic stations (Terelak, Kwarecki & Rakusa-Suszczewski, 1978) and real orbital space flight conditions on the ISS (Alfano, 2018).

2.3.3.5. Spacecraft and orbital stations

Knowing from the past possibility of travelling to the Moon,[30] the latest research focuses on the possibility of using this planet as a "stopover" on the way to Mars. This is pointed out by M. Heppener (2020), who emphasises that this requires cooperation between three important scientific fields, namely physics, physiology and psychology, which must minimise the obstacles to long-range space travel (cosmic radiation and weightlessness), but also develop new technologies (planetary terraforming, nuclear propulsion or human hibernation). This has been confirmed by other authors that we are entering a new era of exploration of ever further regions of the Cosmos, which must take account of a basic fact, that the cosmic environment and other planets are significantly beyond the possibilities of adaptation, both genetic, epigenetic and psychological, determined by terrestrial evolution of species (Shock, Abood, Shelhamer, 2018). The awareness of the colonisation of such planets as the Moon and Mars poses new challenges to habitat designers, which in turn should be analogues of the Earth's environment (Kozicka, 2008).

Stress factors, which are completely new for the human species, and which do not occur in earthly conditions, are, on the one hand, enormous accelerations (excess of stimulation) and, on the other hand, weightlessness, which effectively change the sensations of many sensory impressions and balance. While the acceleration that occurs during take-offs of a space vehicle or the braking that precedes landings takes not very long, weightlessness in long term space flight reduces or even deactivates a number of stimuli that are important for human physiology and social functioning (Terelak, 2016). Both of these factors are aversive in nature, as shown in Fig. 8.

---

30 The last man to walk on the surface of the moon for 73 hours was the late American astronaut Eugene Cernan, a crew member of the Apollo 17 mission from 6th to 19th December 1972. He published his memoirs in 1999 in the form of a book under the symptomatic title "The Last Man on the Moon" (Cernan & Dvis, 2000).

104   Data Sources Concerning The Effects of Sensory Deprivation

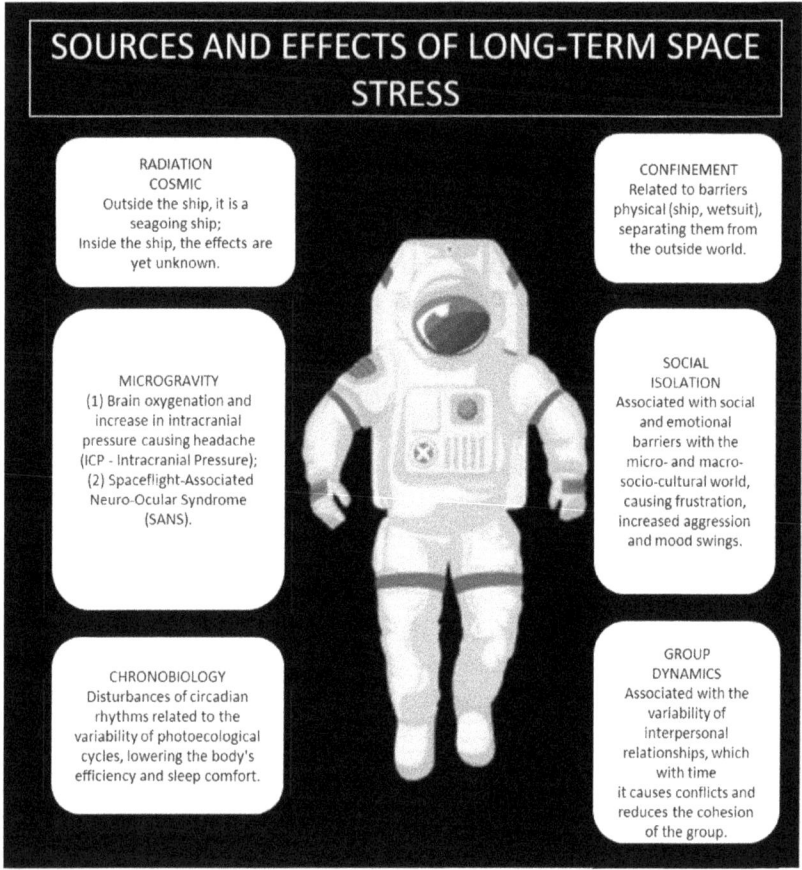

Fig.8. The main stressors affecting astronauts during long-term space flights
Source: own elaboration.

For example, it should be noted at the outset that the experiences of weightlessness are the most recent ones for the human species in the whole phylogenetic development, as they involve a release from earthly attraction, thus cutting off many of the somato-gravitational stimuli that are important for the functioning of the species. Although, with the preservation of certain physical conditions (speed and angle of ascent), in the middle of the 20th century, it is possible to induce the phenomenon of weightlessness inside an aeroplane performing a parabolic flight following the Kepler curve for a short period of time (several dozen seconds), medical and psychological problems appear in the situation of adaptation to work under weightlessness conditions in space flights (Tafforin, 1996). The extremism of this situation is related to such factors as: lack of stimuli registration by appropriate gravitational receptors located in the inner ear; lack of static and dynamic loads of the musculoskeletal system; elimination of hydrostatic pressure of blood and tissue fluids.

Since the state of weightlessness is not experienced in Earth's biogenesis, knowledge of the biological and psychological consequences of prolonged exposure to the state of weightlessness is still incomplete, even though man has been in space for several decades at different times.

We omit the detailed discussion of the medical effects of weightlessness, such as: decrease in circulating fluids, decrease in blood density, escape of calcium from bones, muscle atrophy, clearly impaired ability to physically exercise, as they are described in the available literature (West, 2000). We will only draw attention to the psychological effects associated with reduced working capacity and subjective discomfort. For example, the inability of adequate gravitational receptors to register stimuli leads to sensory illusions, spatial disorientation and response from the vestibular ear (so-called motor disease). Lack of customary load on the musculoskeletal system leads to a significant disorder of visual-motor coordination due to muscle atrophy. The situation of weightlessness as a completely new one causes psychological stress, associated with the need to change not only motor habits, but also previous preferences such as sleeping position. Depending on the length of time a person has been weightless, two phases of adaptation are discerned: (1) Acute stress phase, accompanying the first space flight, generally lasting several hours, accompanied by symptoms of discomfort, anxiety, disorientation (relativity of the terms "up to and down" or "left to right"); (2) Relative adaptation phase, when the cosmonaut becomes euphoric after several hours or days (Friederici & Levelt, 1987). From a psychological point of view, the transfer of motor habits and visual-motor coordination necessary to perform manual operation is a difficult problem. Similar, although to a lesser extent, problems arise with adaptation to reduced gravity on other planets (the Moon or Mars). For example, on the Moon, selenonauts who were perfectly trained in the Earth's conditions, in the conditions of reduced gravity on the Moon, performed only 30% of their work in relation to Earth's habits. Even greater differences occurred during the repair work carried out outside the spacecraft in open space (ISS data). However, the problem of changing the structure of action in conditions of changed gravitation remains open. The literature on the subject draws attention to the wide variety of reactions in conditions of weightlessness: sensory, motor, emotional and vegetative. There are four approaches to their description: (a) the pragmatic approach – describing the degree of deterioration of general well-being and reduced ability to work. The proponents of this approach dichotomise weightlessness responses by distinguishing those that increase or decrease work ability and by scaling their intensity (strength); (b) functional-physiological approach – describes the extent and degree of physiological changes accompanying the deterioration of general well-being and work ability. There is sometimes talk about the so-called physiological cost accompanying "adaptation" to the state of weightlessness; (c) the self-observer's approach – it is the creation of the so-called internal point of view on the basis of self-observation data, images and spatial illusions. An example of a description of not only weightlessness but also other factors of space flight from the perspective of the self-observer is e.g. the book of the Polish cosmonaut Mirosław Hermaszewski emphasising the "importance" of the phenomenon of weightlessness, wittily entitled "The Weight of Weightlessness: The Cosmonaut Pilot's Tale" (2017). This data, although very interesting, as it is not available to the external observer, is of little scientific importance; (d) the external observer's approach – it draws attention to the data on the observation of behavioural reactions and emotional and motor behaviour in the state of weightlessness, which are the basis of the research being designed.

For example, research conducted so far on spacecraft and orbital stations has resulted in a paradigm of individual differences in emotional and motor responses (e.g. active vs. passive), which is important for the methodology of future research, and this translates into work efficiency indicators. Among other things, it has been found that a significant proportion of subjects in weightlessness conditions have a longer processing time for digital signals, visual-motor coordination is deteriorating, probably due to insufficient balance of antagonistic muscle effort, especially horizontal or loop movements (fast and rhythmic movements). There has also been a decline in the sense of hand location during the "writing test." Details of significant localisation errors can be found in the upward or downward tilting of the hand. Localisation errors were also observed within the visual system. For example, in the initial contact with weightlessness, a person does not see the ambiguity of figures (e.g. weight vs. head profiles), which disappears after the adaptation period. Studies under real-life conditions of long-term weightlessness of space flight prove that the experimental data described relates to the initial period of adaptation to the state of weightlessness. After this period, no motor disorders that prevented astronauts from performing the operator's work were observed, except for the extension of the time of performing the task (Convertino, 2007).

Based on the results of experimental research and previous experience of orbital flight, attention should be drawn to the fact that a psychological analysis of human behaviour under weightlessness conditions cannot be carried out only in terms of lack of gravitational strength. This is because we are dealing here with a complex phenomenon of disturbing the processes of spatial analysis, which usually consists of changes in the function of the organ of balance, a reduction in cortical reception and so-called deep feeling. Space experiments to date bear witness to the high plasticity and compensatory capacity of the human nervous system as a species and as a human individual. However, physiological adaptation at many levels of the organisation, especially in long-term spaceflight, encounters a fundamental obstacle, which is the lack of gravity. This, in turn, hinders the redistribution of bodily fluids in the body (see Fig. 5), which leads to cerebral over-oxygenation, caused by blood in the head and ischemia of the lower parts of the body, which in turn leads to many physiological disorders, which are the subject of cosmic physiology studies (Buckey, 2006). The lack of gravity may also be an obstacle to long-lasting space missions, which involve returning to Earth only after some generations. A disturbed mechanism of cell aggregation in the state of weightlessness makes fertilised cells (experiment on hen eggs) die after a few weeks because they do not have the directional information necessary to produce the so-called primary streak, which is the criterion for determining the left vs. right side and the top vs. bottom[31] (Terelak, 2016).

The first tests under real space flight conditions were conducted remotely on ships: Mercury 6–9, Vostok 1-2 and Gemini 6–9, which are considered to be the most serious sources of data on human performance in a spacecraft situation. An example is the own research carried out by means of a specially constructed questionnaire for the Orbital Assembly "Salut 6 – Soyuz 29 Vostok," under the cryptonym "Oprosnik" (Questionnaire) – a testing tool to monitor the well-being of cosmonauts before the takeoff, during the flight and the stay at the space station and after completion of the space flight (Terelak, Kobos, Kozerenko, 1978). According to the instructions, cosmonauts completed the questionnaire eight times: (1) 24 hours before the take-off (A1); (2) directly before the take-off (A2); (3) during the first encirclement of the Earth (B1); (4) half through the space flight (B2); (5) during the space flight after completion of the task (B3); (6) directly after landing (C1); (7) 24 hours after landing (C2); and (8) one week after landing. In cases 1, 2, 6, 7, and 8, cosmonauts answered a set of nine questions, while in cases 3, 4, 5, they answered a set of 51 questions, assessing their feelings

---

31   This knowledge is essential for future space expeditions into the far reaches of the Solar System and beyond.

using a scale from 0 to 10.[32] Figure 9 presents the results of two astronauts differing in their level of anxiety: (1) timid vs. (2) brave in three initial phases of the space mission.

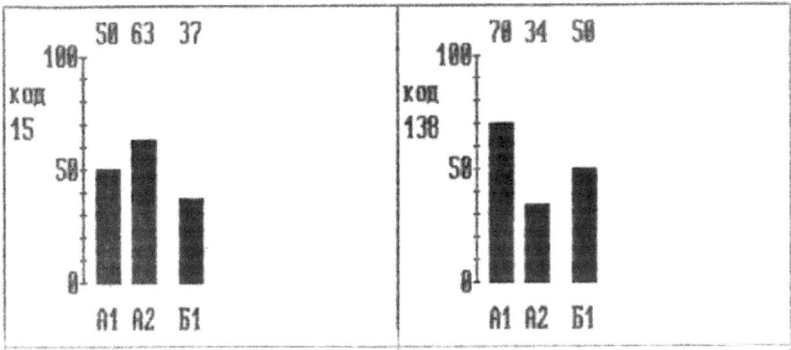

Fig. 9. The results of two astronauts differing in their level of anxiety: (1 – cod: 15) timid vs. (2 – cod: 138) brave in the 3 initial phases of the space mission: A1 – 24 hours before the take-off; A2 – directly before the take-off; B1 – during the first encirclement of the Earth
Source: own elaboration based on Kozerenko, Radkowski, Terelak et al. 1991.

The research objective of the questionnaire was to collect information on 9 aspects of functioning during the space flight, namely: (1) collecting opinions to be used for individualisation of the comfort of working and dwelling at the space station; (2) collecting information on the possibility to maintain the circadian sleep/activity cycles; (3) gathering cosmonauts' opinions on the conditions required to optimise the internal habitat of the space station; (4) collecting data on the ergonomic use of space station interior; (5) assessment of the usefulness of ground working habits in zero gravity conditions from the perspective of improving the ground training; (6) collection of data on the relationships between the international crew members and relationship with the ground flight control team; (7) gathering information on the functioning of the body in zero gravity conditions; (8) the assessment of the

---

32 The questionnaire was elaborated in Polish language at the Military Institute of Aviation Medicine in Warsaw by Jan Terelak and Zdzisław Kobos (Terelak, 2013) and adapted to Russian language by Olga Kozerenko of the Institute of Biomedical Problems of the Russian Academy of Sciences in Moscow in two versions: a pen-and-paper version and an electronic version for IBM computers (Kozerenko et al. 1991).

organization of active recreation; (9) the assessment of the usefulness of the on-board medicine box. In line with the instructions, cosmonauts assessed numerous aspects of well-being in relation to individual stages of the flight using dichotomic answer categories. Thus, in order to assess current well-being, subjects were to mark points corresponding to the following categories: satisfactory/unsatisfactory, absent-minded/concentrated, inhibited/energetic, wavering/decided, relaxed/tense. Subjective assessment of working conditions was based on the following categories: time devoted to work (little/satisfactory), workplace (uncomfortable/comfortable), freedom of movements (extremely restricted/unrestricted), working regimen (too demanding/ideal), lighting (too dim/comfortable), temperature (comfortable/too hot), humidity (comfortable/too dry), noise (low/too noisy). The assessment of recreation included the following attributes: place to sleep (extremely uncomfortable/comfortable), place for recreation (extremely limited/comfortable), time for recreation (extremely short/excessive). The last category, i.e. "time for recreation" was supplemented by an additional instruction: "Assuming total recreation time to be 100%," estimate the percentage of time devoted to: reading, listening to music, watching video, talking to family and friends, playing parlour games, sleeping, resting passively, lazing around." The analysis of the answers of the studied cosmonauts to the preferences regarding rest and entertainment showed that apart from maintaining the Homo ludens attitude, the Earth's preferences towards the cosmic ones turned out to be not very accurate. This is also confirmed by S.J., Pell & F. Mueller (2016), who, when checking the taxonomy of types of games and performances to be used during space flights, found, inter alia, a discrepancy between the preferences previously declared and during space flight. A similar inconsistency in the field of terrestrial and cosmic sensations of taste and smell was found, among others, by J. Kubiczkow during a biomedical experiment conducted by the first Polish cosmonaut Mirosław Hermaszewski on board the SALYUT 6 station using an electro-gustometer (Kubiczek, & Skibniewski, 1979).

The completion of tasks in line with the predefined schedule was assessed using the following categories: associated with mental fatigue (slightly/significantly), associated with physical fatigue (slightly/significantly), associated with haste (slight/significant), associated with interest (slight/ significant). Assessment of motor functions included the following aspects: speed of motor reactions (very restricted/unproblematic), precision of movements (very limited/very good), hand coordination (very limited/very good). The quality of sleep was assessed on the basis of the following categories: time until falling asleep (too long/very short), getting enough sleep (not enough/enough), dreaming (no dreams/frequent dreaming). The assessment of locomotion was based on the following categories: moving around the spaceship cabin (with many difficulties/with no difficulties), speed of moving around (with many difficulties/ with no difficulties). Thinking and decision making was assessed on the basis of the following categories: understanding information from on-board sources (with many difficulties/with no difficulties), assessment of the safety of current situation (with many difficulties/ with no difficulties), occurrence of unpredicted situations (very rare/very frequent), feeling confused by unexpected situations (very frequently/never), necessity to continuously adapt to varying fl ight conditions (very rare/very frequent), illusions of various kinds (never/very frequent). Categories used to assess the appetite included: appetite (too weak/too strong), taste of food (not tasty/ very tasty). After completion of the surveys, answers were sent to a database of limited graphical analysis allowing to simultaneously present the data from not more than four cosmonauts. The results of the studies conducted in a series of space experiments using the presented Cosmonaut Physical State Questionnaire in 12 international space crews at space stations Salyut 6, Salyut 7 and Mir were discussed in another paper (Kozerenko, Radkowski, Terelak et al. 1991), highlighting the usefulness of the obtained results in the analysis of the mechanisms of adaptation at diff erent stages of the space fl ight as well as in space ergonomics recommendations to improve the dwelling and functioning conditions for cosmonauts, particularly in long-lasting space missions (Terelak, 2004). A new phenomenon in the literature on the subject, called "technostress,"[33] which also occurs in space ergonomics, is pointed out by, among others, M. Tarafdar, C.L., Cooper, J-F. Stich (2019). The study of the "technostress" phenomenon began with an analysis of the literature on the relationship between the high rate of development of computerisation of technical devices, based on various conceptual bases, and the discomfort of their use. According to the authors, the principles of designing information systems should be changed in such a manner as to move from technostress to eustress, e.g. conducting future design studies in interdisciplinary teams, taking into account the so-called human factor.

Below in figure 10 we present the content identification of 20 measurement points of "technostress" related to the space mission Salut 6 – Soyuz 29 Vostok, with a narrative from the position of internal observer of Polish cosmonaut M. Hermaszewski,[34] concerning emotional states as stress correlates.

---

33 Technostress – defined as the stress experienced by individuals due to the use of complex information systems which, by analysing huge data packets on-line, provide the operator of technical devices with directive indicators or autopilot feature – is a new preventive trend in ergonomic research.

34 I would like to thank the previously researched as a candidate for a space flight, and for many years my friend General Mirosław Hermaszewski for consent to the publication of his space research results.

Experimental aspect of data sources 111

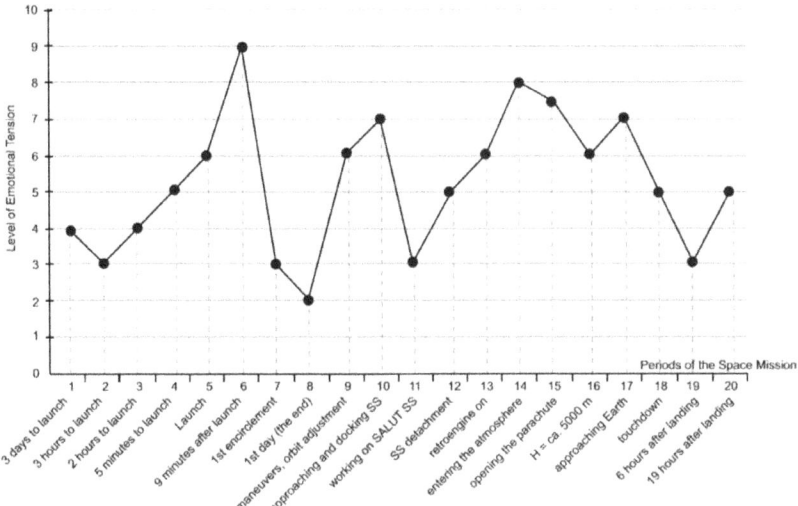

Fig. 10. Changes in Emotional Tension during the Space Mission Soyuz 30 – Salyut 6 in 1978. Subjective feelings by General Mirosław Hermaszewski
Source: Hermaszewski, 2013. p. 7
Legend: (1) three days to launch ("It's so close, hope nothing bad happens or they would cancel the launch"); (2) three hours to launch("This is it – a bath, a meal, my favourite chicken soup, spacesuit on"); (3) two hours to launch ("Spaceship cabin, systems check-up"); (4) 5 minutes to launch ("Work completed, waiting, time drags by"); (5) Launch ("So it started, we're going up, much work, little time"); *(6) nine minutes after launch ("Jettisoning, the orbit, zero gravity, I'm hanging, I don't believe it's real");* (7) first encirclement ("Colourful, beautiful, fantastic, am I hanging upside down?"); (8) first day – the end ("Work has knocked me out, let's get some sleep"); (9) manoeuvres, orbit adjustment ("No time to look at Earth"); *(10) approaching and docking Space Station ("It's hard, but we must make it");* (11) working on SALUT SS ("It's cool, but something haunts sometimes"); (12) Space Station detachment ("Pity we're going back"); (13) retroengine on ("Pray it doesn't starve"); (14) entering the atmosphere ("Look at all this burning, beautiful and horrible, awful vibrations"); (15) opening the parachute ("Pray the rag works (9.6 km)"); (16) H = ca. 5000 m ("A solid shock – gravity load 6.5 g, it's not Earth yet"); *(17) approaching Earth ("Can't see anything, is it going to be soft?");* (18) touchdown ("Soft, sure! – hit, dragging, jolts, no doubt it's the Earth"); (19) six hours after landing ("Relax – pool, sauna, no journalists, champagne, no gabbing"); (20) 19 hours after landing ("The Kremlin, decoration, I am dead tired but have to talk").[35]

---

35 Detailed biographical data of Cosmonaut Mirosław Hermaszewski against the background of the socio-political situation in Poland in the 80–90s of the last century are presented by N. Bażel, P. Gracak & P. Sławęcki (2021).

As shown in Figure 10, three phases of orbital flight are most stressful: (6) – nine minutes after launching and entering the Earth's orbit; (10) – moment of docking to the Salut-6 Space Station; (17) – approaching landing on the way back to Earth ("Jettisoning, orbit, zero gravity, I'm hanging, I don't believe it's true"). These three flight phases were evaluated similarly by all the spacemen in other Salut and MIR missions, while in other assessments of their emotional states they were quite differentiated. These differences were significant especially between "novice" and "experienced" cosmonauts (Kozerenko, Radkowski, Terelak et al., 1991).[36]

Already at the turn of the 20th and 21st century N. Kanas et al. (2000) signalled that the results of simulation tests on Earth and initial experience of real space missions suggest that problems in multinational crews may affect the interaction between crew members and mission control personnel, especially in the long term. The authors examined 5 astronauts, 8 cosmonauts and 42 American and 16 Russian ground mission control personnel who participated in the Shuttle / Mir space programme. The respondents filled in a questionnaire consisting of three scales: (1) Scale of mood profiling, (2) Scale of crew environment, (3) Scale of mission control environment. In order to examine the average differences, ANOVA variance analysis was used: astronaut vs. astronaut, crew member vs. ground personnel. Among other things, it was found that the Americans scored higher in terms of vigour and work pressure, and the Russians scored higher in terms of managerial control, task orientation, physical comfort, self-knowledge and leader support. Those who control the mission scored higher than the crew members in terms of four measures of dysphoric emotion (related to mood swings). There have been significant interaction effects for the subscale measuring leadership support, expression and independence, with US astronauts receiving the lowest score of all three subscale benchmarks. In addition, the research carried out by A.D. Kelly and N. Kanas (1992) shows, among other things, that the structure, content and styles of negotiations with "Earth" in the international ISS crews are changing significantly: for example, the filtering of information by the crew and the selectivity of the contact person. It has also been found that cosmonauts belonging to the same national group dominate

---

36 Similar narrative studies (astronaut notebooks) and psychological questionnaires were conducted at the ISS under the direction of J. Stuster (2010). This report describes an analysis of the content of the narrative, using methods elaborated in previous analogue research, which provided quantitative data to allow the taxonomy of behavioural problems associated with long-range space missions.

communication during flights, which negatively affects the emotional state of the crew. This is partly confirmed by comparative studies from the communication of the three international mission crews on ISS with Mission Control, carried out by V. Gushin et al. (2005), who, on the basis of a quantitative analysis of the dynamics of the content of the conversation within the group and between the crew and the Mission Control (MC), concluded, among other things, that in most cases the crew commander, as the formal leader, dominates (they speak more often than other crew members). In addition, an increase in communication between astronauts and their national MCs, often seeking cultural support there, has been noted in international crews, which means that social isolation is seen as a stress factor, as evidenced by written narratives of astronauts during real space flight (cf. e.g. Lebedev, 1988).

Among other things, the research carried out to date on space crews shows that long-term social isolation, in addition to the other stress factors discussed, has an adverse effect on mood, taste, aesthetic and emotional preferences, generates stress, and the extrapolation of the conclusions from the research carried out to date in ground habits and in the actual conditions of spacecraft and stations orbiting the Earth is still incomplete. However, experience from long-term flights in orbit stations (e.g. MIR, ISS) has provided evidence that the longer a space mission lasts, the more important it is to know the effects of sensory deprivation and social isolation to understand the neuro-brain mechanisms that control behaviour in this extreme monotony of life. In general, monitoring of psychological factors starts immediately in the selection process and in the preparation phase, but especially during the flight, especially in extreme situations such as equipment failures inside and outside the station, lack of fuel and water reserves, etc. It is also important to monitor crew support. Among other things, it has been found that the following crew members cope better with space isolation: Those, who are "tough," persistent, well-trained, task-oriented and do not count solely on psychological support. According to many authors, this requires rethinking the accuracy of the predictive testing methodology related to the selection of candidates for a long-term space mission, on the one hand, and testing ways of support, on the other (Baranov, et al., 2001; De la Torre, et al., 2012; Bernd, van Baarsen, 2020).

To summarise the conclusions of experimental studies carried out in laboratories and under real space flight conditions, one should conclude that they are one of the more important sources of data on human functioning in this type of a situation, as they do not require far-reaching extrapolation of results, which are generally burdened with *ex post* interpretation. Let us recall that the record of multiple space-time stays over the course of five launches is 878 days, 11 hours and 30 minutes and belongs to a Russian – Gennady Padalka (as of 17 April 2020), while the longest human being's single-time stay in space also belongs to a Russian – Valery Polakov, who stayed on board the Mir space station for 437.7 days in 1994–1995.[37]

D.L. Baisden et al. (2008) presented the "Visions for Space Exploration" initiated by NASA in 2004, indicating reference projects for the ISS and the Lunar Base and the expedition to Mars. These include, among other things, issues such as e.g.: (1) medical minimisation of risk factors for diseases which cannot be effectively treated during the mission; (2) minimisation of physiological effects of weightlessness (in space) and microgravity (on other planets) through the use of artificial gravity (e.g. spinning in a rotating spacecraft or local centrifuge; (3) prevention of psychiatric and psychological decompression; (4) autonomous medical care system (e.g. online presence of a nano-robot surgeon in the Mars mission crew).

In the conclusion of discussing the problems of cosmic medicine and psychology, according to J-P de Vera & J. Seckbach eds. (2013) it is worth turning attention to the astrobiological aspect of future research on the exploration of the physical environment of the cosmos in terms of insight into the nature of planets and their ability to sustain life. This research is primarily the domain of astronomy, biology, chemistry, geology, planetology, physics and is concerned with describing: the atmosphere, solar radiation, magnetism, tectonics, mineral composition, availability of liquid water and interaction with life in general. Not without significance is the dynamic development of new technologies providing tools for this research (Beaty et al., 2018). In addition to ground-based laboratory experiments and those placed on orbiting satellites (e.g. LEO, FOTON, BIOPAN) and on the ISS (e.g. EXPOSE-E in the Columbus module, EXPOSE-R in the Zvezda module and various platforms in the KIBO module), currently an exposure platform based on the Moon, and in the future on Mars, are envisaged. For example, *The ExoMars 2022 mission* will have two Mars rovers that will land on Mars (e.g. the European Rosalind Franklin rover) and travel across the surface of Mars in search of signs of well-preserved organic material, and the Russian surface platform (Kazachok), which will depict landing sites and carry out long-term climate and atmosphere monitoring.

Despite the fact that long-term space flights have highlighted the great adaptability of humans to this type of situations, space psychology still treats isolation

---

37 Fifty cosmonauts of different nationalities spent between 333 and 878 days in cosmic isolation.

as an extremely complex stress-related situation and is interested in research carried out in real space flights and in the conditions of Antarctic stations, which have so far provided the most important data explaining the psychological mechanism of the adaptive behaviour of individuals and entire crews of space habitats

2.3.3.6. Long-term Antarctic expeditions as an analogue of Martian habitat

Scientific research in the polar regions involves staying in a small isolated task force under extreme conditions. The initial experiences in this area are related to the reports of the crew of the[38] Belgian Antarctic Expedition, organised by a Belgian naval officer and polar explorer Adrien de Gerlache on the ship "Belgica" (the name comes from the Roman province of Gallia Belgica), trapped in the Antarctic ice for 15 months between 1898 and1899. It was the first wintering-over in the history of mankind in the Antarctic ice (de Gerlache, Arctowski & Rakusa-Suszczedwski, 2016). The next Antarctic wintering-over took place mainly in scientific and meteorological stations in Antarctica and the Antarctic, as the only place on Earth that is not permanently inhabited by humans, and thus isolated from the rest of the world by a geographical barrier, which is treated by many researchers as a natural analogue of interplanetary flights (Terelak, Kwarecki, Rakusa-Suszczewski, 1978).

When discussing various deprivation (isolation) situations, we omitted the data on the Antarctic isolation situation. The characteristics of the stress factors of the Antarctic environment suggest that Antarctic isolation is a particular case of social isolation, since it includes both some elements of perceptual deprivation, monotony, imprisonment and classical experimental isolation, mainly linked to the reduction of information from the outside world and the failure to meet many basic needs. Thus, we treat Antarctic isolation situations as a stressful situation both in terms of physical (climatic-geographical conditions, light and dark conditions, etc.), as well as psychological (lack of gratification of numerous needs) and social characteristics (functioning in a small group of people for quite a long time).

Experienced polar explorers from New Zealand – A. J. W. Taylor & M. M. Brown (1994), in the course of summing up several decades of psychological

---

38 The Belgian crew also included the Norwegian Roald Amundsen (polar explorer, first conqueror of the South Pole) and two Poles: Henryk Arctowski (geographer, geophysicist, geologist, meteorologist, glaciologist) and Antoni Bolesław Dobrowolski (geophysicist, meteorologist).

research in Antarctica, believe that their main advantage is that they are empirical and reproducible under comparable conditions of isolation and threat, and that they offer solutions that maximise human performance and pose questions about the limits of performance, contributing to the body of knowledge about man and the extreme environment. In turn, H. Ursin et al. (1991) underline that placing emphasis in Antarctic research on the efficiency of functioning in extreme conditions makes it possible to treat it as an analogue of future long-term interplanetary flights (cf. also Ursin, Etienne, & Collet, 1990). A series of empirical data collected on the subject seems to support this general thesis (Palinkas, 1986; Suedfeld, 2012). However, the data has different scientific values. Therefore, we will pay attention mainly to research conducted systematically by psychologists and psychiatrists in Antarctic research stations and to their development trends, signalling only the data from observation made by medical doctors participating in pioneering polar expeditions.

Despite the research discussed so far, carried out in various space analogues (e.g. shelters, isolation chambers, submarines, etc.), most of the research on long-term social isolation was carried out in Antarctica, as evidenced, inter alia, by an extensive literature review carried out by G. M. Sandal, G. R. Leon & L. Palinkas (2006). There are currently 40 year-round research stations in the whole Antarctica Continent and in sub-Antarctic regions, belonging to 28 different countries, with wintering-over groups of 10 to 250 polar explorers (these include male only and mixed crews, including family crews, as in the case of Chile), staying in social isolation for a period from several months to three years. For the sake of order, we should remember that the pioneers of systematic psychological research who examined 85 men wintering-over in Antarctica were C.S. Mullin, H. Connery & F. Wouters (1958), who, for the first time in psychological literature, identified the syndrome of the adaptation symptoms of winter-over participants, called "Antarctic fugue," alias "big-eye," which consisted of symptoms such as: distraction, difficulty in focusing attention, deterioration in situational awareness, irritation, etc., which appeared after a few months in Antarctic isolation. Subsequent studies carried out in the 1970s and 1980s supplemented this syndrome with other symptoms observed during the studies, such as depression, hostility, mood swings, sleep disorders, etc., which remained in the literature on the subject as "Winter-Over Syndrome") (Terelak, 2021).

An extensive programme of medical and psychological research is connected with the International Geophysical Year (IGY) 1957–58. Although the IGY scientific programme did not officially cover these fields, the research was undertaken on the initiative of individual researchers or scientific centres (mainly military ones). The research was conducted mainly by American psychologists. For example, C.S. Mullin & H.J. Connery (1959), forming a psychiatric and psychological team, conducted research in the "Ellsworth" station during the Antarctic winter (1957–1958). In addition, C.S. Mullin (1960) and J.H. Rohrer (1961), as psychologists, visited many American Antarctic stations, doing interviews on the stress factors of the Antarctic environment and problems of individual and group adaptation. W.M. Smith & M.B. Jones (1962) developed methods for the selection of scientific personnel going to Antarctica and studied group structure and group interactions in isolation. F. McGuire & S. Tolchin (1961) carried out test studies, observation logs and medical research in the "South Pole" station in the context of adaptation to life in a small isolated task force.g. Palmai (1963a) started psychological research in a small Australian scientific station on the island of Macquarie on subjective complaints about discomfort related to physical and psychosocial factors.

Systematic psychological research was only initiated by American psychologists in 1962 under the auspices of the Naval Bureau of Medicine and Surgery. This research was conducted continuously between 1963 and 1971 on seven Antarctic stations: "South Pole," "Byrd," "Eights," "Hallett," "Palmer," "Plateau," "McMurdo." The psychophysiological research started under the auspices of the British Antarctic Survey after 1960. A detailed programme and discussion of the research results is presented, inter alia, by O.G. Edholm (1965).[39]

Medical and psychological research at that time in Antarctic stations in other countries, such as France, New Zealand, Japan and the USSR, was conducted sporadically. Other countries, however, such as Argentina and Chile, have not carried out scientific psychological research in their fairly numerous Antarctic stations. Polish medical and psychological research has been conducted since 1977, i.e. since the establishment of the H. Arctowski PAS station on King George Island (South Shetland).

Conclusions from the polar research conducted so far suggest, among other things, the creation of appropriately equipped physiological and psychological laboratories in research stations in the Antarctic region in order to expand the scope of the research conducted so far and increase its scientific value, which was to contribute to the creation of a common methodological platform and discussion on the results obtained. The scope of the issues examined so far, the

---

39 At the time, only 12 countries were members of the *Scientific Committee on Antarctic Research* (SCAR), which authorised them to start scientific research in the Antarctic. Poland became the 13th member of this international organisation only in 1961 (cf. Rakusa-Suszczewski, 1973).

research methods and techniques used, as well as cultural differences, give these demands the highest priority and justify further research from the perspective of long-term interplanetary flights.

O. Wilson (1965), in the course of analysing the problems of physiological and psychological research conducted in Antarctic stations until 1963 from the perspective of human biology, distinguishes the following issues: adaptation to cold, circadian rhythms, workload, food, metabolic and hormonal changes, cardiovascular functions, dependence of human functioning on weather conditions, microclimate, ambient temperature, winds, etc. These studies were usually carried out in very difficult conditions with the use of available equipment. Also, studies concerning adaptation to life in small isolated groups were conducted mainly based on medical observations. The programme of systematic psychological research, developed by American psychologists and carried out between 1963 and 1971 in seven Antarctic stations, is basically the only basis for the separation of the main psychological research problems. At the same time, it is quite representative of almost all the research carried out in this corner of the world, by researchers from New Zealand and Australia who have participated in this programme or have used its patterns. Later Antarctic research was actually only a continuation of this programme. E.K.E. Gunderson (1973), in an attempt to summarise the research carried out in the years 1963–1973, identified the following groups of problems: a) selection of station personnel and development of research techniques; b) emotional adaptation (individual and group); c) polarisation of a small task force in isolation.

a) Selection of Antarctic personnel – derives from the experience of occupational psychology, interested in individual differences in coping with stress of employees working in extreme stress conditions (Voas, Zedekar, 1963). Antarctic scientific stations provide an opportunity to check the prognostic accuracy of the applied methods of choice and selection, not only of future polar explorers. An Antarctic "laboratory," according to the E.K.E. Gunderson (1966a,b), has an advantage over others in that, apart from the extreme physical conditions of the environment, total social isolation, characterised by two important characteristics, is present: (a) it is not possible to evacuate the crew or provide external assistance during the Antarctic winter; (b) it is not possible to withdraw from living in a small group during long periods of isolation.

Volunteers who winter-over in the Antarctic are not typical representatives of the general population, as they are primarily selected for good physical health. In addition, they are usually people with a high level of motivation, with specialist education and a profession that is useful for specific mission objectives. The average age of American and Soviet polar explorers in the first period of Antarctic exploration varied between 20 and 45 years, but most people were about 35 years old. There is, however, a tendency, for example, in American and Soviet Antarctic stations, to shift age limits down, especially for scientific researchers (cf. Gunderson, 1964b, Ilyin et al., 1969). The composition of American expeditions has been quite typical for years. This is especially true of the high percentage of military personnel in all expeditions (especially from the US Navy), who usually represent such professions as: transporters, constructors, maintenance personnel, medical personnel (including psychologists), communicators, cooks. Civilian personnel usually secure administrative and scientific activities (cf. Gunderson, 1964a). On the other hand, almost all other countries involve civil personnel, who already have some experience in polar or mountaineering work, in their Antarctic expeditions (Taylor, 1969, 1973). However, the average member of the expedition (apart from the station managers) usually has no experience in the field of resistance to cold and social isolation.

Professional competence the prime initial criterion for selecting Antarctic personnel. An Antarctic group is made up of specialists of various professions. They also differ in terms of education, culture and personality. This creates many psychological problems during winter isolation, affecting both group morale and performance. As the American experience shows, the dichotomous division into military and civilian personnel with two equal leaders and two subgroup organisations with different psychosocial structures is a potential source of interpersonal conflicts (Palinkas et al., 2000).

The very general characteristics of the members of Antarctic expeditions presented above show, among other things, that the basic aim of selection programmes is, first and foremost, to develop selection techniques with good predictive accuracy. The development of selection criteria is usually preceded by an analysis of the organisational objectives of individual expeditions. Thus, for example, in the first Antarctic expeditions, attention was paid mainly to individual effectiveness, and thus the competence and professional skills of individual participants of the expedition[40] were taken into account. However, the assessment of suitability for isolation conditions only according to the professional criterion proved to be too complicated, because in a situation of isolation of a small group

---

40 An example of a personal approach to selection research are J. Rivolier, C. Bachelard, G. Cazes (1991), who used a space simulator for psychological and medical research of candidates for French Antarctic expeditions, in which they evaluate: the cognitive style, attention, emotionality, fear, defensiveness, global behaviour, etc.

of people, each of its members has to perform numerous tasks, usually going beyond one professional specialisation (Gundersan and Nelson, 1962a).

Successive research on the criterion of adaptation to Antarctic isolation situations has provided data confirming that effective adaptation behaviours are the result of the interaction of the following three components: (a) task-oriented motivation, (b) emotional balance, (c) ability to live together in a group. The development of methods and techniques for the measurement of these components is the subject of a number of studies (cf. Gunderson and Nelson, 1966; Gunderson and Ryman, 1971: and others), most of which are focused on the selection of astronauts and cosmonauts planned for interplanetary flights (Jones & Annes, 1983; Santy, 1994; Galarza & Holland, 1999).

The psychologists of all countries that have their Antarctic stations and send polar expeditions every year must solve the problems of validating selection techniques. However, there is a different predictive value of the techniques used and of the criteria used to assess the adaptation of a group that is significantly diversified from the social standpoint. It is therefore worth discussing in more detail the independent variables used in selection studies by American psychiatrists and psychologists for adapting to long-term situations of isolation based on the ancient Platonic paradigm of individual differences between people, dividing them into those who are capable of withstanding long periods of social isolation and those who have difficulties with it. The differences between people who do better or worse can be tested on several levels: cognitive, psychophysiological and biological one. For methodological reasons, selection procedures tend to focus on one of the following levels

b) Adaptation to social isolation at Antarctic stations – was first assessed on the basis of psychiatric biographical inventories on mental health status, occupational status, interests and hobbies, family and upbringing environment, etc. In contrast, data on the level of functioning of the person at the Antarctic station was collected on the basis of assessments by station managers. The results of the biographical questionnaire were then correlated with the evaluations of Antarctic station managers as an external criterion for the accuracy of the diagnosis. The first analyses of this kind were carried out by American psychologists on two small (15–40 people) and one large Antarctic stations (about 100 people) during the IGY (cf. Gunderson & Nelson, 1965a). The study was then continued at the large "McMurdo Sound" base (approximately 250 people) over a two-year period, based on two independent assessments by station managers. Among other things, it was found that, of the many biographical variables, only work experience and general social maturity corresponded with adaptation criteria (manager assessments). The focus was therefore on the validation of biographical inventories over the period of subsequent five years. The study was conducted at all US Antarctic stations, separately for military and civilian personnel. The adaptation criterion included three categories: emotional constancy, task motivation and socialisation. The correlation coefficients between the various categories of the biographical inventory and the adaptation criteria were so low that the authors suggest little utility of the biographical inventory for predicting adaptation under Antarctic station conditions. Rather, they believe that it can be used to help interpret the results obtained using personality inventories.

Personality traits such as autonomy vs. authoritarianism are also included among the correlates of adaptation by American psychologists. E.g. W.M. Smith (1961) found that independence from others correlates positively with general indicators of adjustment. In turn, B.B. Weybrew, E.L.B. Molish, R.P. Youniss (1961) highlighted such correlates of social adaptation as group structure and leadership styles (democratic vs autocratic). Other variables can also be added to this, such as the degree of extremity of the environment, living conditions, culture of group members, etc. (cf. Gunderson & Mahan, 1968).

We will not discuss in detail the selection issues peculiar to representatives of polar psychology from other countries with their own research stations in the Antarctic, as they are mostly based on American models (this applies to Australian and New Zealand psychologists, for example). Thus, for example, the Australian psychologist G. Palmai (1963) states that the basis of Antarctic selection is primarily a psychiatric interview and personality and intelligence tests. In turn, the New Zealand psychologist A. J. Taylor (1973) adds the narrative method and analysis of letters sent from Antarctica to the country. A detailed description of the selection techniques used by French psychiatrists and psychologists is provided by L. Crocą, J. Rivolier, G. Cazes (1973), who include, among other things: medical and biographical questionnaires, psychiatric interview and intelligence and personality tests (e.g. MMPI, Rorschach, TAT, etc.). In addition, various types of scales specifically elaborated for this purpose are used: adaptation assessments at the station, retrospective assessments after return to the country, observation scales for station doctors, and analysis of medical consultations. Based on the latter techniques, adaptation criteria are developed.

The selection techniques used during the first period of Antarctica's exploration by Soviet and Japanese polar psychologists were primarily psychophysiological tests, although some personality tests and biographical inventories were also used (cf. Bogoslovskij & Kriwo, 1975; Hachisuka, 1971). In turn, the selection techniques used to select Polish Antarctic expeditions include: clinical interview (psychiatric and psychological), MMPI, MPI, Cattell's 16 PF and, experimentally, Temperament Questionnaires (cf. Halter & Terelak, 1980; Kwarecki & Terelak, 1981).

Similar studies to those discussed so far, enriched with preselective adaptive training, were conducted by D.L. Lugg (1980) as part of the scientific project The International Biomedical Expedition to the Antarctic (IBEA), which involved 12 scientists with different specialisations. Initial adaptation to the cold in the form of an intensive training programme, covering ten immersions in a water bath at 15°C for up to one hour, took place at the *Commonwealth Institute of Health*, University of Sydney in October/November 1980. In March 1981, the subjects were transported to the French Dumont d'Urvillem Antarctic station for a ten-week journey across the Antarctica plateau (travelling by motorised sledge and living in tents). During the expedition, participants filled in questionnaires, and had interview sessions to assess their acclimatisation to the cold and were subjected to psychophysiological tests of body performance. The predictive accuracy of this form of recruitment and selection of Antarctic polar explorers is discussed by the authors in a monograph (Rivolier et al., 1988; Rivolier, Bachelard & Cazes, 1991).

Summing up the discussion so far on the prediction of Antarctic personnel selection methods, it should be stated that their accuracy, although far from expected, is generally assessed as sufficient.

Emotional adaptation and motivation – are the primary criteria for predicting adaptation to Antarctic isolation conditions. The second condition is the motivation to stay in difficult conditions.[41] Already the first Antarctic experiences have shown that motivation in individual expedition members is quite diverse. So, for example, according to research by E.K.E. Gunderson (1968) high levels of manifestation of motivation in military personnel prior to an Antarctic expedition correlate negatively with subsequent emotional adjustment to life on a polar station. This relationship is not found in civilian personnel. According to the author, military personnel have quite unrealistic expectations of an Antarctic expedition, which, after encountering extreme conditions, causes a lot of cognitive dissonance, disappointment and negative emotions. For these reasons, among others, research is being conducted into understanding the motives behind the decision to participate in a polar expedition in the context of subsequent emotional maladjustment. So, for example, F.E. Pope & T.A. Rogers (1968) conducted a test among participants in an isolation experiment involving walking a

---

41 Motivational psychologists define it as an intentionally directed emotion (cf. Gasiul, 2007).

200 km stretch in the Alaskan region (over a period of 10 days). They identified the following motives for taking part in this original, yet extremely difficult expedition (ranked from 1 to 5): (1) professional interests, (2) adventure and risk, (3) friend's example, (4) desire to test own fitness, (5) other. The authors found that those who were interested in their research in the experiment described (rank 1) were characterised by lower negative emotion indicators (apathy, depression, loneliness, etc.). In turn, A.J. Taylor (1973) gives the following order of motives for New Zealand polar explorers (questionnaire survey): (1) financial (41%), (2) adventure (27%), (3) prior polar experience (24%), (4) competition with colleagues (20%), (5) desire to work in a close-knit team (18%), (6) risk (17%), (7) separation from family and social environment (10%). Other interesting data on the structure of the motives of Soviet polar explorers are cited by I.P. Volkov, A.L. Matusoy, I.F. Ryabinin (1976). The study was conducted by doctors of the 15th Soviet Antarctic Expedition (SAE) among 209 members of groups wintering-over at 7 Antarctic stations. About 50% of the expedition members were taking part in an expedition for the first time in their lives, others already had experience from two or three previous expeditions. About 4% knew nothing about Antarctic conditions, while 46.9% drew their knowledge from friends, publications and books. The study was of a questionnaire in nature. The questionnaire included eight categories of motives: (1) Financial – 79.9% (e.g. desire to buy a house, car; improving standard of living, etc.); (2) Adventurous – 78.9% (e.g. attractive trip, opportunity to travel abroad, visit other countries and Antarctica, enriching knowledge and experience); (3) Professional – 56.5% (e.g. opportunity to work in unusual conditions), (4) Prestigious – 46.8% (e.g. participation in work in a group with high social prestige); (5) Heroic – 41.6% (opportunity to prove oneself in difficult conditions; (6) Teleological – 33.0% (e.g. a dream since childhood, the only goal in life); (7) Random – 19.0% (change of job, life path change, escape from family problems); (8) Scientific – 19.2 (prospects for scientific work, completion of research, new scientific problems, etc.). As can be seen from this breakdown of motives, two representative categories of motives are conspicuous, namely financial (79.9%), and adventurous (78.9%) ones. Respondents could, of course, give several motives, and, according to the authors, they often had "internal conflicts" in assessing their validity. The structure of individuals' motives depends on their psychological and social situation. For example, there is a characteristic trend among young workers with three dominant motives: financial, adventurous, and heroic ones. These individuals, observed during wintering-over, also exhibit neurotic symptoms and depersonalisation more frequently than others. They often think they made a mistake by going to the Antarctic regions. By correlating the different categories of

motives (questionnaire index) with the assessment of adaptation in Antarctic isolation (rating scales), the authors found, among other things, the following trends (p<0.05–0.01): (1) People with predominantly financial motives are more practical, disciplined, efficient at work, submissive, etc.; (2) People with adventurous motives tend to be abstemious, introverted, cautious, overly concerned with their health, with quite limited insight into their problems; (3) People with predominantly professional motives tend to manifest excessively their discipline, diligence, serviceability, kindness towards others, truthfulness (these people tend to be satisfied with their stay in Antarctica, are easy to manage, do not focus on their own problems but on the task, etc.); (4) The prestige motives are not very effective, because this category of motives does not exist in Antarctic conditions, as everyone has a different formal status – a winter-over participant (however, they are of great importance before the expedition starts); (5). Heroic motives require the possession of high moral qualities and a well-developed imagination (in the actual polar conditions people characterised by this type of motives manifest hypochondriacal symptoms, hypersensitivity, indiscipline in the situation of Antarctic isolation and that is why they are called neurotic-compensatory motives); (6) People for whom a polar expedition is "their only aim in life" are generally truthful, resourceful in life, not very capricious, sacrificing themselves for others. Although, on the other hand, they often present emotional disorders of the hypochondriac type, have a "sneaky" nature and are thick-skinned; (7) Task-related motives of a scientific nature are characterised mainly by the group of scientists (hence such a low percentage indicator – 19.2), who, on the one hand, are independent from the group, and on the other, are attached to task-related work and better adapted to Antarctic conditions than others; (8) Persons who cite "the desire to escape from everyday life" (family, professional, personal) as their motive have unrealistic expectations, as it turns out that they are often transferred to an environment of a small isolated group, disorganising its social life.

Overall, the research presented here suggests that the motives that influence the decision to participate in a polar expedition also have a directional effect on emotional adjustment to Antarctic isolation conditions and job performance. Although financial and adventurous motives are the most typical, the authors believe that the most valuable motives, nevertheless, are the professional and scientific motives of a task-oriented nature combined with financial ones.

However, there are many different motives for causing, among other things, symptoms of emotional maladjustment, as a number of empirical studies have convinced us. One of the first studies on this topic was conducted by C.S. Mullin & H.J. Connery (1959) and J.H. Rohrer (1961), who, while they did not find – even during the most difficult periods of the polar night – the occurrence of psychiatric symptoms, acute depressive states, pronounced symptoms of neuroses or psychosomatic diseases, noted periodically occurring symptoms of unease, general apathy, depressed mood, headaches, sleep disorders, hypersensitivity, withdrawal from action, etc., which make up the syndrome of emotional maladaptation.[42] R.E. Strange & W.J. Klein (1973), after having analysed other studies, found that symptoms of emotional maladaptation of Antarctic"winter-over participants" are usually described in four categories: depression, hostility, sleep disturbance and decreased mental performance. Having analysed papers on the subject up to 1971 they support the following characterisation of the emotional maladaptation of Antarctic "winter-over participants:" (1) Depression – symptoms of depression are recorded during the Antarctic winter in most station personnel as: depression, sluggishness, feelings of hopelessness, etc. (e.g. in 1969/71 these symptoms occurred in approx. 72% of American polar explorers); (2) Hostility – symptoms described most frequently as feelings of irritability and outbursts of aggression (e.g. in 1969/71 they such symptoms were found in 65% of American polar explorers; (3) Sleep disorders – appearing in almost all the polar explorers as: insomnia (big eye) (60%), symptoms of falling asleep late or sleeplessness (most often appearing from the middle of winter until the end of it); (4) Decreased mental performance – manifested in approx. 40% of the "winter-over participants" as: intellectual "slowing down" (inertial nature) and disorders of memory and ability to concentrate attention. The symptoms cited above are known in the literature as the "winter-over syndrome." E.K.E Gunderson (1963), on the basis of a questionnaire study of the crews of two successive IGY expeditions 1959–1961, analysed the presence and dynamics of somatic and emotional symptoms in groups differentiated in their occupational roles (military, N=80 and civilians N=78), which show, among other things, that a number of somatic and emotional symptoms occur in percentage terms more frequently at the end of wintering as: despondency, difficulty falling asleep, ease of irritability, feeling lonely, nervousness and tension, waking up at night, difficulty concentrating, restlessness or worrying, feeling tired during the day, criticising others, and that these symptoms occur more frequently for military personnel.

---

42 According to some researchers, the episodes of psychiatric symptoms that were sometimes found in polar explorers during wintering-over were the result of pre-existing disorders that only manifested themselves with increased force under the conditions of Antarctic isolation. The authors believe that the failure to detect these signs in the initial study is the result of faulty selection (Strange, Klein, 1973).

A number of authors looking for causes of emotional disturbance in Antarctic winter-over participants believe that the increased reduction in cognitive and affective stimulation during the Antarctic winter is associated, among other things, with a near-total reduction of physical activity. We are convinced of this, among other things, by the research of E.K.E. Gunderson (1968), who traced the dynamics of these disturbances as a function of geographical location and the size of Antarctic stations. Comparisons were made between "Byrd" station with a capacity of about 30 people, "South Pole" station – with about 20 people, and the three smallest stations: "Hallett," "Palmer" and "Eights" – accommodating about 10 people. The results suggest that the crews of the smallest stations manifest more symptoms of anxiety and depression and fewer symptoms of sleep disorders compared to the staff of larger stations. Symptoms of hostility occurred at small stations already at the beginning of the wintering-over period (they had a chance to get to know each other faster and get bored with each other). Thus, it can be said that the size of the stations and the structure of the isolated group influence, among other things, the magnitude of emotional stress effects. G. Palmai (1963), taking the complaints of wintering-over participants of Australian Antarctic stations as indicators of somatic and emotional disorders, presented the dynamics of changes in the said disorders in the different quarters of the annual Antarctic isolation. It shows, among other things, that an increase in visits to the doctor's office occurs in the second and third quarters, i.e. during the Antarctic winter and the duration of the polar night. It is characteristic that these symptoms decrease towards the end of the wintering-over period, which is undoubtedly related to the reduction of many stress factors. In turn, E.A. Ilyin, B.B. Egorov (1970), after having analysed the so-called winter-over syndrome at the Soviet "Vostok" station in 1968, draw attention mainly to their non-specific nature. The observed behavioural changes have, according to the authors, a weakening direction, entailing a deterioration of human adaptability. Clinical observations over a one-year period of isolation confirmed the hypothesis that neurasthenia gradually transforms into asthenia. The main symptoms include loss of interest in work and lack of desire to make an effort. While there was no deterioration in mental performance as measured by various psychological tests, greater fatigue and a decrease in motivation for mental work were found during the Antarctic winter. Polar explorers furthermore manifested considerable variability of mood and changeability of character. The weakening direction of changes in polar adaptive behaviour found above is confirmed by A.B. Blaekburn, J.T. Shurley, K. Natani (1973), who, in the course of conducting psychological and clinical research using the MMPI test at the American "Plateau" station, observed an elevation of the so-called neurotic triad and neurasthenic symptoms during the Antarctic winter.

Summarising the clinical and psychological problems of emotional adaptation of Antarctic winter-over participants, it should be stated that. the "winter-over syndrome" does not prevent members of polar expeditions from functioning individually. Therefore, although the symptoms described above may cause subjective discomfort, they cannot be classified as emotional "diseases," but rather treated in terms of the so-called psycho-physiological cost associated with adapting to a situation of long-term social isolation. This thesis is supported by the research of A.J. Taylor (1973), who dealt with the emotional problems of readaptation to a normal situation after returning home from Antarctica. The research conducted in the first month after the participants' return to the country shows, among other things, that the test subjects again developed emotional symptoms similar to the classic "winter-over syndrome," namely: increased visual sensitivity to colours (24%), fear of traffic (24%), intolerance and irritability in social situations (22%), increased sensitivity to sounds (17%), unspecified feelings of anxiety (17%), fear of being away from home for long periods of time (11%), desire to return to the Antarctic station (11%), and trouble engaging in satisfying sexual activity (9%). The symptoms described suggest that we are dealing with secondary social adaptation

Social adaptation – alongside individual *adaptation,* is an important research canon of Antarctic psychology (Stuster, 1997). In our deliberations on the selection of Antarctic personnel, we found, among other things, that high professional diversity makes small groups heterogeneous in terms of culture, psychological needs, values, attitudes and personality traits. These factors, as many studies indicate, significantly influence the polarisation of a small group and therefore social adaptation. So, for example, K. Natani, J.T. Shurley, A.T. Joern (1973) found that interpersonal relationships are first formed between people performing similar work tasks at the station. They have shown that the occupational groups: military administrative-technical, military-artisan and civilian scientific-technical groups are significantly differentiated among themselves in terms of preferences for leisure activities and forms of social interaction. These differences are particularly emphasised in small Antarctic stations, where emphasis is placed on workload, responsibility, education, personal suitability, allocated living space at the station, etc. Personality similarities within subgroups may underlie expectations of individual emotional adjustment within subgroups, while differences in value systems between subgroups may disrupt the social adjustment of the whole isolated group. Thus, social adaptation is determined, on the one hand, by the individual level of emotional adaptation and, on the other, by the effectiveness of the group as a whole and its degree of polarisation. Professional diversity of subgroups can be both a source of satisfaction, related to the sense of one's own usefulness within the overall objectives of the station, and a cause of interpersonal stress. These hypothetical situations are often the basis for distinguishing the configuration of the "best" and the "worst" groups in terms of social adaptation. Experience shows that in a small isolated group, interactions between such separated poles of social behaviour are possible, but they reach a degree of functional interdependence, hindering the degree of polarisation of the whole group. The authors also stress that in a small isolated group, social processes are not as static as task motivation or individual emotional adjustment. Indeed, the assessment of social adaptation depends on the duration of isolation. So, for example, during 8 months of isolation, group stability may never be achieved. A different view is taken by E.K.E Gundersoin & D.H. Ryman (1971), who believe that such a view is due to a lack of empirical data. They suggest the introduction of more "sophisticated" research and data analysis techniques, such as: the use of multiple indicators of social behaviour and multivariate analysis, the use of so-called discrete techniques, and the repetition of measurements over long periods of time with as high a frequency as possible. Especially this last postulate seems to be relevant and taken up in own research.

The first information to emerge on patterns of intra- and intergroup interaction at small Antarctic stations came from studies based on trusting single observers (most often expedition doctors). Based on such data J.H. Rohrer (1961) distinguished three phases of social adaptation that occur independently of group composition and duration of isolation: (1) An initial period of high anxiety correlating positively with the magnitude of individual feelings of threat; (2) A period of reduced anxiety accompanying the experience of generalised depression present in all group members; (3) A final period, involving preparation for return to the country, characterised by an increase in hostile expressions. E.K.E. Guinderson (1966d) believes that cultural and personality differences manifested in typical, occupational subgroups (military vs. civilian) in US polar explorers significantly impeded social adaptation and group integration. For example, it was found that the high variability of intergroup interactions is influenced by the "summer" period. This is explained in more detail by J.T. Shurley (1970), suggesting – on the basis of observations – that there is a heavy workload at an Antarctic station during the "summer" period and a need to change working styles. Moreover, the additional crowding in a small space due to the presence of a team of several people (and sometimes dozens people) is a major discomfort, which is reinforced by transmitted behavioural patterns from "old winter-over participants."

The above data suggests that the variability and direction of social interaction in a small isolated task group differs significantly from an analogous group functioning under "normal" conditions. However, it is not clear what exactly the category of "norm" means in psychology (Terelak, 2012), especially since in social psychology the so-called desired norm as the "cohesiveness" of a group is often understood by analogy to its meaning taken from physics (e.g. cohesiveness of gases, liquids) and is supposed to denote the fact that a group consisting of many members exists as a certain (coherent) whole. This concept was introduced to psychology by Kurt Lewin (1951) with his famous Symbolic Interactionism Theory,[43] who sees the mechanism of shaping social structures on the basis of the meanings of symbols, which appear during the processes of interaction between people, influencing our personality, which is the basis for actively determining our own interactions with the environment. Experimental social psychology has confirmed that cohesive groups influence their members more strongly and are more resilient to difficult situations than non-cohesive groups. Hence, polar psychologists' interest in this aspect of small task group structure stems from the fact that in situations of limited stimulation, "undesirable social interactions" (conflicts, aggressive behaviour, etc.) are counterintuitively adaptive in nature, as they are *per se* the source of additional social stimulation, often facilitating "survival" in long-term isolation. Thus, social adaptation can be viewed both from an interaction content perspective (e.g., group morale) and from a formal perspective (e.g., frequency and strength of interaction, etc.).

A number of studies on the social adaptation of Antarctic station personnel have highlighted various variables that modify the adaptation process. Thus, for example, in one of the first studies in this field by C.S. Mullin & H.J, Connery (1959), attention was turned to such determinants of social adaptation in an Antarctic isolation situation as personality structure, station leadership styles, group cohesiveness, expectations and task attitudes. The authors found, for example, that difficulties in social adaptation were mainly manifested by persons characterised by: intolerance, poor social judgement, lack of discretion, professional incompetence, lack of interest, quarrelsomeness, authoritarian personality, aggressiveness, lack of sense of humour, etc. In addition, P.D. Nelson (1962) believes that the factor that can facilitate or hinder social adaptation is mainly

---

43 Symbolic interactionism is a sociological theory derived from American philosophical pragmatism. It is based on three main theses: (1) People act on the basis of meanings that things have for them; (2) Meanings come from interactions; (3) Meanings are modified by their interpretations, made by people in actual situations.

the group leadership style. A group achieves cohesiveness relatively quickly if it can identify with its formal leader. The emergence of an informal leader makes group integration much more difficult. Citing negative interactions between the leader and the group, he stresses that although the leader was a high-calibre researcher, due to his little experience and preference for "rigid" interpersonal relations and "protectionist" principles, he led to a lack of integration of the group over a long period of isolation. Characteristically, however, the tensions between the leader and the group, while being a source of additional stimulation, had what the authors call a "salutary" effect on the group as a whole. Another factor determining social adaptation in a situation of isolation was pointed out, among others, by R.E. Doli & E.K.E. Gunderson (1971), who believe that group size should be taken into account alongside the occupational status discussed earlier. In their study, they found that the number of polar explorers wintering-over at the station (8–10 or 20–30 people) significantly affects social perception and group coexistence. E.g. irritability in social interactions is more common in small stations (around 10 people) than in larger stations (around 30 people). Autbers explain this with less living space and more limited social interaction. In contrast, there is no difference in aggression rates, which probably depends on the composition of the group, especially from its homogeneity vs heterogeneity.

The latter issue of gender heterogeneity continues to be debated, especially from the perspective of the colonization of Mars. We would like to remind you that the first woman wintering in Antarctica in East Base on Stonington Island (Graham's Land in Antarctica) in 1947–1948 was Canadian Jackie Ronne, who was part of the US Antarctic Expedition men's team. In turn, Silvia Morella de Palma was the first woman to give birth to a child in Antarctica, at the Argentinean Esperanza base on January 7, 1978, as part of a political project motivated by territorial claims. Since then, the number of women wintering in numerous Antarctic bases has steadily increased, although opinions about the sex of winterworms are still controversial. An example may be the opinion of Captain David Elliott, the commander of the Antarctic base in Port Lockroy, which is quoted in his Diarius by the Canadian wintering Jean McNeil (2016, p. 145): "- I do not really approve of the presence of women here (in Antarctica – author's note). /… / – It disturbs the balance that used to be here. Men start competing for women, and then community building is not easy. When I spent the winter here, we formed a unity /… / – We were invincible. "

R.E. Strange & W.J. Klein (1973), citing earlier work by G. Palmai (1963) and their own studies, argue that regardless of the above-mentioned factors modifying social adaptation, a small isolated task group is always characterised by specific adaptation dynamics. The authors list three stages of this adaptation.

In the first stage, the group is open, i.e. each member takes an interest in each individual and the group as a whole. The second stage is characterised by the formation of cliques or factions[44] of two or more people (e.g. professional, recreational, etc. factions). The dividing line between the factions is mainly related to professional similarity or interests. Age and authority also matter here. The third stage is characterised by the merging of the group around a "social core" (the asset). There remain, of course, members who are still isolated from the group (so-called satellites) and peripheral cliques. According to the authors, conflicts arising in the group are also characteristic of the three isolated stages of social adaptation. In stage one, conflicts arise between individual group members or pairs of them. In stage two, conflicts involve members who are isolated from the group or peripheral cliques. In the third stage, conflicts should be sporadic and involve people who break away from the group. A group is never completely stable in any of these three stages. E.g. a group that has already integrated may well regress to the clique stage. The greatest individuals usually balance between belonging to a group and becoming dependent on it. On American stations, it is traditional for factions to form between military and civilian personnel. Conflicts are also characteristic of both groups. Factions of the young against the old also form.

W.M. Boriskin,& S.B. Slewicz (1968) drew attention to a general regularity of small task group polarity that is most often revealed in isolated collectives in the Antarctic. The authors believe that up to four periods can be discerned under Antarctic station conditions. The group is not consolidated in the first period. In the second period, numerous, often random subgroups are formed, mainly with professional interests. In the third – group consolidation around the "asset" takes place, but it is not yet permanent. In the fourth – a breakdown into numerous sub-groups occurs, which, however, are not factional in nature.

Summarising the results of psychological and sociological research carried out in Antarctic stations, we can conclude that they are interesting not only from a pragmatic point of view, e.g. by facilitating the selection of the most effective members of groups of successive polar expeditions, but also from a theoretical point of view. As stated by E.K.E. Gunderson (1973a), the emotional and social functioning of a small task group under conditions of Antarctic isolation can be regarded as a model situation for a natural experiment, especially in terms of the so-called group dynamics.

---

44 These terms are used by the authors in a sociometric sense – as a subgroup of people supporting each other in a larger group.

In the review we have presented so far, we have repeatedly highlighted many variables modifying emotional and social adaptation (e.g. cultural, personality, geographic-climatic, expedition objectives, station size, etc.) that justify the purposefulness of undertaking further research, both in terms of validating selection methods and individual and group adaptation to situations of Antarctic isolation. As can be seen from literature review, such research is successively being carried out by all countries that have their own research stations in the Antarctic, which are currently experiencing a renaissance due to their treatment as an analogue of space research, preparing crews for long-term interplanetary flights and their terrestrial isolation on nearby planets, such as the Moon and Mars, or on planetoids or asteroids. Hence, with regard to the usefulness of the conclusions of laboratory experiments and those conducted in the natural environment of Antarctica (and partly the Arctic) as space habitats, we have drawn attention to the possibility of their application to areas such as: (a) selection for adaptability of future space crews; (b) moderators of resilience to isolation and confinement and random extreme events (e.g., sudden habitat failures, etc.); (c) group dynamics from the perspective of isolation duration (e.g., group cohesion, conflict propensity, informal leadership, etc.); (d) meta-analysis of previous research findings on the contribution of individual and group behavioural traits associated with adaptation and functioning in extreme habitat environments (McFadden et al, 1994, Bishop et al., 2010; Vakoch, ed. 2011).

Although the analysis of human functioning presented in this study mainly concerns a situation characterised by limited stimulation, it seems that, based on the proposed own theoretical model, presented in Fig. 4, the phrase "mutual regulation of stimulation levels" can be used, and the stimulation mechanism can be treated as one of the important mechanisms regulating human adaptive behaviour to conditions deviating from the range of optimal stimulation. It seems that the conceptual approaches signalled above are complementary rather than mutually exclusive, as they emphasise different aspects of the same deprivation situation. However, the usefulness of individual concepts varies. Some, for example, have ignored the issue of the physiological mechanism underlying the observed behavioural differences, others have sought to explain them by examining complex physiological mechanisms, and still others have reduced the differences to the level of neurometabolic functioning. We will address some of these issues in the next section on own research on the psychological cost of functioning in long-term deprivation and social isolation at a small Antarctic station from the perspective of a space analogue.

Part Two: Own Research

3. Own Research into the Functioning of Humans in Antarctic Isolation

The review of research to date on the effects of deprivation/isolation convinces us that these effects depend on the very structure of the situation and on many "internal" factors. One of the important factors modifying the effects of deprivation (isolation) is time. From this point of view, space psychology is increasingly interested in empirical data on human functioning under conditions of long-term isolation.[1] This type of isolation, as we tried to demonstrate earlier, is difficult to arrange under laboratory conditions. Therefore, research conducted in the natural conditions of Antarctic isolation remains of interest to cosmonautics.

3.1. Theoretical basis for the research

A review of research to date suggests, among other things, that "total" sensory deprivation is an extreme case that does not occur in human life. Therefore, there is more interest in research on the effects of prolonged partial sensory deprivation and social isolation as stressful task situations, which can be analysed from different perspectives of psychology: clinical, stress, environmental, and work psychology (organisational aspect), including space psychology (Connors, Harrison & Akins, 1985).

As we have shown in chap. 1, the key problem of environmental psychology, which is the subject of many empirical studies conducted in conditions of Antarctic isolation, is the ability of the individual as well as the task group to adapt to the requirements of the physical and social environment of existence, and especially to describe the multidimensional relations between man and the social environment, including the organisational relationship: leader – group. Adaptive behaviour, especially to extreme environments, is the subject of research to explain: "What do people do in order to use their personal competences and, while largely preserving their individuality, at the same time fit into its environmental requirements?" Antarctic research is also conducted within the paradigm of

---

1 Isolation in real spaceflight conditions, such as in the case of the record-breaking flight of Russian cosmonaut (doctor) Valery Polakov, lasted on board the Mir space station 437.7 days (14 months), but future interplanetary flights may last several or more years, hence the interest of cosmonautics in research on the psychological effects of social isolation is valid.

employee adaptation, which in the literature is sometimes considered in several aspects as: (1) The Person – Job Theory model of employee adaptation (Edwards, 1991), which assumes that people fit into the physical work environment. (2) The Person – Environment adaptation model (Tinsley, 2000), which assumes an adaptation between the person and the physical and social environment; (3) The Person – Organization adaptation model (Kristof, 2006), which assumes two types of adaptation between the person and the organizational environment: supplementary and/or complementary adaptation.

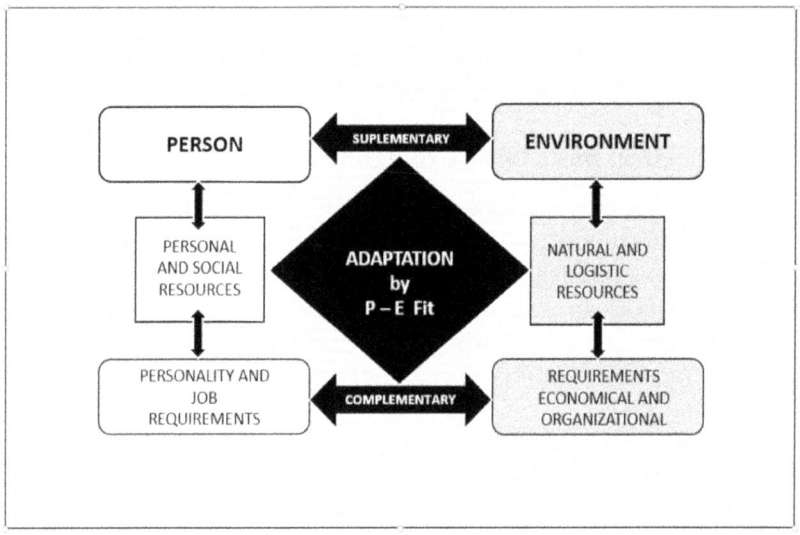

Fig. 11. Types of adaptation P-E
Source: own elaboration of A.L. Kristof. 2006.

The model of human adaptation to the environment presented in Figure 11 takes into account all three of the previously mentioned partial research perspectives: P-J, P-E and P-O: (a) The individual perspective of adaptation, covering such research issues as: learning new behaviours, motivation, personality, perception, training, ability, situation satisfaction, individual decision-making ability, judgement, attitudes, coping with stress (Schneider, Goldstein, Smith, 1995); (b) The group perspective of behaviours concerning such problems as: group dynamics, intergroup communication, attitudes towards management, intergroup conflicts, intergroup behaviour, organisational culture of individual's behaviour under group influence, changes in attitudes towards the group and group members, (Krech, Crutchfield, Ballachey 1962); (c) The systemic perspective, includes such issues as: comparative analysis of value systems, comparison of attitudes, cross-cultural analysis of organisational behaviour, analysis of management styles, etc. (Kristof-Brown, Zimmermanem & Johnson, 2005).[2] One of the first researchers to apply the P-E adaptation concept to Antarctic research was A. Sarris (2006), with an emphasis on the social aspect of Antarctic wintering-over, related to isolation and entrapment. In turn, the relevance of the P-E adaptation theory for assessing deprivation of need and coping with challenges faced by polar explorers wintering-over in Antarctica was assessed on the basis of empirical research under polar station conditions by C. Jaksic (2018). To this end, he conducted two studies. Study 1 used data from personnel currently wintering-over (n=14) in Antarctica, while Study 2 used earlier results from wintering-over participants (n=59). Results from both studies suggest, among other things, that adaptation to Antarctic isolation is positively related to cognitive satisfaction, performance and mood. In contrast, there is no relationship between adaptation to a lack of privacy, but there is instead a positive relationship with the need to belong to a group. The attempts of these two authors to draw on the paradigm of organisational psychology are interesting because there are just a few new theories in polar psychology, compared to other fields of psychology, to explain human adaptation to extreme situations of social isolation and entrapment. The authors furthermore rightly believe that wintering-over groups do not have the status of tourists but of employees, performing specific tasks in which they have to adapt to their new (exotic) workplace in an unusual environment in the first place. Perhaps more comprehensive models can improve our knowledge of ICE polar psychology and allow a better understanding of the relationship between predictive variables (studies of candidates for polar expeditions) and resultant variables (studies under real conditions of Antarctic stations). An

---

2   Some research authors have found that the two independent variables, P-E supplementary and complementary adaptation, can be combined into a single larger theoretical construct (Cable, De Rue, 2002), although they are inconsistent with the original model of Amy L. Kristof (2006), based on the assumption that complementary adaptation stems from supplementary adaptation in the same way that a person's behaviours and actions stem from the values they cultivate, the norms they follow, and the goals they set. Other researchers, such as D.M. Cable & J.R. Edwards (2004), believe that while these two types of adaptation are interrelated, they also have their own independent effects on outcomes relevant to both individuals and organisations.

example of such research is the work of C. Jaksic (2018), in which the author favours a complementary theoretical approach that takes into account various theories that he believes broaden our understanding of polar psychology, which deals with human functioning in unusual places on Earth.

However, it seems that the expectations of the above-mentioned authors regarding the explanatory power of the very detailed P-E adaptation concept as applied to Antarctic and space isolation are too optimistic, as they do not take into account the simple organisational structure at polar stations, which does not exist in other civilisation settings. It is for this reason, among others, that the more general concept of T. Tomaszewski (1978), discussed in chapter 1, adopted in own research, has a more multidimensional nature, typical of environmental psychology. Even if all its segments concerning the perception of the environment (i.e. determining "how it is"), being a synthesis of the relation that occurs between the evaluation of requirements or expectations ("how it should be") and the evaluation of one's capabilities ("how it can be"), do not explain the psychological mechanisms of perception of the Antarctic "new reality," they are at least the basis for an exploration of the transaction that occurs between subjective and objective states of affairs (attributes of the environment, situation, event), which has been studied for several decades at both the narrative (e.g. polar explorers' diaries) and psychometric levels, providing a sufficient basis for their synthesis reflecting a full cognitive representation of the Antarctic environment as an analogue of interplanetary flights. Such a representation is influenced by many mediating variables, such as: self-esteem (aspiration level, achievement level), emotions (positive vs. negative), intelligence, biological clock, etc., which can effectively change both the pattern of demands and the adaptive capacity of a person (Goldstein, 2004; Torbjörn, Göran, & Mats, 2007). Thus, it must be stressed that the narrative of human-environment relations alone is not sufficient to explain these interactions without knowledge of human-environment relations in the psychological sense, which takes into account the awareness of the importance that these relations have for human beings. For this, knowledge of the perception of the attributes of the Antarctic "micro-world" is needed, as well as the perception of other people, of oneself by oneself, of oneself by others, etc. Much detailed information on this subject can be found in the rich literature on the subject, the knowledge of which determines further research perspectives. An attempt to operationalise future research in this regard, using the P-E adaptation concept as an example, is highlighted, among others, by J.R. Edwards et al., (2006) proposing three methodological approaches: (1) Atomistic (correlational) – which involves considering the relevant characteristics of the environment and the person separately and then making a formal comparison.

This approach typically uses an objective measure to describe both elements. (2) Molecular (differential) – which pays attention only to the salient features that differentiate the two elements of the P–E relationship. This approach focuses primarily on the perceived discrepancy at the resource level between the person and the environment. (3) Core (interactional) – which involves matching the characteristics of the environment and the person with mediating variables.

In our own work we adopt the molecular approach – highlighting only the salient features that differentiate the two elements of the P–E relationship and focusing on the perceived discrepancy at the level of resources between the individual and the isolated social environment, which is operationalised as a psychological cost considered at different stages of Antarctic adaptation, and an interactional approach by referring to Tomaszewski's theoretical model, pointing to mechanisms explaining adaptive behaviour in the ICE environment.

3.2. Stress factors in the Antarctic environment

Olar explorers N.R. Deriap, A.Ł. Matusow & L.I. Tikhomirov (1978) argued from the dawn of cosmonautical development that Antarctica, known as the "ice desert," was by its very nature an extreme environment, unsuitable for human life in the long term[3]. Nevertheless, "Terra Incognita" is treated as a natural cosmic laboratory, allowing us to explore the limits of human adaptability to harsh living conditions. Both pioneering medical research, conducted mostly on an occasional basis, and systematic scientific research prove that the Antarctic provides a unique natural laboratory for human sciences to study physiological and psychological adaptation mechanisms, both in terms of physical and psychosocial factors resulting from long-term isolation (Shurley, 1973). In this order we will signal the nature of the various stress factors.

3.2.1. Physical environment of the Antarctic

Human spaceflight was not the only cause of interest in the Antarctic environment. As stated by the Polish biologist and polar explorer S. Rakusa-Suszczewski (1973), the subantarctic and Antarctic waters[4] are extremely rich in fauna

---

3   This view is shared by the winter writer Jean McNeil (2016, p. 212), who wrote in her Antarctic Diary: "After that one day at the base, I knew that we were in a different dimension, that Antarctica is like an extraterrestrial civilization that knows nothing about humanity. People have never lived here, so the land doesn't know us."
4   Antarctica includes only the continent, while the Antarctic also includes all areas (islands, seas) located in the so-called convergence zone. The term is used to describe

(marine mammals, fish, crustaceans, e.g. krill, etc.), while the continent itself possesses various natural riches, and may also constitute a "refrigerator of the Earth," suitable for storing surplus cereals and meat under international control. In addition, the largest continental-sized ice concentration (the volume of ice covering Antarctica is calculated to be about 31 million km) accounts for about 90 per cent of the Earth's freshwater resources.

Physical factors, unlike other environmental factors, can be easily quantified. The most commonly measured parameters of the physical environment include air temperature, humidity, radiation in the visible wave range and wind speed. Human ecology is not interested in the physical environment as such, but in the interaction with the functioning of living organisms, especially humans. The relationship between an organism and its Antarctic environment is quite highly variable. Scientific stations are located on the continent itself and adjacent islands and have different climatic and topographic characteristics, which means that studies conducted at different stations are not always comparable with each other (Collier et al., 1973).

According to E.K.E Gunderson (1974), each Antarctic station should be considered in at least two aspects of the physical environment: internal and external. Characteristics of the internal environment should include the degree of isolation and space available to individuals (e.g. movement around the site, accessibility, visibility, conspicuous natural terrain, layout and interiors of dwellings, individual living patterns, etc.). The degree of isolation, i.e. the physical distance to the nearest human concentrations, and the duration should be considered in interaction with other sources of stress. The external environment includes climatic conditions, latitude, wind speed, atmospheric pressure and altitude. Thus, for example, stations around 60° and 70° S may have a more temperate climate than stations closer to the pole, but they are exposed to hurricane-force winds that increase the cooling of the body.

The discomfort of Antarctic life is adequately reduced by means of the so-called artificial environment,[5] which includes, among other things, attributes such as polar clothing, shelter, food (calorific and vitaminised), tools for work,

---

    the zone where cool surface waters, girdling the continent, meet warmer subantarctic waters. This boundary is very clear and is an effective barrier to living organisms. In the Atlantic and Indian Oceans it runs behind 50° latitude, while in the Pacific Ocean it is shifted further south (according to the Antarctic Atlas, 1966).

5   The artificial environment created for survival in Antarctica is an analogue of the artificial environment in space and on other planets, known as "artificial satellites."

etc. Everything necessary for life (except fresh water, which must be vitaminized because it is a distillate) must be transported from the country (by ship or plane). Hence, the quality of life depends solely on the proper resolution of logistical problems and the affluence of the country of origin.

Since the International Geophysical Year, 81 polar stations have been permanently operating in Antarctica and the Antarctic islands, of which 40 are year-round and 41 are active only in summer. The distribution of stations operating in 2020 is illustrated in Fig. 12.[6]

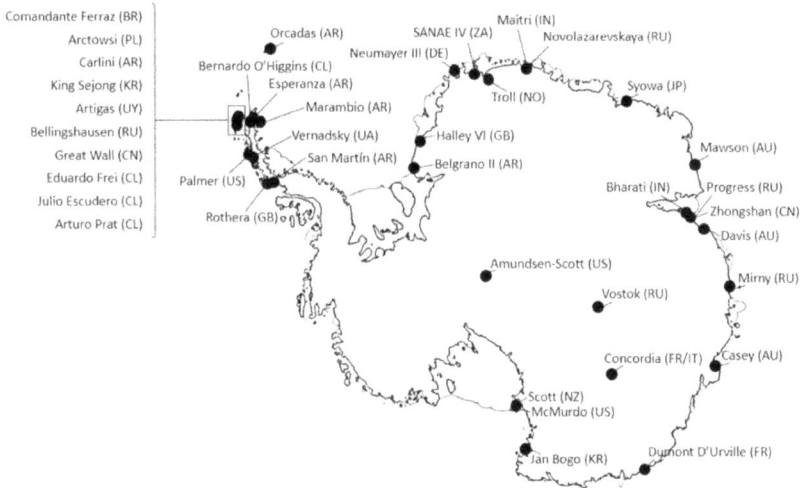

Fig. 12. Location of the 40 Antarctic stations manned during the 2016 winter
Source: elaborated after C. Jaksic, 2018; based on information from COMNAP, 2016.

It is worth noting the terrain and structure of Antarctica (the highest average altitude of all the continents), which influence its climatic conditions. According to data by P.D. Nelson (1963) the average annual temperature in the 1970s on the continent ranged from -1.1°C to -73.3°C.[7] The polar night lasts from two to six

---

6 A list of the distribution of year-round stations in the Antarctic and the continent can be found in Annex 1.
7 The lowest recorded naturally occurring temperature on Earth, -89.2 °C, was found on 21 July 1983 at the Soviet (now Russian) polar station Vostok. Even lower temperatures, as low as -93.2 C, are known from satellite observations. The warmest month is

months. Antarctica, especially the coastal regions of the continent, is the windiest area on the globe. Hurricane winds are caused by powerful storms, temperature differences between the continent and the ocean, and the inclination of the continent towards the sea, followed by the downward movement of descending air masses (known as downwinds). For example, in areas of East Antarctica, winds blow for 220 days a year at speeds in excess of 1.5 m/s, while off the coast of Adela's Land, winds blow for 340 days a year and exceed 23.6 m/s on average. Hurricane gusts can reach up to 90 m/s.

Completely different climatic conditions prevail, for example, inside the Arctic Circle. The centre of East Antarctica close to the geographic pole (the so-called Antarctic plateau zone) is characterised by cloudless weather throughout the year, extremely dry air, minimal precipitation, light winds up to 3 m/s (only occasionally reaching hurricane speeds). The West Antarctic Plateau, located twice as low as the East Antarctic Plateau, is characterised by a slightly milder climate, higher temperatures, precipitation and cloud cover, with frequent winds causing blizzards that can last up to 29 days per month in winter and up to nine days in summer. The coastal climate zone, on the other hand, varies between ice coast belts, glacier shelf areas or oases. Usually high precipitation (mainly in winter) and significant diurnal variations in atmospheric pressure of 20–40 or even 60 milli-bars prevail here (Palmai, 1962).

Summarising the general characteristics of physical stress factors, it can be said that the Antarctic is an environment with extremely harsh living conditions, which threatens to unbalance the internal environment of the human system. Hence, research problems concerning the effect of prolonged exposure to specific climatic conditions on the human system, i.e. acclimatisation problems, were the first to emerge (Dierjap, Matusow & Tikhomirov, 1978). Physiologists, studying these reactions at various levels of system organisation, usually mention three stages in the acclimatisation process: (a) initial – in which contact with new environmental conditions may become the cause of various ailments; (b) gradual adaptation – the progress of which depends, among other things, on the constitutional characteristics of the human being;[8] (c) acclimatisation – that is,

---

January with a minimum temperature of -30 °C, while the coldest month is August, with temperatures dropping to -70 °C. On the coast and subantarctic islands they can reach above 0°C.

8   Numerous syndromes of maladaptive neuroses (so-called meteoneuroses) often develop during this period, often making further residence in the given climate impossible.

relatively permanent adaptation to specific climatic conditions. The reverse process is also gradual and involves two stages: (d) deacclimatisation – which occurs when returning to the conditions of the previous climatic environment; (e) reacclimatisation – which occurs when returning after a long break to stressful climatic conditions (e.g. members of many polar expeditions are characterised by what is known as established acclimatisation).

The pronounced influence of weather on the state of human functioning is of course not only observed in Antarctic conditions. In fact, "meteotropic diseases" are known to exacerbate already existing symptoms in sick people (e.g. hypertension), and also cause some somatic symptoms and reduce mental performance in healthy people who are sensitive to atmospheric changes. The symptoms of a "meteotropic disease" in polar climates are mainly associated with sudden fluctuations in atmospheric pressure. They are characterised by pressure, ringing or tinnitus in the ears (and even temporary hearing loss), headaches or dizziness, a feeling of drowsiness, weakness, a tendency to faint, etc. Although all these conditions are reversible, they are experienced as discomfort that reduces the ability to work. In addition, toothaches, described as "dental hypochondria," often occur during Antarctic winters, which are not related to low temperatures or avitaminosis.[9]

Due to the wide variety of individual stress factors in specific regions of the Antarctic, studies related to acclimatisation are conducted at individual stations. As we have signalled, temperature spikes are a stress factor. For example, the differences in annual mean temperature at US stations located at different latitudes and altitudes are as follows: Byrd (−28.3°C; −28.3°C; 1515 m), South Pole (−49.4°C; 2799 m), Palmer (−6.7°C; 8 m), Plateau (−57.0°C; 3624 m), Eights (−25.0°C; 421 m), Hallett (−15.6°C; 5 m) (Blaekburn, Shurley & Natani, 1973).

Thus, knowledge of the various environmental factors in the area of each station provides the opportunity to supply expedition participants with appropriate equipment and clothing, as well as to conduct relevant research on adaptation mechanisms to the various extreme conditions of the Antarctic environment. With regard to the first problem, a number of studies show that adequate clothing, the provision of living quarters and laboratories, as well as provision of adequate food reduce physical stress factors to a level where functioning in extreme Antarctic environments is possible for relatively long periods (Budd, 1966; Kowalski, 1982).

---

9 The typical scourge of pioneering Antarctic expeditions, scurvy, is now a thing of the past due to developments in pharmacology.

An interesting study on acclimatisation in diverse Antarctic conditions is described by A.F. Rogers (1974), participating in a 12-man expedition (Trans-Antarctic Expedition) which, on the occasion of the International Geophysical Year (1957–58), spent 14 months in the Antarctic making an impressive passage from the "Shackleton" base located on the coast of the Weddell Sea, through the South Pole ("South Pole" base), to the "Scott" base located near the Ross Sea. The participants of the expedition covered a total of about 2900 km in 99 days, in various climatic conditions and at different altitudes (reaching up to 3015 m). During the expedition, individual expedition members carried out research and daily observations on: adaptation to cold, suitability of different types of protective clothing and food kits, circadian activity and sleep disturbance, etc. Among other things it was found that protective clothing protects more easily against cold than against hurricane winds. In addition, quite significant individual differences were found in adaptation to cold, in disturbances of circadian rhythms, as well as individually differentiated resistance to enduring the hardships of the expedition. The author points out that acclimatisation or adaptation to extreme Antarctic conditions is possible, however, it is of a relative nature, because it is connected, on the one hand, with a rather short stay and, on the other hand, with the "artificial" securing of existence.

The characterisation of the physical environment of Antarctica would be incomplete if we omitted such an important factor as the spatial and temporal structure of a polar station's location. Attention should be paid to the often overlooked problem of the variability of circadian and seasonal biological rhythms. In the characterisation of this aspect of the Antarctic environment, the constant change of the time zone and the change of photo-ecological conditions (relation: light-dark) must be taken into account in particular. There is ample evidence that spatial-temporal system change can modify human biological, psychological and social functioning (Harrison, Clearwater, 1991; Mallis, & DeRoshia, 2005). We discuss this issue further in the next chapter.

Antarctica is a good experimental "testing ground" not only for learning about the mechanisms of adaptation to extreme features of the physical environment, but also for learning about the mechanisms of psychosocial functioning of small task forces separated from their natural civilization "base," especially from the perspective of long-term space flight (Vakoch, 2011).

3.2.2. Psychological and social environment

The efficiency of action depending on the characteristics of the emotions depends to a large extent on the situational context varying in degree of difficulty. As an example, a stressful situation has the least disorganising effect on activities that involve reproducing learned skills in a stereotypical manner, while it has the most negative impact on activities whose performance requires inventing new ways of behaving (Terelak, 2019). From this point of view at least, the psychological and social environment of an Antarctic station should be described in terms of stressful situations, because staying in long-term isolation far from one's own civilisation, culture, family, is not natural for a human being, and is therefore characterised by "newness." This prompts often radical changes in the way of life and lifestyle, changes in the value of many physiological and psychological needs, changes in self-activity, many social roles, social habits, attitudes, etc.

Most of the physiological and psychological stress-induced effects of the environment we are discussing are non-specific in nature. It was not possible to measure all environmental factors. It was only possible to use methods of observing the "environmental circumstances" that regularly accompanied changes in work performance, social behaviour or to analyse introspective descriptions of emotional states. However, the narrative approach, despite the attempt to objectify it by using quantitative analysis of narrative content, is still burdened by the influence of the narrator's personality, life history and individual adaptation mechanisms, so that the observational results obtained are too specific and generally not comparable to others (Gunderson & Nelson, 1985a; Ehmann, Altbäcker & Balázs, 2011).

The classics of Antarctic research – E.K.E. Gunderson and P.D. Nelson (1964b) list the following most stressful attributes of a small Antarctic station: (a) Confined isolation, involving geographical and social isolation, emotional isolation with limited psychological space and possibility of escape from such situations; (b) Constant presence of the same people, lack of interpersonal alternatives, awareness of interdependence, etc.; (c) Permanent control of one's emotional impulses and aggressive behaviours, lack of heterosexual objects, inability to reduce anxiety, etc.; (d) Boredom and monotony due to the same physical environment, same faces, similar tasks, lack of diverse external stimulation, etc.; (e) Physical discomforts due to stressful factors of the physical environment, limited food assortment; (f) Heavy workload with minimal health and safety standards; (g) Status limitation, which involves reduction of many social roles and lack of direct financial remuneration. This is confirmed by many other Antarctic researchers (e.g. Mullin, 1960), and others such as G. Palmai (1963a) mention yet other factors of isolation stress: (a) Communication with relatives is severely hampered;[10] (b) Spending most of the time in a closed room ("*encapsulation*");

---

10 Today, in the age of satellite communications and mobile phones, this factor is no

(c) Darkness prevailing for several months of the year; (d) Being "condemned" to stay in a group with a fixed number of people.

The phenomenon of monotony is also related to the external environment, which is diversified very little, since during the polar night only the immediate surroundings are illuminated, while during the polar day the undifferentiated whiteness of the surroundings causes the so-called white darkness.[11] The monotony is also reinforced by the regular repetition of much of the work and ritual of the day[12]. Yet another stress factor is pointed out by W.H. Haythom (1973), who treats Antarctic isolation as a kind of a "micro-climate." Above all, the author stresses that incompatibility in intellectual and cultural demands is a major impediment to the formation of correct interpersonal relations in a small isolated group. According to S.B. Sells (1973) all the stress factors of isolation mentioned so far can be reduced to three basic ones: (1) Separation from one's loved ones and other social groups; (2) Restriction of living space to a specific enclosed area (confinement); (3) Restriction of external stimulation (perceptual deprivation). Aneven more general view of the stress factors of social isolation is proposed by E.Z. Levy, G.E. Ruf & V.H. Thaler (1959), who list four basic factors of isolation: (1) Microcosm – the actual perceived environment of the "isolated" person in terms of the physical structure of the place, the limitations of visual perception, the characteristics of stimuli (quantity, modality and degree of complexity) as well as the time of staying in such conditions; (2) Human psychological characteristics – they are a factor that facilitates or hinders adaptation to the conditions of long-term isolation;[13] (3) Macrocosm, or the so-called "third

---

longer present, although the nostalgia associated with the lack of personal contact with the family remains.

11 White darkness – this phenomenon, caused by the pristine whiteness of the environment in the Antarctic, has not yet been fully explained. It is usually explained by the subsequent reflection of light rays from the white snow surface to the white clouds, and back again. At cloud ceilings of 1,600 m, the frequency of subsequent reflections can reach 180,000/sec, which results in the phenomenon of "absolute whiteness," where the eye accommodation mechanism is in a subliminal situation and the eye behaves as if in complete darkness (after: Machowski, 1959).

12 Wintering Jean McNeil (2016, p. 376) describes the Antarctic monotony in her diary as follows: "It was not about physical isolation or the fact that you can't get out of here, but that each day was planned before it even started. The number of people was limited / ... /. The leisure activities were also limited."

13 So, for example, the study by Altmam et al. (1967) on the so-called territoriality, i.e. activity oriented towards the use of space in order to create bearable living conditions for oneself and to perform optimally, show, among other things, that personality traits

environment," which concerns everything that surrounds a person outside the microcosm and the motivation for staying in it; (4) Communication processes between the macrocosm and the microcosm directly influence the subjective feeling of isolation, which becomes stronger when this communication is impossible or significantly impeded. The approach presented by Levy seems to be very useful in analyses of both "laboratory" and "natural" situations of isolation. However, many authors point out that the most significant factor that definitely distinguishes the situation of laboratory isolation from normal isolation is the "inevitability" of the latter. According to G. Palmai (1963b), although "inevitability" is most troublesome in the situation of Antarctic isolation, it nevertheless fulfils, from the point of view of interpersonal behaviour, a toning task, minimising interpersonal stress, which results, among other things, from the so-called common fate. The psychological consequences of inevitability were studied, among others, in the 1970s by J. Darley & E. Berscheid (1967), who showed that the awareness of the inevitability of spending time in isolation with another person increases that person's positive traits or at least weakens his or her negative traits. In doing so, minimising interpersonal stress is purely rational. This was also confirmed by L. Altman, W.H. Haythorn (1965), who conducted a study on the so-called interpersonal exchange of information in the process of *"self-disclosure."* Nine pairs of men isolated in small rooms for 10 days were studied. They found that isolated pairs of people were more likely to talk to each other about personal issues than pairs of people in the control group. However, post-isolation studies found that subjects talked about themselves less than they could tell a close friend. This means that while isolation conditions are conducive to the exchange of information, they are not always conducive to deeper emotional connections.

A number of researchers believe that excessive caution in establishing lasting emotional connections and excessive emotional control are associated with the

---

such as dominance or affiliation significantly modify the stress effects of social isolation. The authors empirically found that while isolation causes an increase in territorial behaviour in everyone, groups that are congruent in terms of dominance (i.e. where both individuals have shown tendencies towards mutual dominance and control), have progressively divided the space of shared rooms into their own separate areas and have usurped exclusive use of certain objects and places (e.g. at the table). Pairs incompatible in terms of their needs also showed more territorial behaviour (compared to the normal situation) during isolation, but were characterised by a lack of jointly undertaken activities.

so-called zone of privacy,[14] which becomes compromised in isolated group settings. So, for example, D.A. Taylor, L. Wheeler & I. Altman (1968) investigated the interaction between factors such as zone of privacy, degree of reduction in external stimulation and time of expected isolation. The time of expected isolation was manipulated with instructions that announced a four-day isolation for one group and a 20-day isolation for another, without ruling out the possibility of extending the study in both cases (in fact, the study in both groups lasted eight days). It was found, among other things, that subjects expecting a long period of isolation and deprived of external stimulation at the same time scored higher on subjective stress and anxiety levels.

From the review of the stress factors of isolation so far, it can be concluded that social isolation, as it were, reinforces interpersonal stress, deprives one of the possibility of satisfying some basic needs, limits the zone of intimacy, imposes certain interactions with people with whom one has to stay constantly and limits the possibility of social confrontation of one's own assessments and views, etc. From this point of view, among others, one has to agree with J.T. Shurley (1973), who argues that Antarctic isolation can provide a model of a sociocultural laboratory in which people are observed in a micro-culture, disconnected from the sources of their original macro-culture. The analysis of interpersonal behaviour from the point of view of bio-physiological and socio-psychological changes can be useful for experimental social psychology as well as for experimental psychology of emotions.

Evidence for the claim that the Antarctic isolation situation is a difficult situation is also provided indirectly by clinical analyses of so-called adaptive behaviour. Thus, for example, the research that R.E. Strange and W.J. Klein (1973) conducted on emotional and social adaptation to annual Antarctic isolation indicates the so-called psychological cost,[15] which is described in the polar literature as *winter-over syndrome*,[16] characterised by: depression, hostility, irritability, sleep disturbances, reduced mental performance, etc. (Terelak, 2021).

---

14 *Need for privacy* – means the desire to keep certain opinions and emotions exclusively to oneself, in secret from other people, in intimacy (cf. need for intimacy).
15 This is a problem not always appreciated in experimental research. My own observations from a year of social isolation at the Arctowski Antarctic Station of the Polish Academy of Sciences confirm this. Respondents often stated that after a while the systematically conducted psychological examinations felt as a discomfort and a threat to their sense of privacy.
16 Russian authors call this syndrome by a different name, namely "winter-over participant character disorders" (Kielejnikov, 1971) or "polar asthenia" (Ilyin, Poggenpo, 1969).

Although these symptoms do not prevent individual functioning and may disappear after a period of isolation, they sometimes lead to personality disorders, psychoneuroses, psychosomatic illnesses and sometimes psychoses. Another indirect evidence of the stressful nature of long-term isolation is the more frequent use of alcohol than in normal situations. It is a well-known colloquial view that a stressful situation has a particular role in forming a drinking habit. According to, among others, own research, alcohol in small doses reduces mental tension and is a stimulant, while in large doses it is a depressant (Terelak, Koter, 1986). To substantiate the first part of this thesis, we can refer to Alfred Bandura's experiment (1965) in which cats were taught to perform a complex instrumental response for which they received food as a reward. After some time, instead of food, punishment in the form of electric shock was applied. This aversive stimulus completely inhibited the animal's activity. Interestingly, when given a small amount of alcohol during this stressful period, the cat began to perform instrumental actions again. Furthermore, it was observed that after a series of shocks, and therefore in a difficult stressful situation, the cat preferred milk containing 5% alcohol to pure milk. These preferences were the opposite of the normal situation. Although human behaviour is more complex, when one considers that the social ritual of drinking is added, one has to agree with the view presented above that medium doses of alcohol are a powerful reinforcement of the human response, stimulating action and reducing avoidance and escape reactions (Kobayashi, 1969). In general, alcohol makes a stressful situation subjectively more bearable, but sometimes it becomes also dangerous when consumed in excess.[17]

Summarising the considerations made so far on the psychological and social situation of isolated polar explorers, it should be emphasised that the "task-oriented" nature of this difficult situation means that the expedition's basic objectives are generally met. According to E. Aronson (2009), among people who share the same interests, pleasures, hardships and conditions of a particular environment, a sense of a "common destiny" develops – which is much stronger

---

17 For this reason, among others, access to alcohol on Antarctic stations is under the control of the research station manager or doctor. In addition, the realisation that abuse can be a safety hazard (fire is the biggest risk) eliminates potential alcoholics from wintering-over. An example is the incident of eliminating a winter-over participant who abused alcohol during the summer and sending him back to his country at the request of the other winter-over participants for whom he was a danger. The head of the Arctowski station, after consulting a clinical psychologist who was one of the winter-over participants, made this decision without hesitation (cf. Terelak 2021).

than among people who are merely inhabitants of the same planet, region or city. A very interesting suggestion is also made by L. Festinger (2007). When physical reality is characterised by novelty or uncertainty, he argues, people are inclined to rely increasingly on "social reality," i.e. they are more likely to conform to what other people do, not because they fear punishment, but because the behaviour of the group provides them with valuable information to determine what is expected of them. Perhaps this is the reason for obliquely hiding deep antagonisms and controlling one's spontaneous emotional reactions, especially those of a negative nature (the so-called protection of the community). Finally, the research conducted by I.O. Volkov, A.L. Matusoy & I.F. Ryabinin (1976) shows, among other things, that the motivation of polar explorers to perform their primary tasks in isolation is sufficiently high, compared, for example, to other occupational groups, and provides a viable basis for achieving the main mission objectives.

When discussing the psychosocial environment, it is important to consider both the size of the station, the goals set, the degree of comfort, and the size of the isolated group. These differences in the context of stress factors, on the example of a large American "McMurdo" station (250 winter-over participants) and a small one ("Plateau," eight winter-over participants) was pointed out by E.K.E. Gunderson (1968). Furthermore, as reported by N.R. Deriap, A.Ł. Matusow & L.I., al., Tikhomirov (1978) the degree of emotional comfort is different in experimental stations, e.g. at the Argentinian "Esperanza" station, where heterogeneous groups winter-over (even married couples with children), and different in typical stations where homogeneous (male) groups winter-over. Significant differences from the point of view of information deprivation also exist between stations that have no radio communication with the country and those that maintain regular contact with their families.[18] In addition, other variables such as: the average age of the group, prior polar experiences, the role of the group leader and the style of station management, the organisation of life at the station, etc., significantly modify the psychological environment of the polar station.

---

18 In the early days, when it was not possible to use the transponders of geostationary satellites (GEO), some Antarctic stations relied on radio communication, the quality of which was imperfect, especially during the so-called electromagnetic storms wave propagation was considerably disturbed, and the communication itself was one-way. Nowadays, communicators such as satellite TV, the internet and mobile phones have eliminated information isolation.

Among other things, it has emerged from what has been discussed so far that human functioning under conditions of deprivation of important human needs under Antarctic station conditions is characteristic of a stressful situation. We have also shown that the Antarctic is significantly differentiated in terms of the impact of individual stress factors, enough to justify the need for research into human adaptation mechanisms to extreme conditions in its various regions and station types, with particular emphasis on small research stations, to which the Polish H. Arctowski Station belongs.

As we stated earlier, the primary criterion for individual adaptation to Antarctic isolation conditions, besides emotional aspects, is primarily the motivation for being in stressful conditions. Research suggests that motives that influence the decision to participate in a polar expedition are a good predictor of emotional adjustment to Antarctic isolation conditions and job performance (Strange & Klein, 1973). Moreover, the phenomenon of a decrease in the winter-over syndrome at the end of wintering-over is characteristic, which is undoubtedly related to the reduction of many stress factors and the process of adaptation (Ilyin & Egorov, 1970), although the weakening direction of changes in the adaptive behaviour of polar explorers sometimes appears some time after the end of the mission and return to the country. This is undoubtedly a health cost, which was confirmed by A.B. Blaekburn, J.T. Shurley and K. Natani, K. (1973) in the course of research conducted using the MMPI clinical test on the American "Plateau" station, involving the persistence of elevated the so-called neurotic triad and neurasthenic symptoms on the MMPI test after Antarctic isolation.

In summarising up the clinical and psychological problems of emotional and social adaptation of Antarctic winter-over participants, it should be stated that the winter-over syndrome exposed in many studies does not prevent the functioning of members of polar expeditions, although it is certainly a cause of subjective discomfort, and it cannot be classified as an emotional disorder, but rather treated in terms of the so-called specificity of behavioural adaptation to social isolation. This thesis is supported by the research of A.J. Taylor (1973), who dealt with the emotional problems of readaptation to a normal situation after returning home from Antarctica. The research conducted in the first month after the participants' return to the country shows, among other things, that the test subjects again developed emotional symptoms similar to the classic winter-over syndrome, namely: increased visual sensitivity to colours (24%), fear of traffic (24%), intolerance and irritability in social situations (22%), increased sensitivity to sounds (17%), unspecified feelings of anxiety (17%),[19] fear of being away from home for long periods of time (11%), desire to return to the Antarctic station (11%), and trouble engaging in satisfying sexual activity (9%).

A similar thesis can be applied to social adjustment, which has its own specific behavioural dynamics, resulting not only from social isolation, but also from the mediating role of variables such as, for example: professional, cultural diversity, psychological needs, values, attitudes, personality traits, etc., which have a strong influence on the variability and patterns of social interaction (Gunderson, 1966d; Schachter et al., 1968; Natani, Shurley & Joern, 1973). To better illustrate the discussed concept of variability in social interaction K. Natani, J.T. Shurley and A.T. Joern (1973) propose tracing animal husbandry and draw conclusions by analogy. If cattle, horses and sheep were isolated together in a single pasture, they mixed freely with each other and showed widely varying patterns of pasture sharing. However, when a threat appeared, such as a storm, then they banded together in "specialised" groups, gathering in separate parts of the pasture. Consequently, the authors believe that one of the reasons for the reduction in interaction between military and civilian personnel at "South Pole" station in the winter of 1968 may have been the failure of the fresh water and space heating substations. Civilian personnel was then excluded from repair work and melting snow for water as it was considered "less useful." In addition, P.D Nelson (1962) drew attention to yet another factor that can facilitate or hinder social adaptation, namely the role of the group leader and his or her management style. In turn E.K.E. Gunderson and P.D. Nelson (1963), in the course of conducting a questionnaire study[20] at several US Antarctic stations (about 300 were surveyed over two years), at the beginning and end of wintering-over period, found that attitudes changing in the negative direction, e.g. regarding trust in the station's organisation, personal usefulness, boredom, group coexistence, colleague cooperation, group achievements, atmosphere of equality, etc., could be an indicator of social adaptation. The same authors, P.D. Nelson and E.K.E. Gunderson (1964) in their sociometric study highlighted group structure as a variable that modifies social adaptation. A significant number of sociologists believe that the formal group structure is generally permanent and does not change

---

19 In the case of sight and hearing we are dealing here with the phenomenon of habituation to Antarctic conditions and re-adaptation to functioning in domestic conditions.
20 The questionnaire used, "Attitude Study and Group Behavior Description," was constructed by H. Zimmer. By attitude, the authors mean an opinion that contains a valuing and emotional component.

much (Cartwright & Zander, 1968). However, it does not apply to the informal structure of a small task force resulting from a colleague relationship based on liking or common interests.

Summarising the results of psychological research carried out in Antarctic stations, we can say that they are interesting not only from a practical point of view, facilitating, for example, the selection of the most effective members of groups of successive polar expeditions, but also from a theoretical point of view. As stated by T.M. Newcomb (1968), the emotional and social functioning of a small task group under conditions of isolation can be regarded as a model situation for an experiment, especially in terms of the so-called group dynamics in terms of the interpersonal attraction of group members, which is an indicator of informal rather than formal structure.

To conclude this part of our discussion on the mechanisms of adaptation to Antarctic isolation in classical research, four reasons can be mentioned, prompting further research on small task groups from this perspective: (1) Pragmatic – resulting from the fact that every person in everyday life belongs to some small group (e.g. family, professional, social, etc.); (2) Socio-psychological – the pressure of the group on the individual is so great that it is necessary to know the general regularities of the interactions, i.e. of the group on the individual and vice versa; (3) Sociological – a small group is an excellent model for testing theoretical concepts on the basis of observation and experimentation; (4) Ambitious – a small group is a special case of a more general social system, as it is a micro-system, a reflection of larger communities.

In the review we have presented, we have repeatedly highlighted many variables modifying emotional and social adaptation (e.g. cultural, personality, geographic-climatic, expedition objectives, station size, etc.)[21] that justify the purposefulness of undertaking further research, both in terms of validating selection methods and individual and group adaptation to situations of Antarctic isolation as an analogue of space habitats. Such research is successively being carried out by all countries that have their own research stations in the Antarctic.

---

21 Women's adaptation to ICE conditions was not included in the study. Because the group was homogeneous in terms of gender. Gender has been included as a mediatory variable in a small number of studies which, on the one hand, looked for a set of unique female characteristics that might specifically influence coping with stress and, on the other hand, monitored the dynamics of group processes in mixed-gender groups (Burns & Sullivan, 2000).

154   Own Research into the Functioning of Humans in Antarctic

3.3. Assumptions of own research

3.3.1. Aim of the research

Since 26 February 1977, the Polish Academy of Sciences has coordinated scientific research conducted at the H. Arctowski Station, located on King George Island (South Shetland Islands). Preparing the Polish Antarctic expeditions required solving many logistical and theoretical problems in the field of human biology and psychology. Hence, in the scientific and organisational MR-II-16 programme, coordinated by the Institute of Ecology of the Polish Academy of Sciences,[22] "problems of human biology in the Antarctic region" were exposed alongside many topics from various scientific disciplines (cf. Kwarecki & Terelak, 1980). The Military Institute of Aviation Medicine in Warsaw was commissioned to study this issue. The work carried out by the Institute includes both biomedical and psychological research. The main objectives of the research carried out can be put into two groups:

(1) Practical objectives:
The practical objectives related to: (1) Preparation of the procedure for medical and psychological selection tests of members of Antarctic expeditions, both "summer" and "wintering-over" groups; (2) Medical protection of the Antarctic expedition in the area of the Arctowski station for the period of its operation (as well as the A. Dobrowolski station in the Bunger Oasis on the Antarctica continent); (3) Assessment of the suitability of protective clothing, living conditions and sanitary-hygienic condition, as well as prophylactic activities in this area during successive expeditions.
(2) Scientific objectives
The scientific objectives included primarily: (1) Development of health standards and criteria for predictive accuracy of applied medical and psychological selection methods; (2) Studies of biological rhythms (circadian and seasonal) in polar conditions; (3) Physiological and psychological characteristics of human functioning in a situation of long-term social isolation.

Thus, the primary objectives of Polish medical and psychological research are mainly related to the "recognition of the environment" of the Antarctic, considered to be the most extreme for life on Earth. The research objectives formulated this way are substantiated, among other things, by the written data testifying to

---

22 The history of the legal status of the Antarctic and the participation of Poles in scientific research in this region of the world is presented by S. Rakusa-Suszczewski (1973).

numerous symptoms of emotional and social adjustment disorders to the situation of long-term Antarctic isolation. The "winter-over syndrome" discussed earlier is exacerbated at different times during the Antarctic winter and is characterised by quite significant individual differences. Hence the attempt to clarify situational and personality mechanisms and conclusions useful for the selection and staffing of future polar expeditions and their efficacy in the stressful Antarctic environment. Moreover, the use of the natural conditions of Antarctica for research purposes was of great cognitive significance for the explanation of mechanisms of human functioning in extreme situations in the conditions of the space flight of the Polish cosmonaut Mirosław Hermaszewski, who prepared for the flight together with Piotr Klimuk on the "Soyuz 30" ship. A medical and psychological team from the Military Institute of Aviation Medicine in Warsaw was responsible for his fitness, health and psychological preparation. Hence, additional cognitive objectives emerged, correlating with Antarctic research, understood as an analogue of space research, which included: (1) Psychological mechanisms of human functioning in extreme situations (social isolation, monotony); (2) Psychological cost of adaptation to extreme situations, especially long-term ones; (3) Dynamics of small task groups in situations of isolation; (4) Criteria of psychological selection of people so-called resistant to stress (cosmonauts, pilots, polar explorers, professional athletes, directors, surgeons, etc.).[23]

These objectives seem to justify conducting research in human biology and psychology under natural experimental conditions in the Antarctic.

3.3.2. The issues of own research

The primary problem of psychological research at the H. Arctowski Polish Antarctic Station was "human functioning in an extreme situation." The detailed subject matter was carried out in two stages: (1) The first stage was a reconnaissance one and dealt mainly with the problems of selection of members of Polish

---

23 These goals were realised by MIAM (Military Institute of Aviation Medicine) also in further subsequent 1986–1991 within the frames of the Central Programme of Basic Research 01.20 "Guarantee of psychological reliability of crew members in long-duration space flights," implemented within the framework of the intergovernmental "Interkosmos" programme and the Space Biology Section within the framework of the international cooperation of the Space Research Centre of the Polish Academy of Sciences. Space issues also included unique psychological experiments on the space station on the redistribution of work and rest time and on behavioural changes in biorhythms and stress management (Terelak, 2017).

Polar Expeditions and elaboration of physical and mental health standards for groups staying in polar regions. A discussion of this issue is the subject of separate reports (cf. Kwarecki, Terelak, 1980; Halter, Terelak, 1980); (2) Stage two – carried out by a psychologist, a member of the group wintering at H. Arctowki Station.

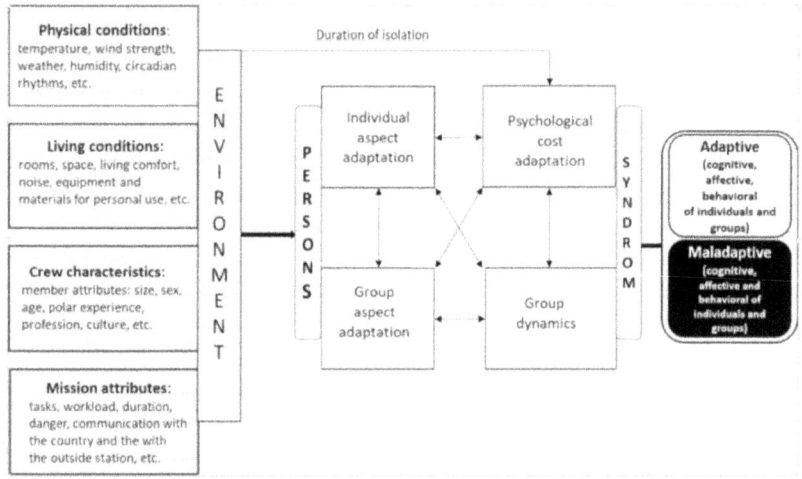

Fig. 13. The transaction model of the psychological cost of adaptation to the Antarctic environment and social "winter-over" isolation
Source: own elaboration.

The research covered two main groups of problems: (1) The dynamics of emotional and social adjustment of a small task group under conditions of one-year Antarctic isolation: (2) Adaptation to changing physical extreme conditions: climatic and photo-ecological ones.

With regard to the first group of problems, reference should be made to S.B. Sells' (1973) assessment of the usefulness of classical Antarctic research from the perspective of its suitability for long-term space missions; in performing a taxonomy of the literature on the subject, he identified three groups of research problems: (1) situational determinants of adaptation; (2) small task group dynamics; (3) individual personality structure. Sells' suggestions prompt the search for a conceptual framework against which to structure the subject matter of individual and group adaptation to situations of Antarctic isolation. It seems that the environmental concept of T. Tomaszewski (1978) described in Chapter 1, which considers four levels of interaction that occur between a person (individual and group) and the environment (situation) in which they find themselves, provides a theoretical basis for the psychological description of such interactions from three perspectives: (1) stimulative, (2) functional and (3) spatial-temporal ones.

(1) The stimulative perspective
For example, when considering the situation of Antarctic isolation in terms of stimulation, two research questions seem to be of interest: (a) How do the specific features of the Antarctic isolation situation (e.g. perceptual deprivation and social isolation) affect the emotional adaptation of winter-over participants? (b) How do non-specific features of the stimulative system (e.g. elements of sensory deprivation, monotony of the environment, restriction of movement, etc.) affect the dynamics and type of human activity?

(2) Functional perspective
Treating the situation as a functional system, we want to draw attention to the task-oriented nature of Antarctic isolation, manifested, among other things, in the directed course of activities based on anticipation of both the final situation (the goal of the expedition) and the activities themselves that are to lead to this situation (the specific programme to be carried out by individual "winter-over participants"). Thus, further important research questions emerge regarding the motivational (values) and personality (capabilities) determinants of emotional adjustment: (c) To what an extent and which personality traits determine emotional adjustment to situations of long-term isolation? (d) To what an extent does deprivation of certain physiological and psychological needs affect adaptive behaviour? (e) To what an extent does the perception of isolation and group situations change and to what an extent do these changes affect emotional adjustment? A number of data cited in chap. 2 on these questions convinces us of the fact that research on this topic has not led to clear conclusions about the personality determinants of "good" adaptation. One of the reasons for this is, among other things, the correlational nature of research in this area (cf. Gundersoin, 1966a).

Another group of research questions is related to social adaptation, and thus to considering the situation as an arrangement of social positions within a group, which includes both the personal elements of the situation (a small task group) and its object elements (features of Antarctic isolation). In chapter 2 we stated, among other things, that a small group with a quite diverse structure is characteristic of a task isolation situation, generating new research problems in terms of: group perception, personal and object elements of the group (e.g. rules of life, isolation characteristics, etc.). There is data to suggest that people who are in a situation of task isolation for a long time change the way they perceive and evaluate the group (cf. Gunderson & Nelson, 1963; Ciosek, 1977). The intensity and direction of these changes is still an open problem, as they depend on both the subjective and objective elements of the isolation situation. Hence, among other things, the study of the personal elements of the social situation, i.e. the structure of small groups, is essential for describing the process of social adaptation to situations of social isolation. As shown by classical psychological research on the social adaptation of Antarctic "winter-over participants," the determinants of this adaptation were sought, among other things, in such variables as personality traits, socio-occupational status, education, work performance, attitudes, etc. (cf. Palinkas, 2003). However, it seems that an equally important problem, requiring detailed research, is the dynamics of interpersonal attractiveness determining the social positions of individual members in the informal group structure. These positions, framed within a specific subjective sociometric structure, can provide a wealth of interesting data on the social adaptation of individual members and the group as a whole (Newcomb, 1968).

An important problem, related to social adaptation, is the informal structure of the small group based on the "relation of liking each other," i.e. on interpersonal attraction. In chapter 2.3.2.5 we cited a number of data showing that this emotional aspect of group members' attitudes towards each other changes somewhat with the duration of Antarctic isolation (cf. e.g. Strange & Klein, 1973). This raises a number of questions: (f) How stable is the interpersonal attraction characteristic for the sociometric structure of a small isolated group?; (g) What positions occupied in the sociometric structure correspond to social adaptation?; (h) What informal position in the sociometric structure is occupied by the group leader with the highest position in the formal structure.

The problems highlighted are of course not exhaustive of all issues related to social adaptation. We have only pointed out that when treating a situation of group isolation as a social arrangement, it is necessary to ask on the one hand about the individual position of the member of the small group in the formal and informal structure, and on the other hand – about all other conditions resulting from specific social interactions.

(3) Spatial-temporal perspective

The "so far specified aspects of human-environmental systems" (stimulatory, functional and social) are modified effectively by the spatial-temporal characteristics of the situation. Thus, for example, the duration of isolation may change both the stimulus value associated with the perception of a monotonous physical and social environment, as well as the hierarchies of values and opportunities for individual or social functioning, associated with the deprivation of many needs, the change of social, professional and family roles, etc. Thus, the physical distance itself (about 18,000 km from the country), as well as the variable duration of isolation, affect the dynamics of emotional and social adaptation.

The issue of biological rhythms (circadian and seasonal) needs to be addressed separately. In many Antarctic studies, especially psychological ones, this problem is generally overlooked or it mainly concerns with physiological "sleep patterns" (cf. Natani & Shurley, 1975; Mallis & DeRoshia, 2005). Meanwhile, the internal synchronisation of biological rhythms is, according to the findings of the World Health Organisation, a condition for "health in general." The absence of some external physical synchronisers (light-dark) or the change in their rhythm during the polar day and night, as well as the absence of many basic social synchronisers (e.g. work-rest cycle, organisation of life, etc.) cause, among other things, the so-called phenomenon of drifting circadian rhythms, which are not without effects on human performance and well-being (Torbjörn, Göran & Mats, 2007).

In summing up the presentation of our own research problems, it should be noted that not all possible questions have been exhausted.

4. Methods and Organisation of The Study

4.1. Characteristics of the study site

Antarctica is a unique natural laboratory for the study of human adaptation to living for some time in the most ectreme environment on Earth (Bradbury, 2002; Schiermeier, 2004). Scientific research in these extreme conditions dates back to the 1960s, when annual Antarctic scientific expeditions began, conducting systematic observations on human adaptive capacity from the perspectives of polar medicine and psychology, and later also space medicine (Mullin, 1960; Palmai, 1963a; Steel, 2005; Palinkas & Suedfeld, 2008).The site of the research was the Henryk Arctowski Antarctic Station[1] of the Polish Academy of Sciences, located on King George Island (aka Waterloo), which is part of the South Shetland Islands (62° 09' 34" S, 58° 28' 15" W).

The characteristics of the Antarctic environment are typical of the various research stations and generally include a description of the physical and psychological sources of stress, the perception of which in terms of discomfort is modified by the personality of the winter-over participants. An attempt to list non-specific elements of extreme environments characterised by long-term isolation, confinement and extreme conditions (ICE – Isolated, Confined and Extreme Environments) in which small, heterogeneous crews must operate was undertaken by N. Smith et al. (2019), who performed a meta-analysis of 284 articles[2] in this field. According to the authors, diagnosing the personal traits that make someone suitable for these types of extreme situations can be helpful in predicting their performance. Personal values are conceptualised as motivational, social, intellectual and emotional attributes of the human individual. Functionally researched for decades, the personal values of winter-over participants set some standard diagnostic criteria needed in the psychological selection of crews, and draw attention to interpersonal differences in adaptability to ICE conditions. Focusing on personal values, they found, among other things, that

---

1 The station's patron Henryk Arctowski, along with Antoni Bolesław Dobrowolski, participated in the pioneering Antarctic expedition on the "Belgica" ship, headed by Adrien de Gerlache (de Gerlache, Arctowski, & Rakusa-Suszczedwski, 2016).
2 However, not all of them were of a scientific nature, because after removing duplicates, consulting the authors and obtaining additional full-text reviews, there were 16 articles left for the meta-analysis.

personal attributes such as benevolence, self-control, high achievement and universalism varied according to the type of methodological approach used by the article authors. It was also noted that personal values change over the course of ICE and that large differences between individuals in personal values can have an adverse effect on group functioning.

The characteristics of the research site will begin with a description of the physical environment of the location of the Henryk Arctowski Polish Antarctic Station of the Polish Academy of Sciences in Antarctica.

4.1.1. Physical environment

A detailed characerisation of South Shetland is given by I.M. Simonov (1965). According to him, the climate of South Shetland is a humid, maritime climate with little change in temperature throughout the year. The sky is mostly covered with low clouds, poor visibility, snow and rain is common. The proximity of the Antarctic front results in very intense cyclonic activity. Large and frequent fluctuations in atmospheric pressure are also characteristic. The southern hemisphere, and therefore Antarctica, receives 7 percent more solar energy than the northern hemisphere. Hence, temperatures in the Antarctic Peninsula and South Shetland are relatively moderate. Due to the location of the Arctowski Station on King George Island, the average monthly humidity value for winter and other seasons is the same: about 80%. The maximum value of the average monthly wind speed is in April (9.7 m/s), the minimum in May (5.4 m/s). However, frequent winds (so-called downbursts)[3] blowing in a very short time (e.g. from 0 m/s to 30 m/s) are characteristic.[4] The highest hurricane wind value at the station was around 50 m/s in 1979.

---

3   Downburst winds, characteristic of the Antarctic coast, slip off glaciers and gain great speed.
4   The following excerpts from the Antarctic Diary can serve as an illustration of their devastating power: "When the four of us were returning to dinner after the net was finished cleaning, a sudden, very strong gust suddenly struck us. All four were cut off their feet and thrown to the ground several meters from the route of the march. Buckets filled with fish and future exhibits "flew" much further. ... Since I had nothing to grab onto, it threw me in the air over a half-meter-long shaft filled with soil and threw me into the freezing water. I don't know how it happened, but I pierced a thin layer of ice and found myself in the water"(Terelak, 2021, pp. 155–156). Such winds are the cause of air crashes in the Antarctic region, such as the crash of the Ik-14 plane in 1979 at the Soviet "Molodioznaya" station, in which three crew members and four polar passengers were killed (Terelak, 2021, p. 64).

Summarising the above meteorological data in the area of the Arctowskiego station, it should be stated that both fluctuations of atmospheric pressure and high speeds of winds or sudden gusts are perceived by the staff of the station as great discomfort. In addition, hurricane winds pose a major threat to the lives of people in the open area of the station at that time ("flying" heavy objects, roofs being torn off, falls, etc.) (cf. Atlas of the Antarctic, 1966).

Thus, although the climate in the area of the Arctowski station is less severe than in the areas of polar stations located in the interior of the Antarctica continent, due to its specificity (high relative humidity, frequent fluctuations in atmospheric pressure and high cyclonic activity), staying at the station requires acclimatisation, and some climatic factors are subjectively felt as discomfort throughout the year.

4.1.2. Logistical problems

(1) *Transport* – both of people and equipment of the Third Antarctic Expedition of the Polish Academy of Sciences took place by sea and lasted a month. The school ship WSM in Gdynia m/s "Garnuszewski" was equipped with air-conditioned five-person cabins. The route led through the Kiel Canal with a short stopover of a few hours in Kiel and through Buenos Aires (two-day stopover). The stops were dictated by the need to replenish provisions (fruit) and fuel. However, transport from the ship to King George Island was by special amphibious vehicles. The winter-over participants, with their personal luggage, were delivered ashore the same day the ship entered Ezcurra Fjord.

It took several days to transport the scientific, economic and technical equipment. After unloading, the ship took on board a group of 19 winter-over participants from the 2nd Antarctic Expedition of the Polish Academy of Sciences and headed back home. Two groups remained at the Arctowski Station: "winter-over participants" – 20 people and "summer people" – 50 persons. From the first day of their arrival at the station, the "winter-over participants" became its hosts with material responsibility for the property, the achievement of the objectives of the Third Expedition[5] and the protection of the Antarctic ecosystem (Bargagli, et al. 2008; Tin, et al., 2009).

---

5   A detailed description of the transport from Poland to the Arctowski Station and the unloading from the ship to land is presented in the author's diary of the 3rd Antarctic Expedition of the Polish Academy of Sciences (Terelak, 2021).

(2) *Supplies* – Supplies were entirely imported from the country. The food was stored in two refrigerated containers throughout the year, which guaranteed the quality of the food throughout the year. For example, fruit was able to be stored for six months and dairy products for nine months. The food supply met both high standards of quality and quantity as well as flavour.

I Technical supplies included repair and construction materials, spare parts for all technical equipment at the station, as well as fuel for vehicles, boats and power plants. Both quality and quantity guaranteed the station's self-sufficiency for more than a year.

I The provision of scientific and research equipment was the responsibility of individual academics, depending on their scientific programmes and research capabilities. This equipment was not part of the station's permanent equipment and was taken home after an expedition or wintering-over. However, medical equipment, pharmaceuticals, ambulatory and operating theatre equipment have been and continue to be replenished along with new expeditions. In addition, the station was equipped with tourist, sports and cultural-educational equipment (library, film library, photographic equipment, radios, tape recorders, adapters, records and common room games).

I It should be stated that the overall equipment of the Arctowski station guarantees its complete independence for more than one year. It also provides basic living, working and leisure comfort.

(3) *Accommodation* – during the summer period it concerned 70 people, members of the Third Antarctic Expedition of the Polish Academy of Science, including 20 people from the wintering-over group.

As shown in Fig. 14, the station is situated on a flat site on which both the utility and technical facilities, the laboratory rooms (which are also living quarters during the summer) and the main accommodation building have been built. All residential buildings are heated with electric radiators. The comfort of summer living spaces is minimal. The central accommodation block, dedicated to the winter-over participants, is a "T" shaped building, nicknamed by the winter-over participants as "airplane." A schematic of the cold storage building is shown in Figure 14.

Characteristics of the study site 165

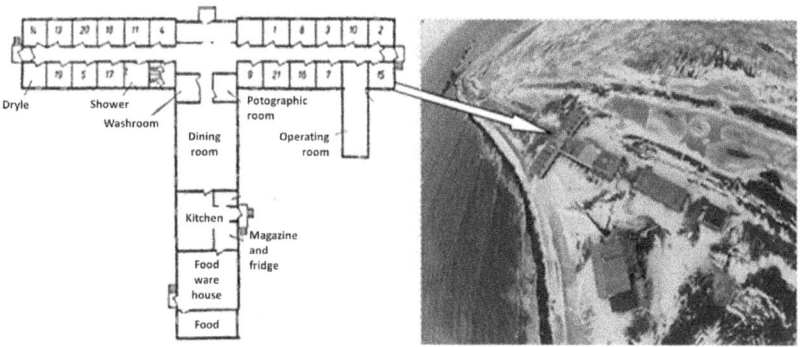

Fig. 14. Residential "Winter-over" room, the so-called "Airplane" at H. Arctowski Station
Source: own elaboration.

The "airplane" contains all the living quarters of the "winter-over participants," sanitary facilities, a dining room (which is also a common room), an infirmary, an operating theatre, an office, a drying room, as well as a kitchen with utility facilities and a handy food store.

Concentration of all rooms is extremely important, especially during the Antarctic winter when going outside is difficult and transport is often not possible. However, such concentration in the event of a fire poses a great danger of the station being destroyed.

The individual living quarters measure 2.20 X 2.00 m and are equipped with basic furniture: bed, wardrobe, table or desk, chair, bookcase and lighting, electric heating, radio or tape recorder.

From a psychological point of view, the living space in the "airplane" is very limited, which "winter-over participants" subjectively feel as a considerable discomfort. For example, the high acoustics of the rooms do not allow for a social life, especially at night.[6] In addition, this design of living quarters does not guarantee privacy (intimacy).

Space heating can be individually controlled and provides basic thermal comfort, although there is a temperature difference depending on the height of the

---

6 In connection with sleep disorders, many people listen to music and, in addition, three people after night duty at the meteorological and the power station sleep before noon every day.

measurement.[7] The humidity of the rooms, on the other hand, was regulated by opening the windows inwards, which is extremely important in the high frequency of hurricane winds.

The washroom, showers and kitchen had running cold and hot water for 20 people. The kitchen with electric and gas cookers guaranteed the possibility of cooking meals in any condition. The dining room was used by 20 people at a time. It served as a common room, reading room and cinema room at the same time. On the other hand, definitely not enough space for various types of recreational and community activities was available. In addition, such rooms as the cloakroom and drying room made it much easier to maintain adequate sanitary and hygienic conditions. The photographic darkroom served as a laboratory and a hobby corner. Sanitary facilities available for the summer group were located outside the "airplane". During the winter the scientific laboratories were also open, equipped with the appropriate infrastructure: two biological laboratories and a geophysical laboratory, as well as a meteorological laboratory, a radio centre, a power station, workshops and a greenhouse.

In general, both the supplies, the equipment of the station and the living, working and resting conditions guaranteed a high level of comfort for the 20 or so "winter-over participants" over a period of more than a year. However, the continuity of the station's operation requires the successive replenishment of equipment and overhaul. This is done by successive Antarctic expeditions of the Polish Academy of Sciences.

4.1.3. Psychological environment

(1) *Stimulating aspect* – The station environment can be seen as a complex system of stimuli that interact with humans and shape their overall level of activation. Two distinct periods should be mentioned here: summer and winter. During the summer period (December–March), the station environment is very diverse in terms of visual and auditory stimuli. The station, located on a narrow strip of Admiralty Bay waterfront, is in two-thirds surrounded by a range of hills. On the opposite side of the bay, the glacier domes and the Keller and Hennequin Hills are visible. The white-aquamarine-blue front of the glacier and the

---

7   E.g. the temperature measured at a height of 10 cm from the floor was 8°C on average, while at a height of 1.10 m it was already 21°C, which was caused by the fact that the "airplane" is placed on concrete pillars with free air flow, which, although significantly cooled the floors, prevented snowdrifts from covering them. After the warming treatments in the following years, the temperature was 15 and 22°C respectively.

Characteristics of the study site 167

brownish-brown rocks and dark blue (depending on the sunshine) of the sea form a colourful backdrop to the area around the station only on sunny days. Some of the hills (to the right of the station) are covered with mosses (shades of green) and lichens (shades of yellow-orange), which in summer creates an interesting mosaic. In addition, at a distance of about 500 metres from the station, on the hills, nests of Adélie, Gentoo and Antarctic penguins can be found, numbering about 50 thousand specimens. The nearby colony of penguins and sea elephants is a source of quite an intense dosage of visual, auditory and olfactory stimulation (pungent smell of excrement) during the summer.

It is not without significance that in 1978/79 two "Mi-2" helicopters operated at the station, which were used quite frequently for cartographic research and as a means of transport. However, in the summer period, an additional source of stimulation were visits of foreign tourist ships (several times the station was visited by the "Lindblad Explorer" and "Discoverer"), fishing vessels: both Polish (e.g. "Jowisz," "Carina," "Włócznik," "Taurus") and foreign (e.g. "Elwira Eisenschneider" and "Erich Weiner") as well as Polish ("Prof. Siedlecki") and foreign scientific vessels ("Bahia Aguire" and "Hero"), as well as two French yachts ("Isatis" and "DamienII") (cf. Appendix No. 2). Thus, it can be said that the psychological environment of the stations varied considerably in terms of stimulation during the summer. This was also facilitated by the organisation of all scientific and tourist expeditions outside the station area during this time.

The psychological environment changed dramatically during the winter period, i.e. from April to December. Complete isolation from the external environment and significant perceptual deprivation exacerbated by the monotony of the environment prevailed.[8]

---

8   Currently, the Arctowski Station no longer meets the criteria for perceptual deprivation and social isolation due to the fact that, compared to the 1980s when there were three year-round scientific stations on King George Island and at a considerable distance, making personal contact impossible, now there are a dozen stations in a large cluster. The current number of scientific stations on King George Island open all year round is: Arctowski – Poland (1977), Artigas – Uruguay (1984), Bellingshausen – Russia (1968), Carlini – Argentina (1953), Escudero – Chile (1994), Ferraz – Brazil (1984), Frei – Chile – (1969), King Sejong – South Korea (1988), Machu Picchu – Peru (1989), Great Wall – China (1985) (Source: Gryziak, 2009).

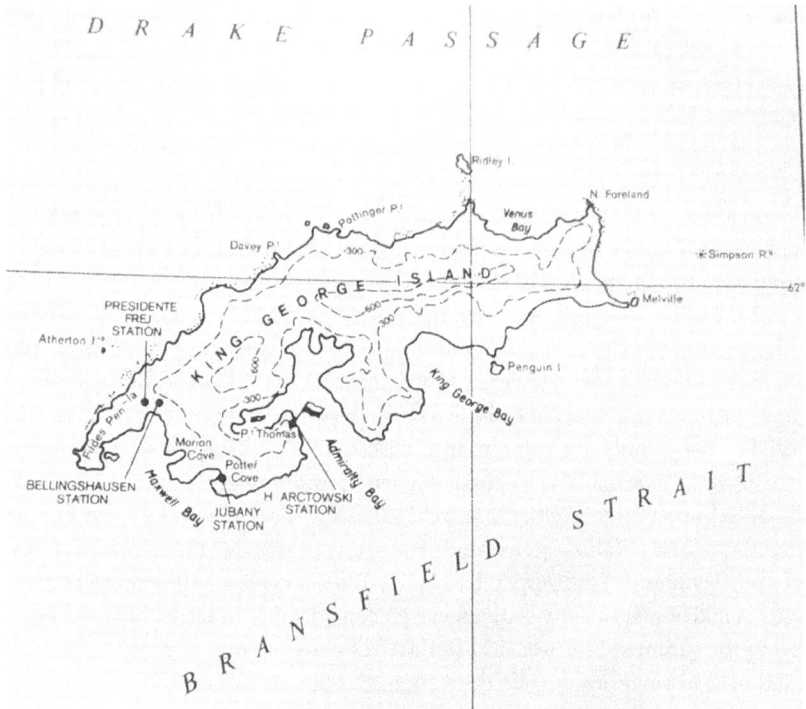

Fig. 15. The "Arctowski" station and other scientific stations on the island of King George in the 1980s (The coordinates of the Arctowski station: 62° 09' 34" S, 58° 28' 15" W
Source: Atlas of Antarctica.

A group of 21 people stayed on the station during this period. Much of the room outside the "airplane" was locked permanently. There have been virtually no visits. Penguins and sea elephants leave the station area and only occasionally do Weddell seals, South American sea lions or Sea Leopards appear on the shore, or crabeater seals staying on ice floes. These animals make virtually no noise compared to elephants or penguins during the mating season.

A thick layer of snow, frequent high-velocity winds and violent gusts mean that free movement outside the station area is not possible, except in the closest areas of the "airplane." In case of favourable weather conditions, outdoor traffic takes place within 500 metres of the station (walking, skiing). The only intense source of stimulation outdoors are sometimes hurricanes, but this is perceived as "white noise" after a while. Thus, the psychological space during the Antarctic winter narrows considerably, encompassing mainly living quarters and workplaces (laboratories, workshops).

In these monotonous conditions, growlers (small ice crumbs) floating in Admiralty Bay and icebergs entering Bramsfield Strait from the ocean, taking on various fanciful shapes and shades of blue and white, are an important source of visual stimulation. This provides a great aesthetic experience and is a source of additional stimulation, as evidenced by the fact that they are eagerly photographed. In addition, during the long winter nights, a 24-hour lighthouse (with its cyclical bright light), the greenhouse illuminating the area around the "airplane," and the string of light bulbs on the way to the meteorological laboratory, the biological laboratory, the power station and the workshops can be considered additional sources of visual stimulation.

Summarising the stimulating aspect of the psychological environment of the station, it can be said that during the eight months of wintering-over, the station environment is characterised by monotony of its surroundings, perceptual deprivation, elements of confinement (in the sense of confining life to a small space) and, strict social isolation.

(2) *Functional aspect* – refers to the changed value system and the possibilities of "winter-over participants" during a period of long-term social isolation. Firstly, separation from one's own family, as well as cultural and professional environment for about 15 months was a major emotional discomfort. With the onset of the Antarctic winter, it was impossible to send or receive correspondence. Besides, letters sent during the summer often arrived at the station only after a year. Contact with the family was only possible by radio (one-way conversation: sender – over, receiver – over… etc.) twice a month on average.[9] However, the presence of third parties (radio operators at the Antarctic station and on the fishing vessels) meant only a formal exchange of the most important information – not satisfying emotionally. In cases of absolute necessity, telex communications could be used. The so-called fishermen's press reached the station daily via a telex bulletin entitled "The Sailor and Fisherman's Voice" (Polish: Głos Marynarza i Rybaka) and brief "Polish Press Agency information." However, the former contained mainly problems of interest to seafarers, while the latter was too concise. Information deprivation was evidenced, among other things, by

---

9   Sometimes in bad weather, which caused radio interference, there was no communication even for several weeks.

the fact that often old magazines brought by fishing or cruise ships during the summer were studied several times throughout the period of isolation.

Secondly – "winter-over participants" are limited in many physiological (e.g. sexual), psychological (e.g. stimulation, intimacy, dominance, etc.) and social needs (e.g. lack of many existing social roles and positions). For these reasons, the primary sources of emotional gratification are being modified.[10]

Thirdly – not only do the hierarchies of previous values change, but so do the capabilities of the polar explorer. Thus, for example, the lack of a "privacy" zone, functioning in undifferentiated social roles, a stable and monotonous landscape, a fixed schedule of the day, spatial imprisonment within the immediate vicinity of the station or in too small living quarters, etc., result in a limitation of human activity and possibilities.

Fourthly, the anticipation of danger and the possibility of loss of life or limb are due, on the one hand, to the awareness that the possibilities for action are limited and, on the other, to the realisation that outside help is impossible in the period of Antatrctic winter. This means, among other things, that the system of values and opportunities means that Antarctic isolation should be considered in terms of extreme stress.

It should also be taken into account that winter-over participants form a small isolated task force that is a kind of macro-community. The individual members are very diverse in terms of culture and personality. These people usually get to know each other during the course of a polar expedition. The only common ground is often the main objective of the expedition, to which all new professional roles and positions in the group are subordinated. Remaining "face to face" with each other for long periods of isolation forces one to accept new social norms, which in some stages of adaptation are the source of many interpersonal conflicts (cf. Mills, 1967).

(3) The spatial-temporal aspect

The living space in the immediate and distant vicinity of the Arctowski station is considerably restricted, and movement around the station is associated with many difficulties and hazards. Thus, for example, any departure from the station area must be reported and must follow a predetermined route with the assistance of others. In particular, moving around the dome of the glacier is

---

10 Many polar researchers have noted that the high position of the cook in the group is due in part to the rise of the significance of food as a new and important source of emotional gratification. A similar role is played, for example, by drinking alcohol, aggressive behaviour, sexual verbalism, etc. (cf. e.g. Mullin, 1960).

very life-threatening due to the existence of deep crevasses (from several to several tens of metres). For a group of five biologists, on the other hand, the living space was extended by the possibility of using a fishing boat for their scientific research. However, again, their expeditions were limited to Ezcurra Fjord and Admiralty Bay. The rather large and abrupt change in the weather (frequent hurricane gusts) and the dangerous ice floes and growlers during the winter period mean that scientific research conducted on the water involves a high risk for life.

The distance separating the Arctowski station from the country is about 20 000 km, which often causes loneliness. In addition, living space, especially on the 'airplane,' is very limited and subjectively experienced as psychological discomfort. Finally, the psychological space narrows considerably due to the prolonged stay in the same group, group composition and lack of privacy conditions.

A separate issue is the altered temporal characteristics of behaviour in relation to "normal" operating conditions in the country. This involves, among other things, the crossing of time zones several times and altered natural lighting conditions, especially during the long winter nights or the so-called white nights.[11] With regard to the first problem, the difference in local time is 5 hours. Thus, there is a change in photo-ecological (so-called primary) and social synchronisers, which is the cause of disruption of circadian biological rhythms and sleep. In addition, the absence of the four seasons is also associated with disruption of seasonal rhythms. The change in circadian rhythms is associated with altered characteristics of the light-dark cycles. Theoretical calculations by J. Speil (1979 – unpublished data) illustrating the proportions of day and night in the area of the Arctowski station presented in Fig. 16 taking into account two parameters: sunrises and sunsets and the so-called civil twilight.

---

[11] The so-called the white night, or summer solstice, in the southern hemisphere falls more or less around Christmas, and then the boundary between day and night becomes virtually blurred, which is associated with a disturbance of circadian rhythms. two and three in the morning. We went to sleep and the darkness did not come. There were minor accidents at work, which meant that symptoms of fatigue and disturbed circadian rhythms appeared" (Terelak, 2021, p. 48).

172   Methods and Organisation of The Study

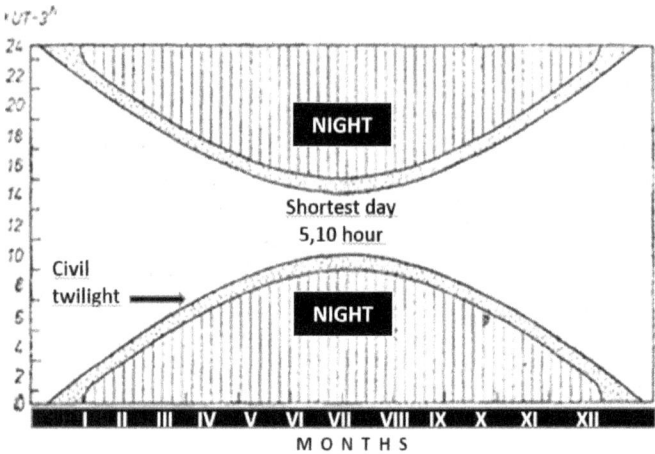

Fig. 16. Day and night at the Arctowski station (62° 10'X and 58* 28'W; according to local time = UT – 3h)

Source: own elaboration of Speil, 1979; data not published.

As can be seen from Figure 16 above, the graph takes into account two parameters: sunrise and sunset and the so-called civil twilight.[12] The characteristics of the astronomical cycles of brightness vs. darkness and the civil twilight bands differ considerably from the cycles prevailing in the Polish latitudes and civilisation synchronisers, which in turn affect the disturbance of circadian rhythms.

In concluding on the characteristics of the research site so far, it must be said that both the physical and psychological environments are predominantly characterised by "novelty." Among other things, the lack of ontogenetic experience makes it necessary to take account of the so-called adaptive behaviours, i.e. the need to develop new patterns of physiological, psychological and social functioning in a new, atypical situation of isolation.

---

12 Civil twilight is defined by the moment when the sun is 6 degrees below the horizon or the moment from which visual perception is impaired, i.e. from which artificial lighting must be used. Astronomical twilight occurs when the sun is 18 degrees below the horizon. From a visual perception point of view, we are more interested in civil twilight, which is also modified by the amount of cloud cover. With a cloudless sky at the end of civil twilight, the illumination of a horizontal surface equals to 0.1 lux.

4.2. Selection of individuals for the study and characteristics of the group

4.2.1. Choice and selection process

Recruitment of candidates to participate in the Antarctic expedition was a two-stage process. The first stage of recruitment (select in) involved an assessment of professional suitability according to the primary objectives of the expedition, carried out by the organisers of the subsequent Antarctic expeditions on the basis of document analysis and an interview. The second stage of selection (select out) included medical and psychological examinations conducted at the Military Institute of Aviation Medicine in Warsaw. Medical examinations were conducted by a multi-specialist team of doctors including a general practitioner, surgeon, ENT specialist, ophthalmologist, neurologist, dentist, anthropologist and radiologist. In justified cases, additional electroencephalographic tests, exercise capacity tests on a cycloergometer and analytical tests (blood and urine) were used. In contrast, psychological examinations were conducted by a psychiatrist (clinical interview) and a psychologist (biographical interview and personality inventories). All participants of the Antarctic expedition were included in the study, although more detailed tests and strict criteria for assessing suitability were applied to the "winter-over participant" group.[13] After qualifying 20 candidates for the group of "winterers," six months before the start of the 3rd Antarctic Expedition of the Polish Academy of Sciences, they were subjected to interpersonal training during a month-long integration stay at the Scientific Field Station of the Institute of Experimental Biology. M. Nencki PAN in Mikołajki.

As we have tried to demonstrate earlier from the literature on the subject, the choice and selection of Antarctic personnel are among the most difficult methodological problems, not only of polar psychology (cf. Hannum, 2008). However, both psychological and economic considerations justify the need for recruitment and selection studies, as well as research into criteria for predicting behaviour under extreme stress (Harris, 1991). By necessity, the first stage of this type of research is negative in nature, consisting mainly in "catching" mental prepathology rather than "selecting" people so-called

---

13 In Antarctica, the names of the residing groups are derived from the names of the seasons. And so, in relation to winter hunters, holidaymakers were "fleeting guests" who usually returned to the country after the Antarctic summer, while winter workers were the station's owners all year round.

resistant to the extreme situation of social isolation. The next stage requires thorough familiarity with the stress factors of the polar station environment, as well as the specificity of long-term natural isolation in a small task force. This is exactly what the research carried out for many years at Antarctic stations during the wintering-over period was intended to achieve, which is also referred to in own research presented here.

4.2.2. Characteristics of the persons studied

Research at the Arctowski station included only "winter-over participants" during the entire period of the Third Expedition, i.e. from the initial research in the country, through cyclic research at the station and the final research on the way back home. The characterisation of winter-over participants presented here concerns 21 individuals, with complete results from only 20 individuals.[14]

4.2.2.1. Biodate

The whole group has already been formally divided in the country into two parts: technical (11 people) and scientific (10 people). As such, the group was significantly diverse in terms of professional background, education, culture, family status, etc. Its characteristics, developed on the basis of a structured biographical interview, are presented in Table 1.[15]

---

[14] The reason for this was that one "winter-over participant" had to be sent back home with the returning "summer" group with a clinical diagnosis of "acute depressive conditions." In his place, after 5 months of the 3rd Antarctic Expedition of the Polish Academy of Sciences, a 21st participant of wintering-over arrived at the station.

[15] In this table we have quoted only some data from the biographical interview, including the following information: (1) record number, (2) age, (3) scientific experience, (4) education, (5) marital status, (6) marital status, (7) number of children, (8) involvement in child-rearing, (9) preferred type of child-rearing, (10) physical activity, (11) social activity, (12) intellectual activity, (13) size of place of residence, (14) social background, (15) parents' education, (16) size of place of birth, (17) number of siblings, (18) birth order, (19) type of upbringing in the family, (20) power structure in the family, (21) satisfaction with current occupation, (22) criminal record, (23) polar interests before the expedition, (24) motivation to go to Antarctic wintering-over.

Selection of individuals for the study and characteristics 175

Table 1. Demographic characteristics of the "Winter-over" of the Third Antarctic Expedition of the Polish Academy of Sciences

Age	Occupation	Education	Family status	Children	Place of residence
38	cook	wed professional	married	2	above 100 thousand
52	technician	engineers	married	2	above 100 thousand
44	bricklayer	wed. professional	married	-	above 100 thousand
33	technician	avg. professional	married	3	above 25 thousand.
49	mechanic	wed. professional	married	1	above 100 thousand
44	meteorologist	higher	married	2	above 500 thousand
33	electrician	diam. professional	married	2	above 2.5 thousand
44	radio operator	wed. professional	married	1	above 100 thousand
28	radio operator	higher	married	1	above 500 thousand
44	meteorologist	higher	married	-	above 500 thousand
38	fisherman	wed. professional	married	3	above 2.5 thousand
38	surgeon doctor	higher	married	2	above 500 thousand
30	chemist	higher	married	-	above 500 thousand
40	biologist	higher	married	2	above 500 thousand
24	biologist	higher	married	-	above 500 thousand
29	geographer	higher	single	-	above 25 thousand
36	psychologist	higher	married	1	above 500 thousand
35	microbiologist	higher	married	1	above 500 thousand
39	biologist	higher	married	1	above 500 thousand
24	biologist	higher	married	-	above 500 thousand
23	physicist	higher	single	-	above 100 thousand

As can be seen from Table 1, it should be stated that the separated groups: technical and scientific one, differ in terms of age, education, social status, work experience etc. The mean age of the entire group is 34.7 years.. The scientific group is significantly younger (28.9 years) than the technical group (40.6 years). Aside from that, the scientific group is more homogeneous, in terms of social status, education, although on the other hand more diverse in terms of family status. Thus, it can be concluded that the studied group of polar explorers was heterogeneous in terms of biographical variables. This is characteristic of the small task forces involved in the Antarctic expeditions of the 1970–80s (cf. Gunderson, 1964b).

4.2.2.2. Motivation

As for the structure of motives for deciding to participate in an Antarctic expedition, it included the following categories: (1) Financial: a) improvement of financial position, b) desire to purchase a car, house, etc., c) improvement of general standard of living; (2) Adventurous: a) attractive travel, b) opportunity to travel outside the country, c) visiting Antarctica and other countries, d) enrichment of knowledge and life experience; (3) Professional: opportunity to work in inaccessible conditions; (4) Prestigious: participation in the work of a group with high social recognition; (5) Heroic: opportunity to test one's character and physical endurance in difficult, extreme conditions; (6) Teleological: a) childhood dreams of achieving something great, b) participation in a polar expedition as the sole purpose of my life; (7) Fortune: a) the need to change my job, b) the need to change my life path, c) escape from family or personal problems; (8) Scientific: a) attractive prospects for scientific work, b) completing already started research with interesting materials, c) participation in an important scientific programme, d) opportunity to explore new scientific problems. The ranked indicators of motive importance in the military and civilian groups and in general are shown in Table 2.

Table 2. The ranks of the categories of "Wintermen" motifs determining participation in the Third Antarctic Expedition of the Polish Academy of Sciences

	Categories of motives	Technical group	Scientific group	In total
(1)	Financial	1	3	1
(2)	Action	2	2	2
(3)	Professional	3	5	3
(4)	Prestige	5	4	4
(5)	Heroic	4	6	5
(6)	Teleological	7	6	6
(7)	Random	8	7	7
(8)	Science	6	1	8

As Table 2 shows, although financial motives are important in both subgroups, the scientific group is more homogeneous in terms of anticipation of scientific benefits. This is confirmed by the psychological interview data, as out of 10 people in the scientific group, four people were preparing their tenure dissertations from research conducted at the station, while three people – PhDs. Thus, we can say that the scientific group was characterised by a more defined task motivation as compared to the technical group. In addition, adventure motifs are highly ranked in both groups, which may be related to the personality structure of the respondents. The feeling of social prestige resulting from participation in the expedition of the Polish Academy of Sciences is also not without significance. The hierarchy of motives in the technical subgroup is fairly typical of analogous Soviet Antarctic expeditions (Volkov, Matusoy & Ryabinin, 1976) or American ones (Gunderson & Mahan, 1966).

4.2.2.3. Personality

Questionnaire surveys included: Eysenck's Personality Inventory (EPI), Cattell's Sixteen Personality Factor Questionnaire (16PF) and Minnesota Multiphasic Personality Inventory (MMPI), as well as Srelau's Temperament Questionnaire.

178 Methods and Organisation of The Study

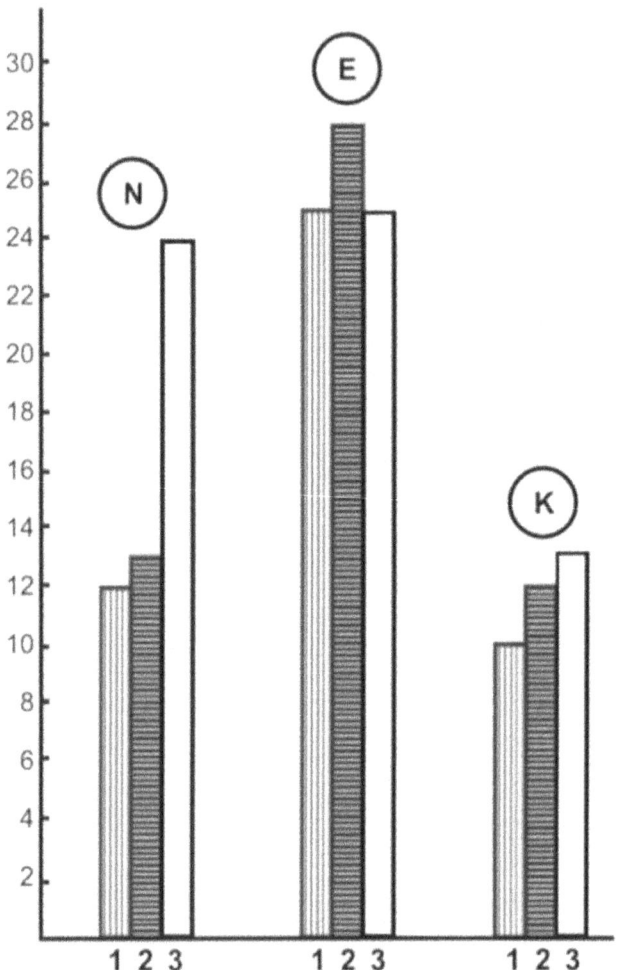

Fig. 17. Averaged personality profile in EPI of the Polish candidates for Antarctic Expeditions PAS in 1976–1985 (N= 250)
Source: own elaboration.
Legend: N – neuroticism; E – extraversion; K – the key of lies; 1 – winter-over; 2 – summer visitors; 3 – average Poles.

The data obtained on the basis of the EPI suggest that the studied group of Antarctic "winter-over participants" is characterised by low neuroticism scores and a higher proportion of extraverts, which is undoubtedly caused by the natural preselection of candidates for polar expeditions. This is confirmed by studies using other personality questionnaires.

Thus, the averaged personality profile in Cattell's 16 PF of the members of the 3rd Antarctic Expedition PAS is shown in Fig. 18.

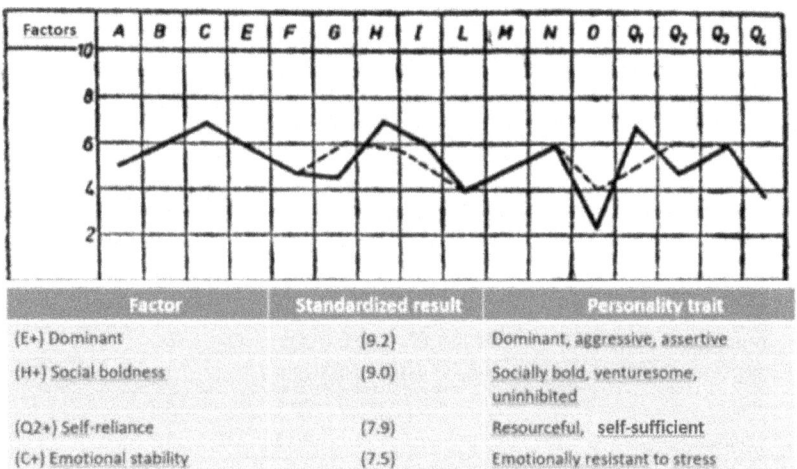

Fig. 18. Averaged personality profile in 16 PF Cattell Winter-over of the 3rd Antarctic Expedition PAS
Source: own elaboration.
Legend: -------- winter-over. - - - - - - summer group

As shown in Fig. 18, Cattell's personality profile[16] of Antarctic wintering-over candidates can be characterised as follows: A+ cyclothymia, B+ high intelligence,

---

16 The individual factors of the Cattell questionnaire are: A: cyclothymia–schizothymia; B: high–low intelligence; C: emotional maturity–neuroticism; E: dominance–submissiveness; F: surgence–desurgence; G: high–low superego; H: psychological resilience–no psychological resilience; I: sensitivity–no sensitivity; L: excessive suspicion–no suspicion; M: unconventionality–conventionality; N: rationality–simplicity; O: depressive self-doubt–self-confidence; Q1: radicalism–conservatism; Q2: self-sufficiency–group dependence; Q3: high self-esteem–low self-esteem; Q4: high ergic tension–low ergic tension.

C+ emotional maturity, E+ dominance, F+ surgency, G+/- medium superego, H+ high mental toughness, I+ sensitivity, L- trust, M+ unconventionality, N+ rationalism, O- self-confidence, Q1+ radicalism, Q2+/- partial group dependence, Q3+ high self-esteem, Q4- low ergic tension. This profile, in terms of low neuroticism, is confirmed by the results of the Eysenck's EPI personality test in terms of the absence of the so-called Neurotic Triad in Cattell's 16PF, characterised by: schizothymia (A-), low self-esteem (F-) and low psychological resilience (H-). Detailed results by technical and scientific group and their interpretations are presented in Table 3.

Table 3. Personality characteristics of winter-over according to the 16PF Cattell of the III Antarctic Expedition of the Polish Academy of Sciences

Study groups/Scales	A	B	C	E	F	G	H	I	L
Techies	17.3	12.9	26.1	15.3	18.0	28.6	24.0	20.7	18.5
Scientists	15.9	17.0	29.0	21.1	22.7	22.8	27.8	23.0	19.3
All together	16.6	14.9	27.5	18.2	20.3	25.7	25.9	21.8	18.9
Study groups/Scales	**M**	**N**	**O**	**Q1**	**Q2**	**Q3**	**Q4**		
Techies	8.0	23.8	9.0	24.4	17.6	30.1	7.6		
Scientists	13.6	23.3	6.4	25.9	23.1	25.9	10.2		
All together	10.8	23.5	7.7	25.1	20.3	28.0	8.9		

As Table 3 shows, the personality characteristics present themselves in a slightly different way if we consider the division into subgroups: technical and scientific ones.

Firstly, the scientific group has a higher proportion of people with a predominance of extraverted traits compared to the technical group. The latter also has a slightly higher proportion of people with marked neurotic tendencies. Secondly, the Cattell's "B" factors (the level of so-called academic intelligence) are conditioned by differences in the level of education between the two groups.

A more detailed analysis of other factors in Cattell's 16 PF shows, among other things, that the scientific group, compared to the technical group, is characterised by: greater self-confidence, dominance tendencies, high activity, unconventionality (factor E, M), sociability (factor I), radicalism characteristic of intellectuals and broad interests (factor Q1).

In turn, the clinical personality characteristics based on the MMPI tests, are shown in Figure 19.

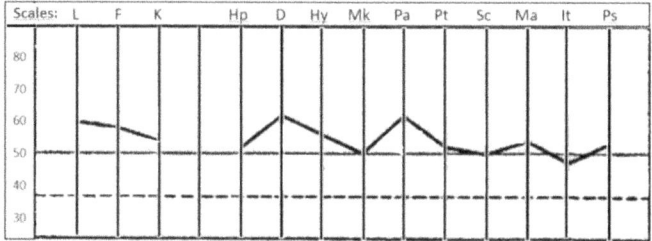

Fig. 19. Personality profile structure of candidates for Winter-over of the 3rd Polish Antarctic Expedition tested MMPI
Source: own elaboration.

As can be seen from this Figure 19, from a psychopathological point of view, the "winter-over participant" group is characterised by good personality defence mechanisms (cf. control keys L, F, K,) and no personality disorder tendencies (all clinical scales are normal). Noteworthy is the absence of the so-called Neurotic Valley, as evidenced by, among other things, low levels on the Hypochondria (Hp) and Hysteria (Hi) scales and the absence of depression (D). These scores correlate with low neuroticism and high extraversion on the Eysenck EPI and low social introversion (It) score on the MMPI.

The phenomenon of dissimulation tendencies is characteristic, the indicator of which is the so-called Gough's simulation vs. dissimulation index, which is calculated from the relationship of the two control scales of the MMPI: F–K and is −9, 3). This means that one has to take into account the possible falsification of results towards showing oneself in a better light. This phenomenon is quite typical for small task groups, characterised by high task motivation, such as in spaceflight candidates.[17]

Based on the Strelau Temperament Survey (KTS) we can say that polar explorers are characterised by relatively low emotional reactivity (indicator P – strength of the excitement process) and significant mobility of nervous processes (Rl). This is consistent with our initial hypothesis, saying, among other things, that low-reactive people with a high need for stimulation show interest in stressful situations (e.g. mountaineers, athletes), or practice occupations that expose them to stress (e.g. polar explorers, pilots, cosmonauts). Assuming a

---

17 I obtained similar results in the MMPI for 60 researched candidates for the Polish cosmonaut (cf. Terelak & Błoszczyński, 1978).

random distribution of factors influencing the decision to participate in an Antarctic expedition, it can be assumed that individual differences in stimulus demand are a significant influence. The results presented in the KTS support the above suggestions that there is a higher proportion of low-reactive individuals among those who prefer the harsh situations of Antarctic isolation.

Based on the literature on the subject, a rule was adopted to exclude candidates with psychopathic and psychotic traits from the expedition (psychiatric history and scores on the MMPI, e.g. "Neurotic Valley: Hypochondria, Depression, Hysteria, and Psychasthenia), neurotic with high levels of introversion (Eysenck's MPI and some of the factors in Cattell's 16 PF, e.g., the "Neurotic Triad" of factors A-F-H"), and with low intelligence (the "B" factor in Cattell's 16PF). The "simulation vs. dissimulation" ratio in the MMPI clinical test was also evaluated based on control factors F–K. These results were analysed in the context of data obtained from a structured biographical interview and motivation.

The recruitment studies discussed above were conducted 2–3 months before the subsequent Antarctic expedition set off. In relation to the "summer" group, the research was one-off, while in relation to the "winter-over participants" – it was longitudinal – as the subsequent part of the research was conducted with the same methods by a clinical psychologist – a participant of the 3rd Scientific Expedition of the Polish Academy of Sciences to the Arctowski Station during the annual period.

4.2.3. Tasks of the members of the wintering-over group

Since the group we are describing is primarily task-oriented, the description presented so far would be incomplete if we omitted the scope and types of tasks that were associated with participation in the Third Antarctic Expedition of the Polish Academy of Sciences.

4.2.3.1. Responsibilities of the technical group

As can be seen from Table 1 presented earlier, the group of "winter-over participants" was formally divided in the country into two subgroups: technical and scientific. This division was based on the specific tasks that the expedition organisers set for the individual participants. Thus, the technical group as a whole was responsible for the maintenance of the station, the operation and ongoing repair of all technical equipment located at the station. All the work was managed by the technical deputy, an engineer by training. The tasks of the individual members of the technical group were primarily related to their professional specialisation. However, it should be mentioned that not everyone was equally burdened by the responsibilities of their profession. Among the busiest workloads were the cook, meteorologists and radio operators.

The cook prepared three hot meals a day.[18] During the entire period of his stay at the Station he had only a few days off work. Although he had a helper (each wintering-over participant was required to be on duty in the kitchen), the cook's work still lasted about 10 hours a day.

Meteorologists (2 persons) were on alternate 24-hour duty at the weather station. Their duties included conducting meteorological measurements and observations on a 3-hour cycle, preparing weather forecasts on the basis of their own research and meteorological data received by teletype from other meteorological stations in the Antarctic region. These forecasts were systematically sent by radio to the Antarctic fishing grounds of the Polish fishing fleet. In addition, they conducted scientific research as part of an international meteorological programme, the results of which were successively sent back home. The arduousness of their work consisted in the need to carry out frequent measurements outdoors, regardless of the day or season, and of the weather conditions (e.g. cyclones, snow drifts, low temperatures, etc.). Moreover, they practically slept only every other night.

Radio operators (2 persons) were periodically on duty around the clock. Their duties included maintaining constant radio and telex communications with the country, other Antarctic stations and Polish fishing vessels in Antarctic waters. They were also responsible for communication within the station (telephone network), for communication with expeditions outside the station (radio communication with helicopters during the summer, communication with a fishing boat carrying out scientific research for most of the year, communication with expeditions on foot on glaciers, etc.), and for maintenance and repair of the radio station, antenna field, teletypewriters, walkie-talkies, telephones, etc.

The skipper of the fishing boat had a very risky job in the waters of Ezcurra Fjord and Admiralty Bay, related to the handling of marine biology research. Working in very difficult conditions (low water and air temperatures, frequent hurricane force winds, ice fields and icebergs, the presence of whales, orcas, etc.) and being aware of the low effectiveness of any possible rescue (in water temperatures approaching −2°C one can only survive for a few minutes) required exceptional resistance to physical stress and the stress of danger. In addition, in his spare time on the water, the skipper also performed the duties of the tailor.

---

18 Although his contract of employment stated 1 hot meal a day, he also prepared an afternoon snack "with a cake."

The chief electrician was also the maintenance man for the power plant (oil-fired generators). This responsibility was mainly related to the knowledge that the lack of electricity practically disorganises life on the station completely.

Water equipment driver, operating amphibious crossing and rescue vehicles, was also a maintenance technician and, in his spare time, the cinema operator.

Locksmith – the locksmith's duties included making all the necessary parts needed to repair technical equipment and operating the finished parts warehouse, as well as operating two refrigerated containers and working in the greenhouse.

The carpenter repaired furniture, made new furniture, and was also responsible for burning rubbish and storing (until it was dumped) non-flammable waste.

In addition to the above-mentioned tasks, the members of the technical group were under the general obligation to perform periodic round-the-clock duty in the power station, and – together with the scientific group – to perform all-day duty in the kitchen every dozen or so days (waitering, washing dishes, peeling potatoes, washing floors in the kitchen, corridors and dining room, cleaning bathrooms and toilets, ironing, taking out the rubbish, etc.).

Station Manager did not fulfil other duties, as his responsibilities were primarily to coordinate the work of the two groups. He also had full responsibility for the implementation of the expedition programme, life and work safety. The manager was also a research ichthyologist and was therefore included in the scientific group.

4.2.3.2. Responsibilities of the scientific group

Biologists – constituted the largest group among scientists. The scientific projects they carried out were on subantarctic marine biology, much of which involved dangerous work on the water.

Ichthyologist – conducted research on the description of new fish species living in cold Antarctic waters. He conducted these studies in addition to his primary duties as head of the Arctowski Station as part of the Third Antarctic Expedition of the Polish Academy of Sciences.

Chemist – worked with biologists, taking water samples in various areas of Admiralty Bay and Ezeurra Fjord.

Parasitologist – he had a rather arduous job, because in addition to preserving the material he collected (parasites of various Antarctic animals), once a month he and his companion (responsible for safety) went on a dozen-kilometre long scientific expedition along the coast of Admiralty Bay to observe seal life.

Geographer – conducted seismological research and astronomical observations. He sent the data he compiled on an ongoing basis by telex to the country.

Geophysicist – carried out electromagnetic field measurements and repaired the station's electronic equipment.

Surgeon – in addition to his primary duties of treatment, prevention and hygiene, he conducted scientific research in the field of physiology on human adaptation to cold (in a monthly cycle) and chronobiology (drifting circadian biorhythms).

Psychologist – carried out a project from a space psychology perspective on psychological reactions to long-term social isolation. In addition, he was responsible for organising recreational and sporting activities as well as controlling the work of the kitchen duty staff and hygiene in the social areas.

In addition to collecting materials, the researchers, with their own laboratories, conducted preliminary analyses and compilation of results and were involved in ongoing clean-up work at the station.

4.3. Measurement techniques and organisation of surveys

The choice of appropriate research techniques in the field of space psychology was conditioned by several important considerations, due in part to the natural environment in which the research was conducted. Firstly, the exploratory nature of the research justified the use of various methods and techniques, covering the widest possible area of interest. Secondly, the choice of some methods was determined by data from the literature on the subject. Thirdly, the choice of questionnaire methods was determined by longitudinal studies conducted with high frequency (e.g. once every fortnight), which sought to capture the dynamics of adaptive behaviour throughout the period of isolation against the situational context. Fourthly – the use of experimental and some sociometric techniques was possible because of the situation of the psychologist, who was one of the members of a small isolated group on Arctowski Station. Fifthly, the use of measurement techniques should not negatively affect social interaction or iatrogenically affect emotional adjustment. The application of the techniques of situational experimentation, for ethical reasons, was limited to practice alarms under the cryptonyms of "power plant failure"[19] and "communication system failure" planned in the country by the technical management of

---

19 Similar alert exercises are routinely conducted at other Antarctic stations. A description of one of these alert exercises at the British Rothera Research Station located in Adelaide on Margaret Bay, managed by the British Antarctic Survey (BAS), is described by Canadian winterist Jean McNeil (2016, p. 347): "The generator was turned off. Suddenly the light went out. Candles were forbidden in the database. /.../ I put on a down jacket and went outside. This might be the only chance to see the base plunged in darkness. "

the Polish Academy of Sciences, which were a natural opportunity to observe the psychological behaviour of the station's crew in a threatening situation.[20] Early on, similar exercises under the code name "lifeboat alert" in open ocean waters were carried out by the ship's captain m/s A. Garnuszewski on the course to Antarctica. Two situational experiments, with the consent of the participants concerned "susceptibility to suggestion," codenamed: "Shaving the head bald" and "Tanning," the results of which are detailed in the author's Diary (cf. Terelak, 2021, pp. 122–123 and 241–242).

A detailed model of the empirical research conducted at the Arctowski Station is illustrated in Fig. 20.

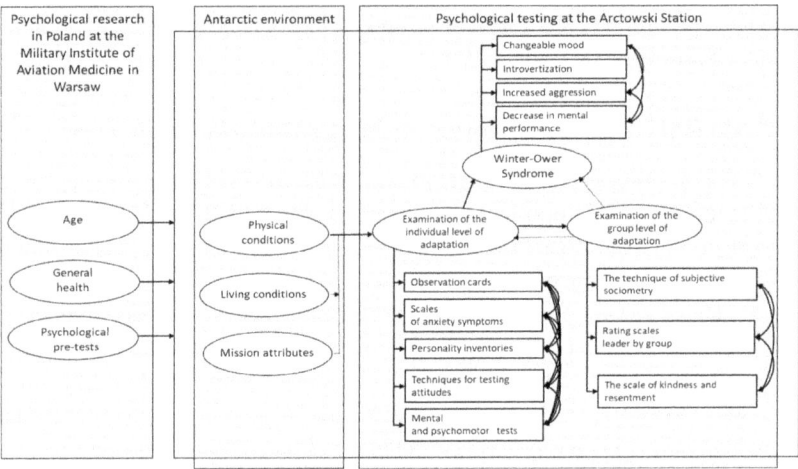

Fig. 20. Structure and schedule of own research

Realising that the research area is an extremely stressful natural situation, in which many real-life variables cannot be precisely controlled, it was decided to

---

20 Similar emergency drills are routinely conducted at other Antarctic stations. A description of one such drills at the British Rothera Research Station located on Adelaide Island on Margaret Bay, managed by the British Antarctic Survey (BAS), is described by Canadian winter-over participant – Jean McNeil (2016, p. 347): "The generator was switched off. Suddenly the light went out. Candles were banned at the base. /.../ I put on my down jacket and went outside. This may have been the only opportunity to see the base plunged into darkness."

use mainly descriptive techniques, concerning both personality, description of the individual's and the group's adaptive behaviour, description of the group's structure and dynamics, description of the situational context. Most generally speaking, the following research methods and techniques were used: (1) psychiatric interview and structured biographical interview; (2) continuous observation (diary, standardised observation sheets); (3) anxiety symptom scales; (4) personality inventories; (5) attitude survey techniques; (6) subjective sociometry techniques; (7) interpersonal attractiveness scales; (8) situational experiment; (9) mental and psychomotor performance tests; medical observation scale of adaptive behaviour.

Some of the above-mentioned demands have been resolved positively. Group surveys were abandoned in favour of individual surveys. In addition, all the completed tests, after a cursory check of how they were filled in, were kept by everyone in their respective rooms for a whole year in a folder prepared in advance. These files were only submitted to the psychologist on the ship on the way back home. The use of surnames was also abandoned in favour of corresponding digital codes, the key to which everyone had in their briefcases. This way, discretion was preserved and the fear that the findings could be used by the psychologist against the subjects was partly reduced. The tests usually lasted for 2–3 days and this term was meticulously observed. To create a "democratic" situation, both the station manager and the psychologist were included in the tests. The multiplicity of tests used and their variety required the development of a precise schedule, which included both one-off tests (psychological selection in-country) and cyclical tests (on the ship on the way to Antarctica, on the station and on the ship on the way back home) at quarterly, 2-month, 1-month and fortnightly intervals.

Summarising the considerations contained in the theoretical and methodical part of this study, we can state, among other things, that an important variable modifying the so-called isolation effects is the duration of Antarctic isolation. Hence, as many points of measurement as possible were included in the research in order to capture the dynamics of these effects, which is an innovative approach in the literature to date.

5. Transactional Characteristics of the Winter-Over Syndrome

In chapter 2.3.2.5 we characterised the situation of Antarctic isolation as a complex, difficult (extreme) situation consisting of both elements of perceptual deprivation (change in intensity and modality of stimuli coming from the environment), elements of social isolation (separation from the previous socio-cultural environment, lack of many important sources of emotional gratification), and elements of imprisonment (significant restriction of physical and psychological space). Defining Antarctic isolation this way as a detailed case of social isolation has its advantages and disadvantages.

The disadvantage (undoubtedly) is the overly broad scope of the term "isolation" making it difficult to operationalise the individual components. However, in research conducted in the natural environment, based on "recognition of the Antarctic environment" rather than "manipulation" of its features, this definition of isolation also has its advantages, as it allows the important stress elements of the Antarctic environment to be highlighted not "as such" but as the site of action of a small isolated task force.

Among the diverse issues of Antarctic research presented in chapter 3.2, we will focus primarily on that part of the problems that relate to individual and group adaptation in the broadest sense, and the "psychological cost" that accompanies adaptation processes to the stressful conditions of a polar station. Conducting research in conditions of long-term social isolation is associated with certain methodological difficulties. Firstly, the choice of research techniques and methods had to be limited so that the research situation itself does not increase stress and cause research refusals. Secondly, uncertainty about whether longitudinal annual psychological examinations on a fortnightly basis would be feasible.

C. Jaksic (2018), in the course of reviewing the literature up to 2016 on adaptation to social isolation in polar station conditions, proves the thesis that adaptation to extreme environment in general (EE) exceeds natural functioning and requires the support of logistics based on human cooperation and new technologies, while J.S. Barnett & J.P Kring (2003) added the crisis management of a small task force, which in turn leads to the conclusion that the extreme environment (EE) of the polar region should be described from a broader perspective: isolated – confined – extreme (ICE). An exemplification of this broader perspective is the winter-over syndrome described by many polar researchers, which is the result of an attempt to adapt to ICE, and it manifests itself in behaviours such as depressive, irritable and aggressive behaviour, periodic decline in mental performance, dissociative changes, sleep disturbances, mood swings, interpersonal conflicts, looking ahead and neglecting responsibilities, etc. The mediating variables responsible for the magnitude and frequency of the appearance of these symptoms, are: (1) Personal characteristics (personality, mental agility, education/profession, interpersonal skills, gender, health, etc.); (2) Social factors (group size, nationality, group cohesion, management styles, mission objectives, access to external support, etc.); (2) Social factors (group size, nationality, group cohesion, management styles, mission objectives, access to external support); (3) Environmental characteristics (climatic factors – varying temperature, humidity, atmospheric pressure and wind strength, as well as photo-ecological factors (related to the duration of day and night) and residential habitat (technical equipment, room comfort, food quality, sanitation, rest rooms, communication with the outside world, on-line access to contacts with family and relatives, etc.).

Citing literature data (Carrere, Evans & Stokols, 1991; Le Scanff, Larue, & Rosnet, 1997; Schmidt et al., 2005; Zimmer et al., 2013 Smith et al., 2019), the considerations in chap. 1.2. Concerning the understanding of the notion of a "situation" and the characteristics of the physical, psychological and social environment of Arctowski's station presented in paragraph 3.2., we adopt, following T. Tomaszewski (1978), four aspects of a situation (stimulative, functional, social and spatial-temporal) as the basis for the classification of the results of own research presented in this chapter. By defining a situation of isolation as a complex system of mutually interacting systems of: stimulation, values and possibilities, social and spatial-temporal, I am aware of the fact that a human being is in the very flow of specific events of such a situation and that any separation of its individual aspects is an artificial procedure and is only of an ordering nature. Hence, situation perception, understood as a set of subjective conditions on which it depends whether certain features, states of affairs and events in a situation of isolation are meaningful to a person, was singled out as a separate problem.

5.1. The stimulative aspect of Antarctic isolation

Treating the situation of Antarctic isolation as a set of stimuli of a specific intensity and modality, we assume that for a human being with a specific need for stimulation, the isolation situation is characterised by "newness." This is due, among other things, to the fact that under normal conditions a person has not acquired the experience needed to function in a situation of prolonged isolation. The need to adapt to annual Antarctic isolation, as a consequence of a decision taken earlier, therefore requires the development of new ways of behaviour. For this reason, we treat the situation of Antarctic isolation, characterised among other things by significant stimulus restrictions compared to the normal (domestic) situation, as an extreme situation.

A detailed description of the isolation situation at the Arctowski station from the point of view of its stimulation value was presented in paragraph 3.1. Referring to this data and the considerations in chapter 1.3. on the need for stimulation as a determinant of adaptation, we present our own observations and research results on the mainly non-specific[1] effects of Antarctic isolation.

The situation of Antarctic isolation understood as a system of stimuli acting in a general, non-specific way regulates the general psychophysical condition, which in a non-linear way determines both the level of performance of specific tasks and the subjective feeling of comfort vs. discomfort (e.g. attractive or boring, differentiated, relatively monotonous situation, etc.). In accordance with this approach and the assumptions of the theory of optimal activation (stimulation), which we have presented in chapter 1.3., a certain course of dynamics of adaptive behaviour should be expected, leading in consequence to a certain level of adaptation, understood as habituation to stimuli typical for the environment of Antarctic isolation. D.W. Fiske and S. Maddi, eds., (1967) suggest that adaptation to situations of limited stimulation is characterised by a lowering of sensitivity thresholds (increased sensory sensitivity), thus facilitating functioning in such a situation (the habituation model of adaptation). A confirmation of this thesis in the situation of Antarctic isolation is provided by observational data presented in detail in another study (Terelak, 2021), describing the facts of decreasing auditory and olfactory sensitivity thresholds during the Antarctic winter (e.g. accurate recognition of wind strength, sensitivity to the smell of exhaust fumes from a long distance, etc.). Thus, the transition from active to passive regulation is likely to be determined by both the degree of discrepancy between the need for stimulation (level of reactivity) and stimulation and the duration of isolation. However, regulation of stimulation at the physiological level is automatic, so the formulation of active and passive regulation should be treated conventionally.[2] However, the concepts of active and passive stimulation

---

1 Specific effects are described in detail in the diary (Terelak, 2021).
2 The term of active stimulation regulation replaces the awkward phrase "stimulation regulation by means of negative feedback." In turn, the concept of passive regulation replaces the term "stimulation regulation by means of positive feedback" (cf. Eliasz, 1985).

regulation better correspond to the actual processes of stimulation regulation taking place at the level of behaviour.

In the present study, the level of reactivity was assumed to be *constant*, as already in the process of selecting members of the Antarctic expedition a higher proportion of low-reactivity individuals was found compared to the Polish population. This is evidenced by the results in Strelau's Temperament Questionnaire (strength of the excitation process), which for members of the 3rd Antarctic Expedition of the Polish Academy of Sciences is 68.15 (N = 66). For the corresponding male cohort, the ratio is 56.00 (N = 1265) (cf: Stawowska, 1973).

The empirically detected fact of a higher proportion of low reactivity individuals in the polar explorers' group and its too small size (N = 20) determined the omission of the issue of individual differences in the tolerance of long-term Antarctic isolation. Assuming the level of demand for stimulation to be relatively constant, we therefore turn our attention to the relationship between the duration of isolation and adaptive behaviour. At the same time, speaking about adaptive behaviour in the stimulus aspect of the isolation situation, we obviously mean its non-specific nature, i.e. the level of activation which, according to D. Hebb (1965), is the basis of directional processes, shaping the level of human activity.

In order to observe the various forms of activity of the polar explorers, a special "Daily activity sheet" was elaborated, including hourly settlement of forms such as: sleeping, working inside the room alone, working outside the room alone, working inside the room in a group, working outside the room in a group, indoor games and plays, outdoor recreation (walking, hiking, skiing), meals, social conversations, reading, watching films, listening to the radio (tape recorder, record player), hobbies (e.g. photography), other activities (e.g. laundry, cleaning, toileting, etc.).[3] The activity indicator in this case was the percentage of time that individuals and the group as a whole spent on a particular type of activity in relation to the "week day." A similar technique to study the quantitative characteristics of sleep-wake cycles was used on Antarctic stations by American psychologists such as R.E. Brooks et al. (1973). The results obtained regarding the dynamics of the daily activity of the winter-over participants on an annual basis are illustrated in Fig. 21.

---

3 The "Daily activity sheet" was filled in by everyone once a month for a period of 7 consecutive days. At the same time, in order to avoid a certain bias, the weeks covered were mobile, i.e. in subsequent months they always included the following week compared to the previous month.

The stimulative aspect of Antarctic isolation 193

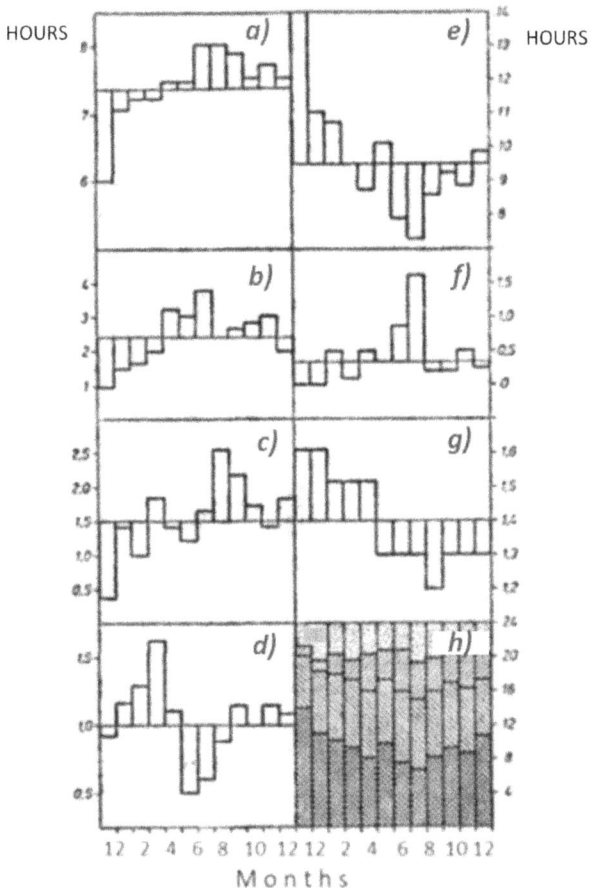

Fig. 21. Dynamics of the daily activity of winter workers over the year
Source: own elaboration.
Legend: Mr – average annual activity; Types of activity: a) sleeping, b) indoor games, c) conversations, d) other activities, e) work outside, f) outdoor group work, g) meals, h) monthly days including (from the bottom of the figure): work – sleep – rest – other activities

Analysing the results presented in Fig. 21, it can be seen that the two periods of isolation differ significantly, from the point of view of stimulating value: "summer" (D/ecember, February, March) and "winter"[4] (remaining months). A more detailed characterisation of these two periods, derived from a percentage scale in surveys conducted fortnightly throughout the year and considering four types of activity: a) task activity, b) pro-social activity, c) social activity and d) autoactivity (reflexivity) is presented in Fig. 22.

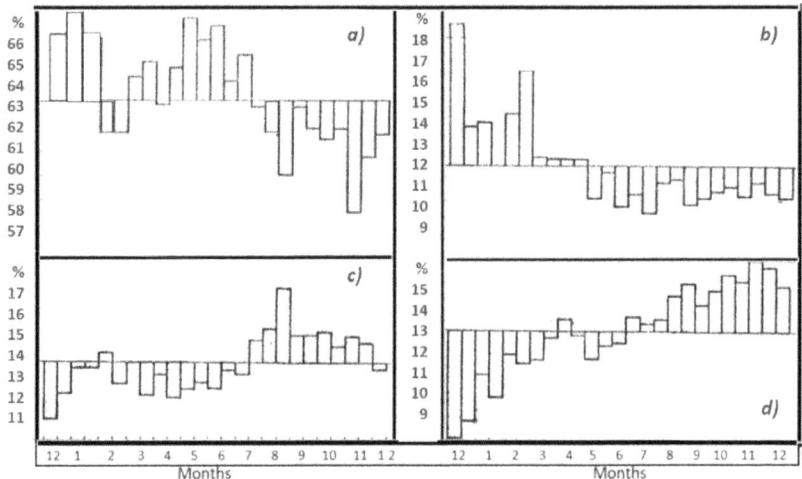

Fig. 22. Fluctuations in polar activity around the average annual social isolation (N = 20)
Source: own elaboration.
Legend: (a) task activity; b) pro-social activity; c) social activity; d) autoactivity – reflextivity.

Fig. 23 shows, among other things, that task activity declines sharply in the months that are poorest in terms of stimulation, i.e. characterised by the total monotony of the environment (endless whiteness, restrictions on movement outside the station area and often outside the living quarters, etc.). Referring to our own model of the stimulation level regulation, shown earlier in Fig. 4, we can assume that the switching of active regulation of excitation and stimulation to a passive form probably results in some form of fatigue of those physiological

---

4 The period of "wintering-over" is conventional and refers to the period of actual isolation from the outside world (i.e. April to December). It therefore does not coincide with the astronomical duration of the Antarctic winter.

mechanisms that counteracted the natural decrease or increase in excitation. In the case of Antarctic isolation, we are dealing with a retuning of regulation from active to passive one, caused by a collapse of stimulation regulation at a given level, which is related to a greater involvement of "higher" regulators (e.g. autoactivity) in the stimulation regulation process than before, as confirmed in Fig. 22. This may also be associated with a fundamental change in activity style, or the abandonment of certain goals previously pursued by "winter-over participants" (cf. Fig. 22a and 22b). The findings on the decline of some forms of winter-over participants' activity along with the duration of social isolation are confirmed, among other things, by the data cited by C.S. Mullin (1960), indicating that in about 40% of American polar explorers, symptoms of intellectual retardation (inertia), impaired memory and ability to focus attention, appeared most frequently during the winter months. The author makes it clear that during the Antarctic winter there is a significant decrease in activation levels in all polar explorers, which is a significant symptom of the "winter-over syndrome." This phenomenon surprised many people because with an excess of free time they were unable to achieve many of their intended goals (e.g. learning foreign languages, writing a series of articles, etc.). This is also confirmed by my own observations in another study (Terelak, 2021).

The non-specific aspect of the "winter-over syndrome" is pointed out by E.A. Ilyin and B.B. Egorov (1970), who analysed the symptoms of maladaptation of "winter-over participants" at the "Vostok" station. They found that the observed behavioural changes had a weakening direction. Their clinical observations during a one-year period of isolation confirmed the thesis that neurasthenia gradually transforms into asthenia, the main symptoms of which were a decrease in interest in work and a lack of desire to exert oneself, that is, in the terminology of D. Hebb (1965), a decline in the "directional function." A similar, weakening direction of changes in the adaptive behaviour of polar explorers is confirmed, among others, by studies by A.B. Blaekburn, J.T. Shurley & K. Natani, K. (1973) who found neurasthenic symptoms at the US "Plateau" station to be inversely related to achievement motivation, job satisfaction and group coexistence.

With regard to the latter, i.e. coexistence in a group, a seemingly paradoxical phenomenon must be stated, as shown in Fig. 22c, that despite the reluctance to coexist in a group, an increase in social contact is marked during these very periods of isolation. This occurs with the aforementioned switch from active to passive stimulation regulation. When there is a reduction in sensitivity to stimuli, a paradoxical need to seek more stimulation may occur, despite deepening apathy during the winter months. Winter-over participants try to overcome this by organising an active social life (a paradoxical increase in the need for stimulation at the behavioural level). Polar explorers, despite their considerable apathy, then decide to have some incidental entertainment, a social gathering, because they expect this activity to stimulate them for a while. In the absence of this minimum stimulation, which is necessary for undertaking activity, people are in danger of sinking into a state of further apathy. This is partly confirmed by my own observations and those of other polar psychologists, who characterise the behaviour of polar explorers in the final 2–3 months of Antarctic isolation by an analogy with the "sea elephant position."[5]

A similar effect, resulting from a significant decrease in stimulation levels and reduced movement, is described by, among others, S.J. Freedman, H.U. Grunebaum & M. Greenblatt (1961) as so-called positive contemplation. It is characterised by the activation of "primary mental processes" (i.e. dreams, memories, fantasising and sometimes even hallucinations). The authors' research shows that the greater the reduction in stimulation and movement, the more this type of mental activity is increased. Obviously, this stress effect in interaction with social isolation was correspondingly smaller.

5.2. The functional aspect of the Antarctic isolation situation

The data presented in the previous paragraph supports the thesis that polar explorers achieving adaptation understood as habituation to conditions characterised by a certain level of stimulation, do not exhaust the whole range of adaptive behaviour, especially when we take into account, on the one hand, the subjective attitude to the attributes of the environment, and on the other hand, the ability to understand the situation of isolation (e.g. from the point of view of its task-relatedness, common fate, etc.). This set of conditions and characteristics of a situation of isolation, involving many restrictions on human values and capabilities, narrows our focus to the issue of individual adaptation. The "winter-over syndrome" described in the literature takes into account the too low dynamics of adaptive behaviour (this is due to the low frequency of research), as well as a too modest repertoire of behaviours, which is often due to the choice of research techniques used (cf. Wilkins, 1973).

A detailed review of the literature on human functioning in Antarctic conditions, has been done by, among others: E.K.E. Gunderson & J.L. Mahan (1968),

---

5   The sea elephant, after a period of foraging, usually spends 2–3 months completely passive, lounging on the beach. Prior to the arrival of their ship, "winter-over participants" await it for 2–3 months, preferring a completely passive manner of "bleeding time off." "The sea elephant position" is described in detail in another study (Terelak, 2021).

A.J. Taylor (1987), I.I. Tkhomirov (1973), C. Tisch, (2005), A. Sarris (2007). A. Gavalas (2011), M. Zimmer et al., (2013), C. Jaksic (2018), which shows, among other things, that the cultural distinctiveness of polar explorers of various nationalities, as well as variables such as the size and structure of the isolated group, motivation and the type of tasks performed, require the implementation of research at different stations, located in different places in the Antarctic and at different times.

In the presented own research, we made two assumptions about emotional adjustment: (1) The "winter-over syndrome" is an adaptive behaviour, so it should be considered in terms of a psychological cost rather than an emotional disorder; (2) Significant inter- and intraindividual differences, conditioned by situation and personality, must be taken into account in emotional adaptation.

5.2.1. Clinical psychology perspective

In analysing adaptive behaviour so far from the point of view of the stimulus value of Antarctic isolation, we have highlighted an important, but not the only, aspect of this situation. Antarctic isolation also brings with it other significant human limitations, such as human values and capabilities. The effects of this adaptation process can appear at the level of both psychopathological symptoms and less drastic symptoms of emotional strain, related to the failure to meet many psychological needs and to perform many social roles, as well as to the subjective perception of discomfort associated with the stressful attributes of the polar station environment.

The primary criterion for assessing the effectiveness of emotional adjustment to long-term Antarctic isolation is usually the absence of psychopathological symptoms (Nardini, Herrmann & Rasmussen, 1962). In the present study, two techniques were used to measure the severity of psychopathological symptoms, namely: Clinical symptom list – M. Siwiak-Kobayashi KLO and Minnesota Multiphasic Personality Inventory – MMPI.

(1) *Clinical Symptom List*[6] – is a tool for assessing the intensity and dynamics of behavioural changes over a short period of time and across a wide range of symptoms, grouped into six classes: anxiety (general anxiety, non-canalised), depression (both psychological and somatic), psychological tension, obsessive (including phobic anxiety symptoms), somatisations and mixed (all not previously classified). The raw scores obtained, converted for each symptom into standardised "z" units, are shown in Fig. 23.

---

6   Siwiak-Kobayashi KLO was used with permission from the author, for which I offer my thanks. For a detailed description of this technique, see M. Siwiak--Kobayashi (1974).

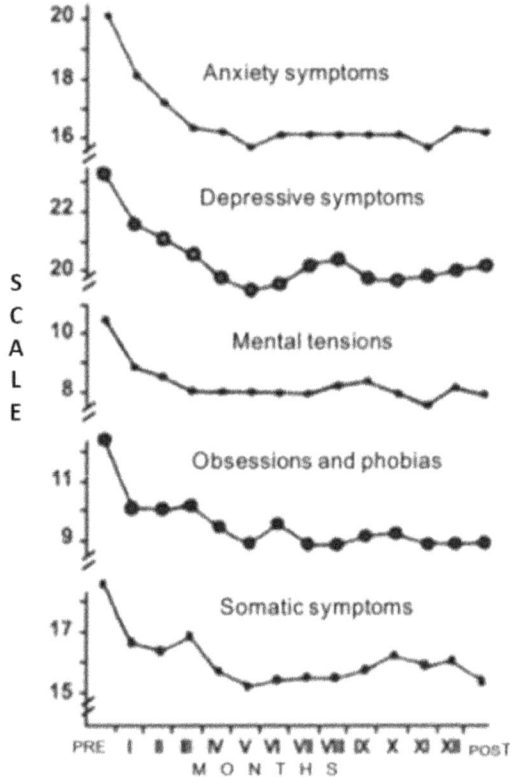

Fig. 23. Clinical distribution of the level of anxiety and its attributes in the annual isolation of wintering-over at the Arctowski Station

Source: own elaboration.
Legend: Pretest. – initial tests; XII, I, II…, – monthly tests; Posttest – tests on the ship on the way back to the country; all differences between Pretest and subsequent tests are significant at p <0,01 and p <0, 05.

As can be seen from Fig. 23 there is a clear period of stabilisation of the adaptation process of the second part of the wintering-over with some increase in the level just before the return to the country, related to the journey and the new challenges of civilisation. Two conclusions can be drawn. Firstly, there was no evidence of psychopathological symptoms of such severity as to prevent adaptation to conditions of Antarctic isolation, although they certainly hindered it. Secondly, it was found that the most difficult period, from the point of view of adaptation, is the initial period of isolation, covering the first 3–4 months of stay at the station. This is not consistent with the views, especially of American psychiatrists and psychologists (e.g. Gunderson, 1968), who consider the middle of wintering-over to be the most difficult period. However, this seems to be more due to the way the surveys were conducted, which were most often carried out three times a year, i.e. at the beginning, middle and end of the wintering-over season (less often quarterly). The absence of many intermediate points of measurement led to generalisations that were inconsistent with the actual dynamics of the symptoms or behaviours under study.

As Figure 23 suggests, each symptom class has a distinctive annual pattern. E.g. if psychological tensions remain more or less the same (relatively high), depressive and somatic symptoms have quite a varied course.[7] Depressive symptoms increase during the initial period of isolation, in the middle of winter-over and towards the end. Somatic symptoms, on the other hand, occur mainly during the initial period of isolation and towards the end of the isolation period, indicating human fatigue (psychophysiological cost), despite the achievement of a relative level of adaptation. Similar trends of psychopathological symptoms are cited among others by E.K.E. Gunderson (1963), who compared the somatic and emotional symptoms of American winter-over participants at the beginning and end of the Antarctic winter.

(1) *Minnesota Multiphasic Personality Inventory* – is one of the more prominent methods used by American clinical psychologists to assess emotional adjustment. The MMPI[8] was used quarterly to diagnose the differential diagnosis of psychiatric nosology in an Antarctic isolation situation. The results are illustrated in Figure 24.

---

7   L.A. Palinkas, M., Cravalho & D. Browner (1995) speak explicitly of the seasonal nature of polar depressions.
8   The MMPI was used with the author's permission in the women's version of the WISKAD-MMPI, a description of which can be found in the study by Z. Płużek (1971).

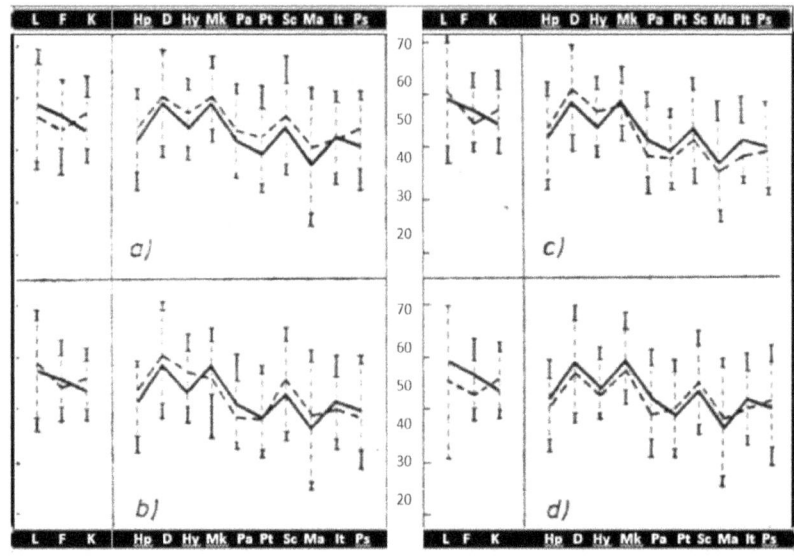

Fig. 24. MMPI profiles as clinical indicators of wintering-over's adaptation to Antarctic isolation

Source: own elaboration.
Legend: Control scales: F – (number of questions that the respondent did not answer), L – (Lie scale), K – (Defensive attitude scale); F – K (simulation index / Gough dissimulation) of the simulation; Clinical scales: Hs – Hypochondria, D – Depression, Hy – Hysteria, Mf – Masculinity / Femininity, Pa – paranoia, Pt – Psychasthenia, Sc – Schizophrenia, Ma – Hypomania, Si – Social Introversion, Ps – Psychopathy; a) 1st quarter, b) 2nd quarter, c) 3rd quarter, d) 4th quarter.

Diagnosis with the MMPI is intended to answer three questions: Is the respondent disturbed? Does the respondent have a personality disorder? Does the respondent have psychosis? As can be seen from the analysis of Fig. 27, MMPI profiles did not significantly differentiate the study group of winter-over participants in terms of emotional adaptation vs maladaptation during long-term Antarctic isolation, suggesting rather good mental health. However, the rather large scatter of results indicates that the study group is not homogeneous in terms of emotional adjustment. In addition, an analysis of the control scales: F – (number of questions not answered), L – (Lying Scale), K – (Defensive Attitudes Scale) in the form of the so-called Gough Index: F – K

= 11.0, indicates a high level of dissimulation[9], which means that the test results are subject to a high level of falsification towards "showing oneself in a better light" and are not suitable for clinical interpretation. The operation of the social approval variable is characteristic of a small isolated task group, as suggested by both polar (e.g. Seymour and Gunderson, 1971) and cosmic (e.g. Terelak and Blochinski, 1978) psychologists[10]. This fact makes it very difficult to study personality using a variety of inventories that do not have a so-called lie key. Right are A.B. Blaekburn, J.T. Shurley & K. Natani (1973) who argue, among other things, that the operation of the social approval variable and the study of personality by questionnaire lead to the fact that some of the most significant personality changes accompanying extreme Antarctic isolation have so far not been sufficiently documented.

---

9 In clinical psychology, it is the attitude of the subject to a psychological test manifested in an attempt to conform to prevailing health and social norms and to conceal real or imagined disorders.
10 A similar phenomenon of dissimulation is found in candidates for space flight, which is generally associated with a high level of task motivation and is used in selection and choice as an indicator of achievement motivation (Terelak, Błoszczyński 1978).

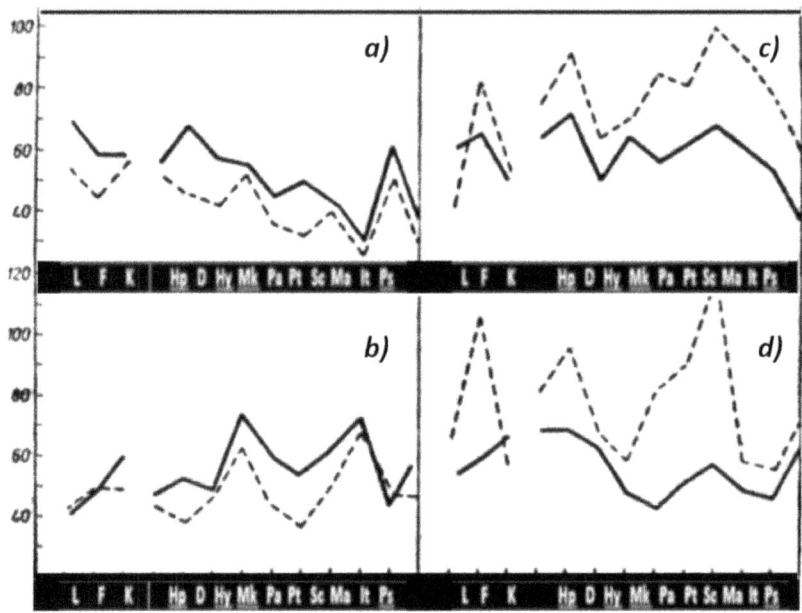

Fig. 25. Examples of the MMPI profile of winter beds well adapted (research a and b) and unadapted (research c and d) to the conditions of the annual Antarctic isolation at the Arctowski station (fig. a and c) and at the Plateau station (fig. b and d)

Source: own elaboration (fig. a and c) and A. B. Blackburn, et al., 1967 (fig. b and d).

After a more detailed analysis of individual clinical personality profiles in the MMPI, individual cases of adapted and maladapted individuals to long-term Antarctic isolation could be distinguished. Examples of such profiles as case studies found in studies at the Polish Arctowski Station and at the US Plateau Station (Blaekburn et al. 1973) are illustrated in Fig. 25.

The personality profiles of two Antarctic winter-over participants, presented for comparative purposes, confirm our conviction that psychological selection based only on a negative criterion, i.e. the absence of psychopathological symptoms, does not augur well for individual adaptation to Antarctic isolation conditions. Figure 25 shows that individuals classified as "maladapted" at Arctowski and Plateau stations were characterised by relatively high levels of neuroticism and introversion. A more detailed analysis of the clinical scales, on the other hand, reveals significant differences between the adapted and the maladapted in the so-called depression triad, which includes scales: Hp-D-Hy, which means, among other things, that maladapted people are characterised by symptoms such as: complaints of psychosomatic disorders (Hp scale), low mood and subjective feeling of depression (D scale), longing for emotional contact, malaise, inhibition of aggression (Hy scale), oversensitivity (Pa scale), generalised symptoms of neurosis (Pt scale); emotional alienation, lack of control over Ego (Sc scale). It is also worth pointing out the differences between Arctowski and Plateau. Thus, for example, at Arctowski, symptoms of hypomania (Ma scale) and social withdrawal, discomfort in contact with others and hypersensitivity (It scale) are more intense, while at Plateau station the highest level was reached by the Sc scale, describing symptoms of emotional alienation. Significant cultural differences in value systems may be the cause of these discrepancies.

The unfavourable changes in the MMPI profiles correlated with the results of other studies, suggesting that these individuals felt the most lack of satisfaction with their tasks during their one-year isolation, repeatedly manifested a desire to return to the country, a subjective sense of discomfort. This is confirmed by research using the Sense of Psychological Wellbeing Scale.

(3) *Mental Comfort/Discomfort Scale* – allowed us to describe a very interesting phenomenon, not previously reported in the literature, showing that the highest level of symptom severity in all categories is found in a study conducted during a voyage to Antarctica. This data clearly shows that due to the anticipation of the stress of Antarctic isolation and the voyage in dangerous subantarctic regions (fog, icebergs), the voyage period, etc. is associated with the greatest psychophysiological cost. This also applies to the first 2–3 months at the station. This is confirmed by studies assessing subjective feelings of psychological discomfort related to Antarctic station stress factors, isolation and monotony, and lack of interpersonal alternatives.

Subjective feelings of psychological discomfort were examined according to detailed instructions.[11] Studies with this test were conducted on a fortnightly cycle. The categories listed in the study and the results obtained are shown in Figure 26.

---

11 "Listed below are various things that sometimes make us anxious, upset, unhappy or sad. For each reason you have a pool of 100 points to express how important it is to you (0 points if it is not important at all and 100 points if it is very important). If there is anything else that is important to you, please add it along with your score at the end of the list."

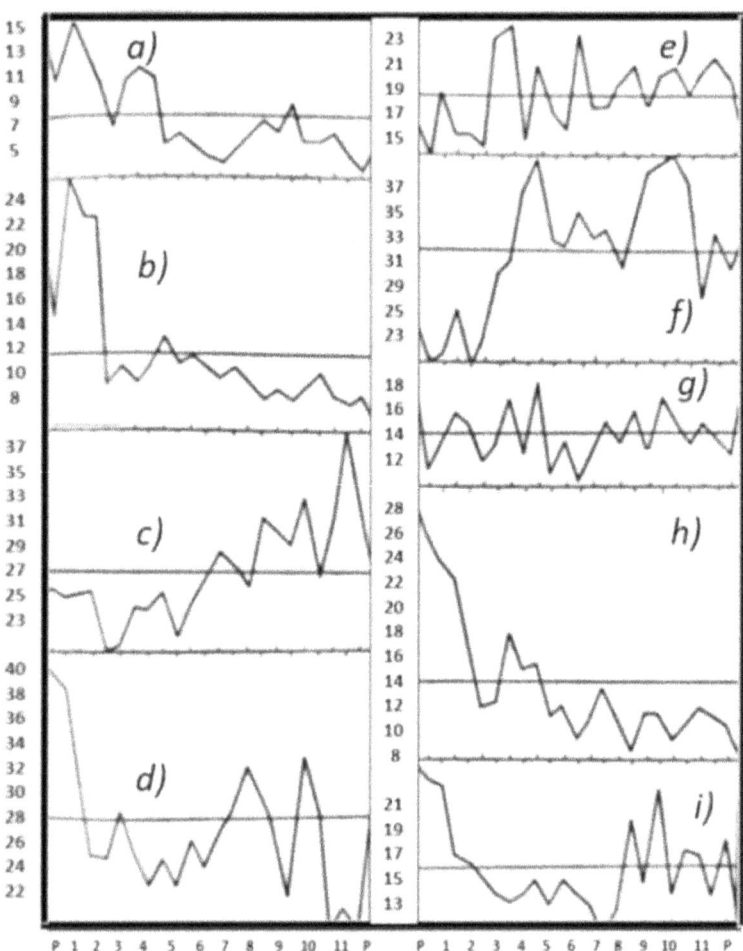

Fig. 26. Dynamics of the subjective discomfort of polar explorers during the annual Antarctic isolation (N = 20)

Source: own elaboration.

Legend: P – pre i post test; a) narrowness of one's own room, b) narrowness of the common room, c) way of managing the station, d) rare communication with the country, e) relations with others, f) own work efficiency, g) lack of someone for intimate conversations, h) lack of time and conditions for privacy, i) the need to adapt to the requirements of others, Mr – annual average for the group.

Figure 26 shows, among other things, that the greatest difficulties in adapting to discomforts such as housing conditions, information deprivation (infrequent telephone communication with the country) occur during the initial period of isolation. Other matters: e.g. the way in which the station is run, relations with others, one's own work performance, etc., are a cause of discomfort throughout the annual period of isolation, making emotional adjustment difficult. Noteworthy is the rather rapid adaptation to the discomforting conditions of the station, concerning: the cramped conditions of the own room (a) and the common room (b) and the lack of time and conditions for privacy (h). An interesting source of discomfort in the last months of isolation is related to the liberal way of managing the station, which means that this period is sometimes regarded as a big factor of threat stress, in which people expect support rather from an authoritarian style of management, giving a greater sense of security (Terelak, 2005). Also noteworthy are two discomfort factors related to interpersonal relationships, which cyclically persist above the average annual discomfort, and concern relationships with others (e) and lack of intimate conversations (g). The fact that not all polar explorers achieve adequate levels of emotional adaptation, despite prior psychological selection, was highlighted earlier when discussing the predictive accuracy of polar expedition member selection methods.

5.2.2. The personality perspective

The achievement of an adequate level of adaptation by winter-over participants may also be the combined effect of long experiences with reduced levels of stimulation, but also certain personality traits. Features such as: *extra- introversion and neuroticism* examined using the Eysenck EPI questionnaire.

5.2.2.1. Dynamics of neuroticism and extraversion

Eysenk's Personality Inventory was used to diagnose extroversion and neuroticism in winter-over participants and to assess so-called lying. The results obtained are shown in Fig. 27.

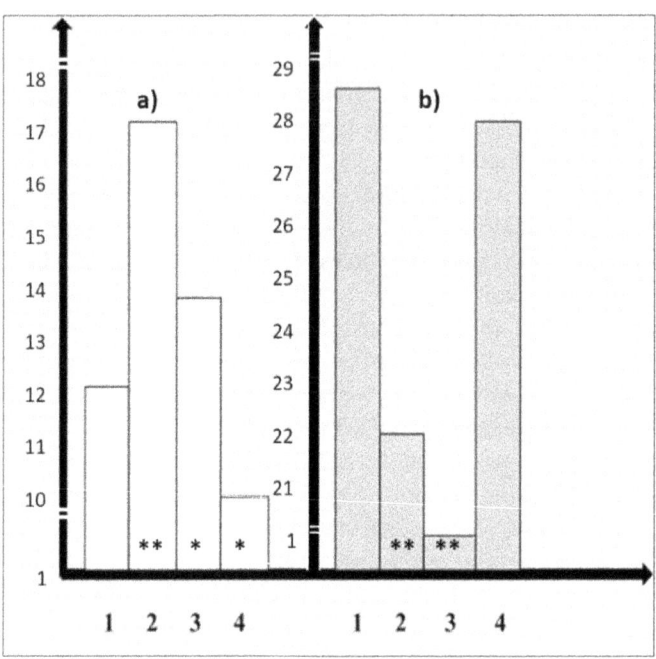

Fig. 27 Neuroticism (a) and extra-introversion (b) measured by Eysenck's EPI in winter-over
Source: own elaboration.
Legend: 1 – country pretest; 2, 3, 4 – beginning, middle and of Antarctic wintering-over; differences between pretest and subsequent tests are significant at * p <0.05; ** p <0.01.

Figure 27 shows, among other things, that the process of adaptation to a situation of isolation is accompanied by a process of introversion and that this adaptation is situational (not very stable), as it returns to its initial level after wintering-over. Moreover, the fluctuations in the neuroticism index suggest, among other things, that this is a difficult situation for 'winter-over participants,' especially during the initial period. Similar effects related to emotional lability, characteristic of neuroticism, along with the duration of isolation, were found, among others, in a 105-day experimental isolation by M. Nicolas et al. (2013), who observed that with the duration of isolation there are fewer and fewer positive emotions, and more and more often negative emotions, including depression.

Our own research compared the results of psychological tests from studies conducted during a one-year wintering-over period at two Polish polar stations on Spitzbergen[12] and King George Island (Terelak & Maciejczyk, 1989). The research was based on the assumption of the existence of adaptive symptoms specific to each research station as well as non-specific, i.e. common to both stations. Specific symptoms are a direct result of differences in the extreme stress nature of the Arctic and Antarctica environments. The first fundamental difference comes down to climatic differences, as "the Arctic is a real ocean, surrounded by continents, and Antarctica is a continent surrounded by oceans" (Pyne, 1986). Hence the Arctic is "full of life," as there are foxes, bears, reindeer, walruses, whales, vegetation and, above all, people, while the Antarctic is not only colder but completely dead, that is, devoid of all attributes of life. In turn, the characteristic non-specific features of the Arctic and Antarctica can be attributed to attributes such as low temperatures and social isolation in scientific stations

Indicators of the level of adaptation were scores on Spielberger's STAI tests (Spielberger, Gorsuch & Lushene, 1970), Eysenck's MPI (1959) and the Buss-Durkee Attitude Inventory (1961). The respondents were members of polar stations of the Polish Academy of Sciences, staying on Spitsbergen (n = 10) and King George Island (n = 21) during the polar winter of 1979. Test surveys took place: before the expedition in the country, at the beginning, middle and end of the wintering-over.

---

12 Polish Polar Station Hornsund Stanisława Siedlecki is a year-round research station on Spitsbergen (in the Svalbar archipelago), located a few hundred meters from the shore of the Bay of White Bears inside the Hornsund fjord. A. Skorupa (2015) presents the full environmental and psychological characteristics of this Polish Arctic station.

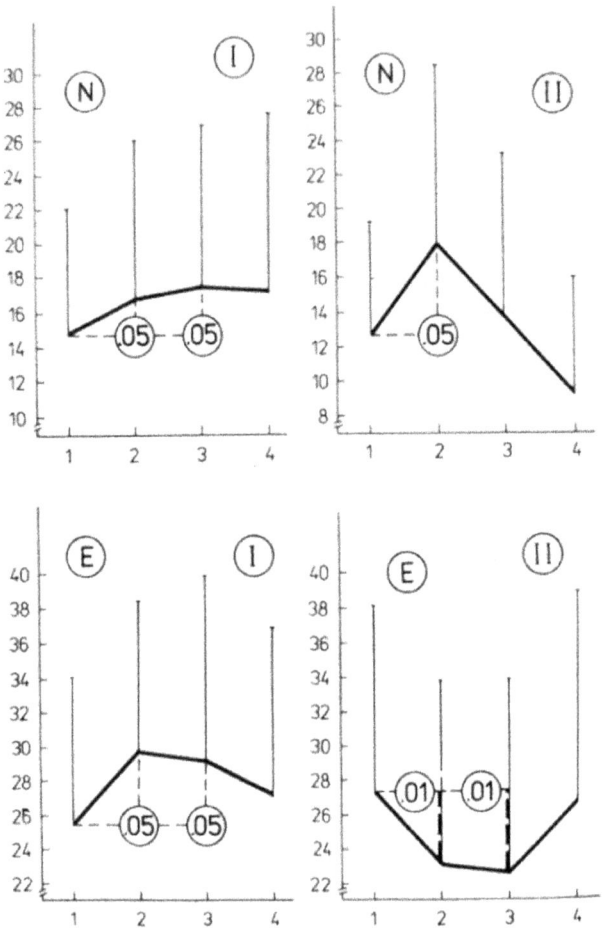

Fig. 28. Neuroticism (N) and extraversion (E) of winter-over parties in isolated Arctic (I) and Antarctic (II) stations

Source: own elaboration.
Legend: 1. Pretest; 2. Pre-winter; 3. Mid-winter; 4. Post-winter.

As can be seen in Fig. 28, adaptation to polar station conditions is characterised by an increase in neuroticism, especially in the early wintering-over phase. However, the differences between a "small" Arctic station and a "medium" Antarctic station can be seen. The elevated levels of neuroticism that persist in the

Arctic until the end of wintering-over are indicative of the more stressful conditions (photo-ecological and social) of the small station there. The differences in extraversion point to another, no less important aspect: they reveal different patterns of social adaptation in "small" and "medium" polar stations. For example, the lack of interpersonal alternatives at a small station on Spitsbergen "forces" social contact with all winter-over participants, despite an increase in neuroticism, which may suggest a link to psychological discomfort. At the "medium" station on King George Island, the pattern of social adjustment is different, characterised by progressive introversion, manifested by social withdrawal. However, regardless of the divergent adaptation patterns at the two polar stations, the situations are stressful.

We pointed out earlier that the polar explorers we studied are characterised by a higher proportion of extraverted individuals, in whom there is a high dependence on stimulation from the environment (cf. Fig. 17). Their achievement of a certain level of adaptation reveals at the same time that they have become accustomed to a certain degree of discrepancy between the expected and actual levels of stimulation. The mechanism of this discrepancy, which is of a discomforting nature, and ways of reducing it are presented in my own model in Figures 4, 5 and 6.

The clinical consequences of the introvertisation process from the perspective of space flight are pointed out by N. A. Kanas (2016), who documents that psychiatric problems occur during space missions, despite the fact that astronaut candidates are screened for predispositions to psychiatric conditions. No problems associated with the most common types of mood and thinking disorders were found (bipolar affective disorder, schizophrenia). Most often, the adaptive response to cosmic isolation is accompanied by symptoms of anxiety or depression in about 5%. A similar percentage of symptoms are seen in people working in analogue space environments, such as submarines and Antarctic bases. Russian psychologists and psychiatrists call the symptoms of introvertisation "asthenisation," which is the result of adaptive reactions to an extreme long-term isolation. These symptoms include: fatigue, irritability, emotional lability, attention difficulties, restlessness, heightened perceptual sensitivity and emotional instability, and they comprise the "winter-over syndrome" (Terelak, 2021). These symptoms are treated with psychoactive drugs, available to crew members during space missions, such as: anti-anxiety drugs (diazepam), antipsychotics (al. haloperidol); painkillers (al. codeine and morphine), sleeping pills (flurazepam and temazepam), stimulants (dextroamphetamine; and promethazine for motion sickness), and psychopharmacology (the effectiveness of some drugs has been studied on the ISS). As stated by S. Eyal & H. Derendorf (2019) although drugs have been used during space missions for more than half a century, further research should be conducted during space flights, due to the specific artificial environment of the spacecraft and possibly other planets or asteroids, taking into account the pharmacokinetics and pharmacodynamics of drugs. In addition, future commercial flights with space tourists, especially interplanetary flights, will require more knowledge from space pharmacology on how to optimise drug therapy. This is supported by, among others, E. Salas, et al. (2015),, who highlight that the studies of the geology, atmosphere and potential for life on Mars may be useful to develop requirement criteria on medical and psychological preventive and therapeutic measures tailored for Martian crews during long-duration space exploration (LDSE) missions.

A more detailed characterisation mediating the role of neuroticism and extra/introversion in the process of adaptation to social isolation is presented in Table 4.

Table 4. Dynamics of changes in Neuroticism and Extraversion measured by EPI in wintering-over (N = 20) in the process of adaptation to social isolation

Study	Neuroticism			Ekstraversion		
	Mean	SD	T-test	Mean	SD	T-test
1	12.61	6.72	2.64*	27.28	11.09	—4.97**
2	17.83	10.72		22.94	10.89	
1	12.61	6.72	0.54	27.28	11.09	—5.78**
3	13.61	9.74		22.55	11.36	
1	12.61	6.72	—2.14*	27.28	11.09	—0.63
4	9.28	6.95		26.61	12.40	
2	17.83	10.72	—3.21**	22.94	10.89	—0.63
3	13.61	9.74		22.55	11.36	
2	17.83	10.72	—4.81**	22.94	10.89	4.19**
4	9.28	6.95		26.61	12.40	
3	13.61	9.74	—3.65**	22.55	11.36	4.17**
4	9.28	6.95		26.61	12.40	

(Legend: 1 — preliminary research in Poland; 2 — research at the beginning; 3 — in the middle of wintering-over; 4 — at the end of Antarctic wintering-over; ** $p < 0.05$; * $p < 0.01$; SD — Standard deviation)

Table 4 shows, among other things, that the process of adaptation to a situation of social isolation is accompanied by a process of introversion or "closing in on oneself," although this process is slower in extroverts than in introverts, which is in line with the suggestion of M. Zuckerman et al. (1968), stating that it is possible that real situations of sensory deprivation may have a high excitation value and thus should be considered as stressful situations. Among other things, this is how they explain the results of their own research on the better tolerance of sensory deprivation situations in extroverts. This is confirmed in Table 4 by variations in the neuroticism index suggesting that sensory deprivation is a stressful situation, especially during the initial wintering-over period. This is consistent with our own observations in the "Antarctic Diary" (Terelak, 2021), in which the described symptoms of introversion, combined with mood swings and apathy, lead to a decrease in task activity, resulting in the abandonment of previously planned tasks. Examples include neglecting periodic research, which must be reported on every quarter, or wasting time watching the same films several times or taking up the position of "sea elephants" in the common room. Another example is the increase in aggressiveness, seen as a new source of stimulation.

5.2.2.2. Dynamics of different types of aggression

The Arnold H. Buss and Ann Durkee Aggression Questionnaire[13] examines the manifestations of various forms of aggressiveness, the annual dynamics of which are illustrated in Figure 29.

---

13   This tool was used in the Polish version elaborated by M. Choynowski (1972) under the name "Moods and Humours" (T-153: abridged textbook for the "Moods and Humours" test by A. H. Buss and Ann Durkee, based on the textbook by M. Choynowski / Ministry of Education and Upbringing).

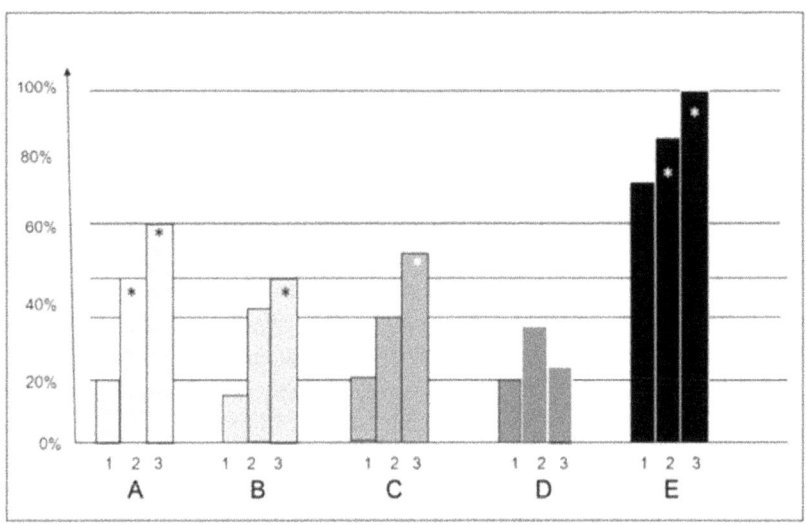

Fig. 29. Dynamics of different types of aggression in wintering-over
Source: own elaboration.
Legend: 1 – initial study, 2 – after half a year, 3 – after a year of Antarctic isolation; A – general aggression, B – physical, C – transferred, D – irritability, E – verbal; *p < 0,05.

Figure 29 shows a clear increase in aggressive behaviour with the duration of social isolation, which is particularly evident in the case of the overall aggression index. One of the suggestions explaining this state of affairs is that of J. Reykowski (1977) who, referring to the problem of the need for stimulation, proposes to treat aggressive behaviour in a situation of stimulation deficit as a source of strong and diverse stimulation. Particularly for low-reactive people, as is the case with winter-over participants, who, as extroverts, feel a great hunger for stimulation, aggression proves to be a good way to raise stimulation levels towards the optimum. This type of aggressive behaviour, having the nature of a situational, temporary adaptation, has nothing in common with other personality predispositions, such as aggressiveness as a defence mechanism, manifested in the form of neurotic-compensatory behaviour (Hraciarek, 2003).

The data quoted above and our own observations (cf. Terelak, 2021) seems to support the thesis that aggressive behaviour may be a learned form of providing oneself with stimulation appearing especially when there is a deficit of stimulation. In an Antarctic isolation situation, such a deficiency is typical, hence any "irritating" factors increase the level of adaptation, which consequently leads to the facilitation of aggressive activities,[14] what A. Sarris, & N. Kirby (2007) call "behavioural norms" at Antarctic stations. Although we address the causal aspect of aggression in the context of individual differences, and this thesis requires further detailed empirical verification, it explains in an interesting way the aggressive behaviour that is a basic component of the "winter-over syndrome."

The comparison of the results presented in Fig. 29 in relation to the symptoms of aggression also shows differences in the patterns of aggressive behavior that are characteristic of small and medium-sized polar stations, as shown in Fig. 30.

---

14 Observations suggest that in the situation of Antarctic isolation, aggressive behaviour was treated with great forbearance and often even with overt social approval ("finally something is happening") (Terelak, 2021).

214 Transactional Characteristics of the Winter-Over Syndrome

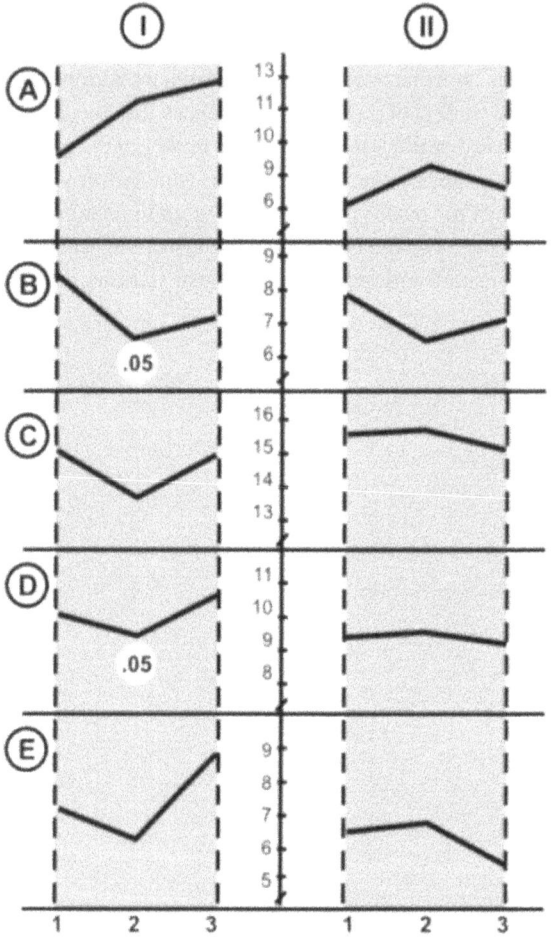

Fig. 30. Level of aggression: physical (A), projection (B), verbal (C), general (D), irritation (E) of winter-over participants in the isolated Arctic (I) and Antarctic (Il) stations
Source: own elaboration.
Legend: 2. Pre-winter; 3. Mid-winter; 4. Post-winter.

Particularly clear differences emerge with regard to the physical aggression index at the small station on Spitsbergen, the level of aggressive behaviour is clearly higher than at the medium station in Antarctica. Differences also appear in the irritation component of aggressive behaviour. Further comparative analyses should also take into account the largest Antarctic station, which is the American McMurdo Station with about 1000 crew members (Oliver, 1991), and standardise the variety of psychological methods used.

Stimulation demand theory (Strelau, 1995) provides one possible explanation for the psychological mechanism of aggressive behaviour in polar isolation conditions, whose adaptive mechanism suggests that in situations of stimulation deficiency, aggression is its medium, as evidenced by the fact that it is socially accepted (cf. Mullin, 1960; Terelak, 2021). This is one of the reasons why the results shown in Fig. 61, are not indicative of disturbed polar behaviour, but reflect specific patterns of adaptation to situations poor in social stimuli, which are particularly characteristic of small and medium-sized polar station crews. This is an interesting analogue of the situation of numerically small space habitat crews.

Another personality trait opposite to aggressiveness, namely humour, was pointed out by J. Brcic et al. (2018) as one of the strategies for coping with stress. The Humour Coping Scale (HCS), based on which five categories of humour can be assessed, was used in the study: (1) affiliative humour, facilitating positive connections with others (Affiliate); (2) humorous perspectives on life (Enhancing); (3) sarcastic (Aggressive); (4) self-discrediting (Self-defeating); (5) stress coping-oriented (Problem-oriented). In the retrospective narratives (memoirs, summaries, interviews, etc.), a content analysis was carried out as a basis for identifying two groups of astronauts studied: (a) an international group of 46 active astronauts and cosmonauts; (b) 20 retired, long-serving male cosmonauts. The results for sample one showed that astronauts were more likely to use humour as a coping strategy during long-duration rather than short space flights. Astronauts who represented the national majority in the crew during the flight were more likely to indicate the use of aggressive humour compared to those representing minorities. Retired cosmonauts tended to mention positive, solution-oriented humour. It should be noted that narrative research is burdened by the variable of social approval. A similar role of humour in dealing with stressful situations was confirmed in the author's own research conducted on young military pilots (Szymanik & Terelak, 2015).

In summarising the stimulatory aspect of the Antarctic isolation situation, we have pointed out that as the possibility of active regulation of stimulation at the level of behaviour is reduced, the subject adjusts their activity to external stimulation. Some forms of adaptive behaviour presented so far support the abovementioned hypothesis.

However, it is clear from Figures 21 and 22, cited earlier, among other things, that the level of adaptation achieved is situational and that many behaviours lose their adaptive significance as the situation changes (e.g. on the ship on the way back home). This demonstrates, among other things, the considerable lability of the physiological mechanisms responsible for satisfying the need for stimulation. These problems are presented in detail by A. Eliasz (1985) and my own own research on the relative stability of some temperament traits under conditions of social isolation

5.2.2.3. Dynamics of personality changes in selected factors of the 2nd degree Cattell's 16PF

In characterising the Antarctic isolation situation as a complex stress situation in paragraph 4.2, we pointed out that it is characterised primarily by its newness. Many hitherto learned behaviours may turn out to be useless in a new situation of isolation or simply take on new values. This suggests that adaptation to the Antarctic situation involves changes in some aspects of personality (preferences, attitudes, needs). An illustration of this thesis can be found in Figure 29, which shows some trends in changes in some personality traits of the subjects tested with Cattell's 16 PF test under the influence of Antarctic isolation.

A formula describing the dimensions of personality measured by Cattell's (1946) 16 PF in the following form: $Z = f(S1T1 + S2T2 + S3T3...)$, where: T is trait; S situation, refers to the definition of personality traits and their aggregation by G.W. Allport (1937), according to which a personality trait is a reflection of innate factors as well as environmental influences. There may be innumerable of them and that is why G.W. Allport created a nomothetic-idiographic classification of the so-called predispositions, differentiating them into those concerning people in general and individuals. The former, *dominant dispositions*, are powerful in virtually every situation, while the latter, *central dispositions*, manifest themselves in a range of adaptive behaviours in different situations. The author also mentioned a third category of traits, consisting of *secondary dispositions*, which are revealed only in certain situations. In turn, R.B. Cattell (1946), using factor analysis, combined the personality traits of individuals into groups with similar predispositions and this way, inspired by Allport's earlier observations, extended the knowledge of the interrelationships of the sixteen personality dimensions identified. By constructing the 16 PFs, he provided a psychometric tool to examine not only the structure of personality, but also the dynamics of its various dimensions in different life situations. Fig. 31 is an example of using this tool to assess the dynamics of adaptive behaviour in a situation of long-term Antarctic isolation.

The functional aspect of the Antarctic isolation situation 217

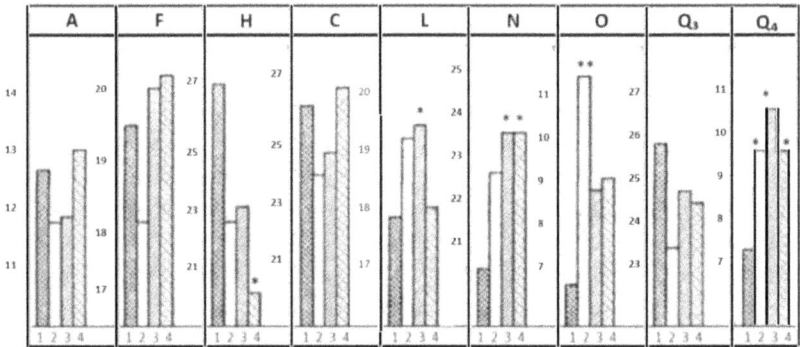

Fig. 31. Dynamics of personality changes of winter-over participants in selected factors of the 2nd degree Cattell's 16PF Questionnaire in conditions of annual Antarctic isolation

Source: own elaboration.
Legend: Second factors in the field of "introversion:" A + (cyclothymia), F + (surgency), H + (resistance to stress), N + (openness to the world), Q_3 + (high self-esteem): Second degree factors in the field of "anxiety:" C + (emotional maturity), L – (confidence), O – (self-confidence), Q_4 – (low ergic tension); 1 – Pretest in the home country, 2 – beginning of wintering-over, 3 – middle of wintering-over, 4 – end of wintering-over; * p <0.05; **p<0.01.

From the analysis of the results shown in Fig. 31, it is evident, among other things, that statistically significant differences between the pretest and subsequent longitudinal studies were obtained in several factors. Thus, for example, differences in the H-factor (p<0.05) between the first and last study suggest that with the duration of isolation a gradual closing in on oneself (situational autism), aloofness and caution in dealing with others occur, as well as lack of commitment to tasks and a decreasing tolerance of others. It is of great interest that the H factor in Cattell's concept is treated as a relatively constant predisposition (central trait), whose physiological basis is found in the autonomic nervous system (the difference in H1- and H4- measurements is determined by the predominance of the sympathetic and parasympathetic systems). Significant reduction in the H factor at the end of isolation characterised by autism compared to significant reduction in neuroticism (cf. Fig. 27) and a decrease in task activity (Fig. 23) can be considered as achieving a specific level of adaptation.

Differences in the L factor (p<0.05 between pre-test and mid-wintering-over), indicating an increase in criticism towards others, distrust, aggressiveness, hostility and psychological tensions, point to a large increase in small task group polarisation in mid-wintering-over. This is confirmed by the results for the N factor (p<0.05 between the initial and subsequent surveys), indicating a gradual establishment of emotional contacts between group members, an increasing understanding of other people, an attention to good social manners, with a simultaneous restraint in words. This is the result of the realisation of the irreversible necessity of a "common fate" (Aronson, 2009). In the light of the above, the findings in the L and N factors are only seemingly contradictory, as they rather indicate an ambivalent attitude of the group members in the process of polarisation and integration.

The differences in the O factor (p<0.01) between the first and second surveys indicate a realisation during the initial wintering-over period that impending long-term social isolation is an objective fact from which there is no escape. This causes, among other things, feelings of loneliness, sadness, self-aggression often caused by regret for a decision taken earlier, worrying about the difficulties ahead, etc.

Finally, the differences in the Q_4 factor (p<0.05) between the baseline study and the others testify to high nervous tension, anxiety, a sense of fatigue in isolation situations, irritability in social situations, and the fact of deprivation of many psychological needs. Similar data for the N, L, Q_4 factors were obtained by, among others, A.J. Taylor (1973), studying New Zealand polar explorers wintering-over at the Antarctic "Scott Base" and "Lake Vanda" stations. The changes described in the Q4 factor draw attention to the fact that many psychological, social and physiological needs are not being met.

In chap. 1. we have pointed out that contemporary psychology, especially Polish psychology (e.g. Tomaszewski, 1975), sees the human being as a complex regulatory system between him/her and the environment. Such an arrangement is usually assigned a set of needs as standards of regulation related to optimising the condons for its own development. However, in a situation of sudden stressful change in the environment, a person has to switch from a system of relatively constant needs (standards) for the individual to a variable (situational) system of preferences required by the constraints of a given moment, situation, specific unit of time. The preference system, as an adaptive strategy variable, depends directly on the current ability to satisfy needs, which is consistent with our model presented in Fig. 1 (in chap. I). It is therefore to be expected that long-term deprivation of a particular need may also change the system of need preferences, as confirmed by our study.

5.2.2.4. Dynamics of mental needs

The individual needs preference inventory Z. Borucki (1978)[15] measures one's attitude towards the fulfilment of the following eight needs: safety, social contact, sexual love, recognition, achievement, dominance, emotional sensations, self-actualisation, knowledge and understanding of the situation and emotional sensations. or cognitive. A detailed description of the questionnaire can be found in a separate paper (Borucki, 1986). This questionnaire was used in the longitudinal study on three occasions: at the beginning of isolation, at the halfway point and at the end. The results of the studies are presented in Fig. 32.

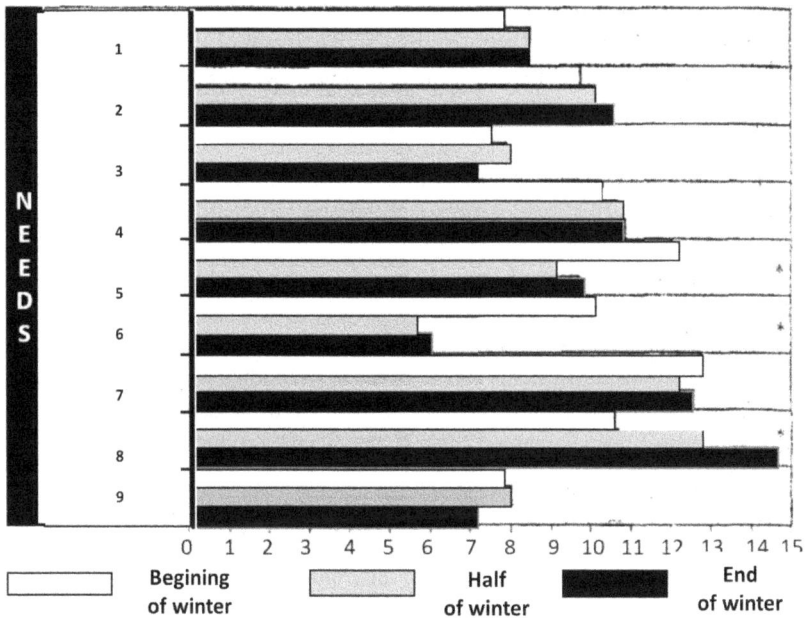

Fig 32. Dynamics of mental needs of winter-over in the situation of long-term Antarctic isolation at the Arctowski station**Endof winter**

Source: own elaboration.
Legend: * p <0.05; 1. Safety, 2. Social contacts, 3. Sexual love, 4. Acievements, 5. Recognitions, 6. Domination, 7. Self realization,8. Cognitive, 9. Emotional impressions

---

15 This questionnaire was elaborated for a survey of seafarers on deep-sea vessels and was used experimentally with the author's permission, for which I express my thanks.

For the purposes of this study, we define need as a subjective feeling of lack (no satisfaction) or desire (expectation) for certain things or situations. Thus, considering need as an objective and subjective fact, we would like to draw attention, following T. Tomaszewski (1979), to man's dependence (in the biological and behavioural sense) on the system of their environment and/or current situation, of which man is a distinguishing element. In the process of socialisation, preferences for specific psychological needs are formed, on the fulfilment of which human wellbeing depends. By analysing the structure of needs presented in Fig. 30, it can be seen that there are clear changes in three needs during the Antarctic isolation: dominance, recognition and cognitive one. Needs for dominance and recognition diminish along with the duration of isolation ($p<0.01$). This is due to the peculiarities of a small isolated task group, remaining without external social pressure (Gunderson and Nelson, 1965b). The interesting phenomenon of a systematic increase in the need for cognition ($p<0.01$) throughout the period of isolation is a peculiar reaction to monotony, manifested in self-stimulation through intellectual activity, such as reading books.

Summing up the discussion of certain adjustments in the structure of needs of an adaptive and situational nature, it should be assumed, following M. Kofta (1977), that we are dealing not so much with a change in the system of needs, but with a changing system of preferences as a reaction to their periodic deprivation. This is supported by previous research conducted in both experimental (Haythorn, 1973) and natural isolation situations (Gunderson, 1969), which indicates that some adjustments can be made to individual hierarchies of needs. E.g. Z. Borucki (1986), using the questionnaire we discussed above to investigate the needs of trainee seafarers (N=39) on a seven-month deep-sea voyage, found at the end of the voyage a slight increase in the need for safety, excitement, self-actualisation, dominance and achievement and a decrease in the need for cognition. These results are quite opposite to those we presented above, perhaps because the isolation situation was quite different (e.g. knowing that there is an SOS rescue system in case of life-threatening emergencies, which is not available during the Antarctic wintering-over period, etc.). However, regardless of the adopted direction of interpretation of the quoted data, it testifies to the fact that becoming aware of one's limitations in terms of values and possibilities in a specific situation leads to the "activation" of such defence mechanisms of the personality which generally[16] change the attitude towards human needs or cause the

---

16 We have used the phrase "generally" because the lack of conditions to satisfy a specific need can also lead to pathological reactions, as described in the diaries of polar doctors.

appearance of substitute behaviours of a compensatory nature. This is supported by data from the literature and by own observations of, for example, responses to sexual deprivation in situations of Antarctic isolation. Japanese psychologist H. Hachisuka (1971) and New Zealand psychologist A.J. Taylor (1973) believe that the phenomenon they describe of a higher than normal frequency of dreams with erotic content in a situation of Antarctic isolation and an intensification of sexual verbalism (of a humorous or even vulgar nature), testify, among other things, to compensatory reactions to a situation of deprivation of sexual need.

Sexual Opinion Survey (SOS) by L. White et al. (1988) as a method of examining of sexual deprivation of needs are well illustrated by my own research on the change in preferences towards sex during one-year Antarctic isolation, using the SOS scale, translated into Polish and modified in terms of the set of rating scales by B. Jaźwiński-Buczyńska (1978).[17] The subjects expressed their attitudes towards such aspects of sex and eroticism as pornography, homosexuality, onanism, group sex, oral sex, etc. Based on the assessments made, two indicators of attitudes towards sex can be assigned to the subject: (1) General attitude direction index – expressing how much the individual approves or disapproves of the sphere defined as sex; (2) Attitude direction index within each sex category.[18] In both indicators, the rule of thumb was that the higher the score, the more positive the attitude towards sex. Results from eight consecutive measurements (in-country, at the Arctowski station and on the ship on the way back home) are shown in Figure 33.

So, for example, O. Nordenskiöld, while wintering-over with his team in Antarctica (the ship "Antarctic" trapped in the ice) on Snow-Hill Island in 1902, noted among other things: "Winter has not yet arrived and tobacco is already in short supply. It is difficult to imagine a man deprived of tobacco in any of the great cities of America, Poland or South Australia; in a man-inhabited cave in Antarctica or on an Antarctic ship – it is a disaster, a reason for crazy hallucinations, for tormenting dream visions. And yet it is so: there is not a hint of tobacco left in the villa 'Under the Penguin' in Hope Bay" (quoted by Forster, 1959, p. 207).

17 In the version used, the scale consists of 21 statements containing content to which the respondent can respond on a 7-point scale (from extreme disapproval to extreme approval).

18 The attitude direction index is the product of multiplying the weighted average for a given sex category by a fixed scale factor. This way, the upper and lower limits of the positive (42 points converted) and negative (6) attitude direction were unified for all sex categories.

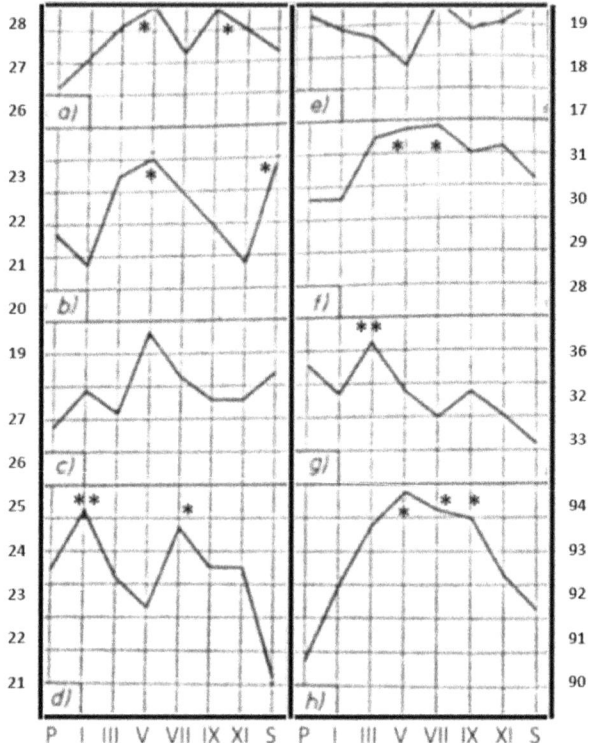

Fig. 33. Dynamics of needs against sex during the annual Antarctic isolation at the Arctowski station

Source: own elaboration.
Legend: P – Pretest; I, II V,… XI – research at Arctowski's station; S – on-board research on the way back home; significance of differences * p <0.05 between the initial and subsequent tests; attitudes: a – pornography, b – masturbation, c – homosexuality, d – sexual perversions, e – erotic fantasising, f – excitement with the opposite sex, g – public eroticism, h – global index.

Analysis of the global index of attitudes towards sex shows that it takes the shape of an inverted "U," i.e. it oscillates towards a positive attitude, reaching a peak at the end of May/beginning of June (middle of wintering-over), and then leans towards a negative attitude, reaching in the last survey (on the ship on the way back to the country) a level close to the initial one. The differences between baseline and climax surveys and between climax and final surveys are statistically significant (p<O.05). This seems to be an extremely interesting effect of

sexual deprivation in an Antarctic isolation situation, not previously reported in the literature. Despite the scarcity of data on this subject, based mainly on observation and interviews, it is fair to say that the direction of attitudes towards sex presented above is at least indirectly reflected by periodically increasing verbal behaviour with sexual content (e.g. sexual humour, telling intimate erotic details, etc.) or masturbatory practices. Data from the literature and our own observations confirm that the above-mentioned sexually-tinged behaviours acquire gratifying values, especially in the mid-winter period, which in our study manifests itself most in positive attitudes towards sex (cf. Fig. 33h). However, with the duration of isolation, the effectiveness of this type of behaviour becomes less and less effective, which consequently leads to a decrease in their "stimulative" value, as evidenced, among other things, by the fact that, for example, sexual verbalism is explicitly "forbidden" in the last period of isolation (cf. Mullin & Connery, 1959; Hachisuka, 1971, Terelak, 2021).

A more detailed analysis of the results allows us to conclude that depending on age (the older, the more neutral the attitude towards sex), education (the higher the education, the more positive the attitude) the approval or disapproval of certain sexual attitudes varies considerably. For example, other variables modifying attitudes towards sexual deprivation were pointed out by G. Palmai (1963a), who in his research showed, among other things, that sexual deprivation in a situation of Antatrctic isolation was better endured by respondents from "happy" marriages and "bachelors" compared to "divorcees" and "newlyweds." This is confirmed by our own narratives, especially those of bachelors and newlyweds[19] (Terelak, 2021).

As the small size of the group does not allow for quantitative correlations to be captured with the above-mentioned variables, we have therefore made a general assumption that the attitudes towards sex presented in the baseline studies are representative of Polish culture. The analysis of the individual categories of attitudes shows that they vary quite considerably during the one-year isolation, E.g. the two peaks in the course of attitudes towards pornography (Fig. 33a) reflect periods when pornographic magazines were popular (statistically significant

---

19 In the diary of the 3rd PAS Expedition to Arctowski Station, I present a case study of a polar explorer who left for an Antarctic wintering-over three months after his getting married. Observations show that he was unhappy with the separation from his sexual partner, which was evidenced by such behaviours as not locking himself in his room, depressive states, wallpapering the walls with pictures of his "beloved," etc. (Terelak, 2021).

differences between baseline and May and September surveys (p<0.05). The dynamics of attitudes towards onanism (statistically significant differences between January and May and the shipboard survey (p<0.05) and between May and November (p<0.05) and towards homosexuality (Fig. 33c – no statistically significant differences) correspond with the data of C.S. Mullin (1960) and H. Hachisuki (1971), who in their study on small Antarctic stations (American and Japanese) found that both erotic dreams, nocturnal pollutions and masturbatory activity appear or increase during periods of inactivity and personal emotional stress, as well as on the way back home (re-enlivening sexual fantasies and imaginations), associated with anticipation of engaging in sexual activity. The authors also found that there is an aura of utmost intimacy and secrecy around homosexual issues. This data is consistent with my own observations based on interviews with some of the winter-over participants[20] at Arctowski Station (Terelak, 2021).

An interesting change in attitudes is also manifested in the category of perverse behaviour (Fig. 33d), as evidenced, among other things, by statistically significant differences between the January survey and the May and ship surveys (p<0.01) and between the March and ship surveys (p<0.05), suggesting that the closer the return to the country, the more attitudes change in a negative direction. Similar trends are seen in attitudes towards public eroticism[21] (statistically significant differences between March and July and ship surveys at p<0,01), which anticipates an imminent return to a "puritanical" attitude towards sex.

Finally, "the excitement of the opposite sex" (Fig. 33h) shows statistically significant differences between pre- and post-test and May, July and September (p<0.05) i.e. the months characterised by the most vigorous sexual verbalism, which is also confirmed by the positive attitude in the same period towards pornography and onanism.[22]

Summing up the discussed dynamics of reactions to sexual deprivation, it should be stated that it is a good illustration for the three stages of the attitude object response proposed by J. Reykowski (1973). Thus, in the first stage, the separation of the object of attitude is associated with an increase in sexual tension. This is indicated by the differences between the pretest and follow-up tests,

---

20 Only some winter-over participants from the scientific group agreed to such interviews.
21 This includes incidents of public exposure for fun (Terelak, 2021).
22 During this period, erotic films screened in the station's "Terelax" cinema were well attended, accompanied by numerous erotic commentaries, as narrated in my Antarctic Diary (Terelak, 2021).

which suggest the fact of increasing interest in sex as sexual tension increases (no such tension was felt in the country, as conditions existed to "reduce" it naturally). In the second stage, an emotional-motivational attitude towards the object of the attitude can be observed, consisting of associating sexual activities (in this case substitute, compensatory activities) with pleasant (positive direction) or unpleasant (negative direction) emotions. However, the situational conditioning of sexual deprivation does not lead to the third stage, i.e. to the formation of a new pattern of behaviour, because, on the one hand, substitute sexual behaviour cannot constitute optimal sexual relaxation on an emotional level and, on the other hand, the awareness of the temporariness of the situation of sexual deprivation does not compel the formation of such a pattern. Hence from Fig. 33 it is clear that preparation for the "old" sexual pattern (an attitude similar to that expressed in the pretest) takes place on the ship on the way back home. In other words, it can be said that in the situation of Antarctic isolation we have described, deprivation of sexual need entailed an attitude shift in the positive direction and, in the situation of anticipation of a return to the original conditions, a preference shift in the opposite direction.

Summarising the discussion of needs, one may think that achieving an adequate level of adaptation to a situation of long-term Antarctic isolation mainly involves the need to change a particular system of preferences while attempting to maintain the system of needs as relatively fixed standards of behavioural regulation. However, it cannot be ruled out that in the case of long-term isolation (such as during interplanetary travel), the system of needs may also change. This is evidenced, among other things, by the change in the need for dominance and recognition shown in Fig. 33, as well as by the data of C.S. Mullin (1960) on the so-called positive effects of Antarctic isolation, to which he includes, among other things: greater internal discipline, greater tolerance of others and patience, and a better understanding of oneself and others. However, the lack of data on the sustainability of these changes does not authorise overly far-reaching conclusions on this. Further research in this area therefore seems to be of great interest.

So, for example, J.G. Corneliussen et al. (2017) highlight that the research into personality traits, personal value structure and a tendency to conflict behaviour in an isolated, confined and extreme environment (ICE) can be helpful in the selection and screening of space crews for long-duration interplanetary missions. In the longitudinal studies (biweekly and monthly) conducted before, during and at the end of the two-year stay of 10 Danish soldiers at the polar station in Greenland, using three personality questionnaires: the NEO PI-R, Triarchic Psychopathy Measure and Portrait Values Questionnaire, supplemented by the structured interview, it was found, among other things, that the subjects well-adapted to ICE conditions were characterised by personality traits such as: conscientiousness, agreeableness and boldness, and by personal values such as: kindness and self-control. The important behavioural attributes also include peaceableness or conflict resolution skills. Similar views on personality predictors useful for crewing long-duration space missions are cited by L.A. Palinkas et al. (2000) based on the research conducted at Antarctic stations. The authors examined, among other things, the influence of personality traits, interpersonal needs in 657 Americans wintering-over in Antarctica in the years 1963–1974, taking into account demographic and environmental variables of the station (crew size and the harshness of the physical environment). They found, among other things, that socio-demographic characteristics, personality traits (low neuroticism, extroversion and conscientiousness, as well as self-control) and station environment attributes together accounted for 9–17% of the variance for predicting adaptation to the ICE. The results of this research were confirmed by A. Kjærgaard et al. (2013) in a longitudinal study conducted on six two-person "Sirius Patrol" Danish military teams[23] using the NEO PI-R questionnaire on personality traits, personal values and self-development, and the Triarchic Psychopathy Measure (TriPM) and the Portrait Values Questionnaire (PVQ) on self-control, boldness, universalism and need for stimulation. The research results presented above are also partly confirmed by the data of M.A.G. Peeters et al. (2006) originating from a meta-analysis of the Big Five personality traits relationships concerning the impact of personality traits such as agreeableness and conscientiousness on the functioning of small task teams (ranging from 106 to 527) in the ICE. Based on these results, the authors made suggestions for the future composition of small task teams, which should consist of members with a similar profile of personality traits in terms of agreeableness and conscientiousness, excluding traits such as introversion, emotional lability and lack of openness to experience. These suggestions are agreed with by, among others, G. R, Leon & N.C. Venables (2015), who examined the relationship of personality with decision-making ability in an emergency situation from the perspective of a space analogue. A two-man Special Forces team performing the task of reaching the North Pole in the shortest possible time was studied. Both subjects were characterised by personality traits such as leadership-dominance, low anxiety

---

23 Sirius Patrol – a special unit that has been in existence for about 70 years, originally formed during World War II to monitor German ship activity in the Baltic Sea area. At present, its main mission is related to maintaining Denmark's sovereignty in an uninhabited national park.

and high achievement, risk-taking. However, they differed significantly in empathy, agreeableness, extroversion, emotional control and value systems preferences. According to the authors, characteristics such as the dominance of both team members on the one hand and incompatibility of other personality traits and value systems on the other, as well as the lack of survival training, were the reasons for the mission failure (aborted tracking).

In concluding our discussion of the personality aspects of adjustment, we should state that the observed changes in certain personality traits, preference systems and attitudes, reassure us that these should be considered in terms of "psychological costs" rather than emotional disturbances, as, for example, some American psychologists do (Mullin, 1960; Gunderson, 1966 and others). Furthermore, the data cited so far support the thesis that individual adaptation should be regarded not so much as a relatively stable state, but rather as a process characterised by specific dynamics that are not unrelated to situational factors characteristic of Antarctic isolation.

5.2.3. Emotional perspective

In characterising the physical, psychological and social environment of Arctowski Station in paragraph 4.2, we found that the characteristics of the elements of the environment and the ability to understand the situation change with the duration of isolation. For these reasons, among others, the situation of annual Antarctic isolation should be regarded as varying considerably in terms of the action of many stress factors. Research into mood levels and mood swings is one example of this.

5.2.3.1. Dynamics of moods

Mood as an emotion which changes in short periods of time and which has its own characteristics, described by means of intensity and dynamics of change, was examined using my own scale, on which the continuum on one side was minus 100 points (great sadness, anger, anxiety), and on the other side – plus 100 points (great joy, cheerfulness, happiness). Each respondent marked their mood over the past two weeks on a scale. In addition, below the scale, it was necessary to indicate by how many plus/minus points the mood changed during the surveyed period (e.g. if by 10 points +/-, i.e. it was rather stable, and if by 50 points both ways, i.e. it was very changeable). Indicators of mood intensity and fluctuation here are group mean scores from each survey, conducted on a two-week cycle through the one-year isolation period and on the ship on the way back home. The results are shown in Figure 34.

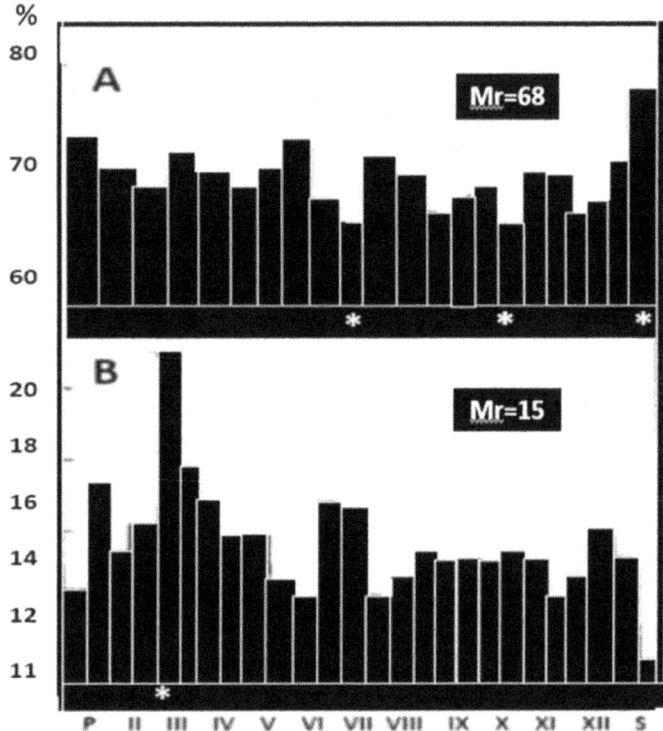

Fig. 34. Variation in the mood of polar explorers during the annual isolation (N = 20)
Source: own elaboration.
Legend: Annual average: A – intensity (68), B – Fluctuation (15); ↔ – the figure indicate the significance level of differences *p <0.05 between the pretest and the next tests; P – pretest; II, III, IV,... XII – monthly tests; S – ship tests.

The data contained in Fig. 34 shows that the level of mood is relatively high, and its changes reflect major events at the Arctowski station. For example, the decreased mood in February compared to previous months is linked to the departure of the 42-member summer team to the country and the start of de facto social isolation. In the following three months, there is a significant improvement in mood, which may be related to undertaking physical work with station's preparation for the Antarctic winter. In April, the lowering of mood is to be associated with a drastic deterioration in weather conditions (strong hurricane winds blew for 26 days, making it impossible to stay outdoor). As the weather got better, an improvement occurred. The significant deterioration of the mood at the end of June (p<0.05) can be associated with the "Antarctic midwinter holiday," which makes every polar explorer aware that in fact the most difficult period of isolation is still yet to come. Throughout the entire Antarctic winter, the mood is, however, lower than the annual average. The mood does definitely change for the better no earlier than on the way back home (p<0.05).

The phenomenon of considerable emotional instability is characteristic for Antarctic isolation. As can be seen from Fig. 35, the range of mood swings in individual months confirms the correctness of treating mood swings as one of the basic symptoms of the winter-over syndrome (Gunderson, 1966a). That is why we were interested in mood components in our research.

The Mood Adjective Check List (Nowlis, 1965)[24] is based on defining mood as a manifestation of a subject's general functioning and orientation, noting the basic patterns of its identification, such as activation level, control level and concentration level, among others. Mood can therefore be triggered by stimuli (cues) associated with these patterns in ontogeny and manifested at affective, cognitive, motivational and motor levels. These reactions may modify or trigger further reactions. Based on the above-mentioned assumptions, one might expect, among other things, that in the situation of Antarctic isolation, with the change in activity patterns of functioning, one should expect more mood variability. Hence, in our study, this technique was applied once a month throughout the annual isolation period and on the ship on the way back home. The theoretical model proposed by Nowlis is convenient for use in situations where an individual's responses are not a simple function of an external stimulus, or where the determinants of response are inaccessible to manipulative control. This technique is therefore suitable for use in natural isolation conditions. The fact that it is measured several times is also due to the understanding of mood structure itself, which is inherently a dynamic structure and refers to the transient tendency of certain personality traits to be revealed or not revealed in certain situations. Although mood can manifest itself in various forms of behaviour, the research technique employed by Nowlis used, as an indicator, the individual's verbal behaviour.[25] The results of the study are illustrated in Fig. 35.

---

24 A cultural adaptation of the Mood Adjective Check List was used in my own research, with the permission of the author, to whom I extend my thanks at this point.
25 Nowlis' Adjective List contains 49 adjectives, based on factor analysis classified into 12 factors, which in turn can be reduced to four second-level factors: activation or deactivation, positive or negative social orientation, positive or negative sign of experienced emotions and level of control. The subject obtains 12 mood indices formed by

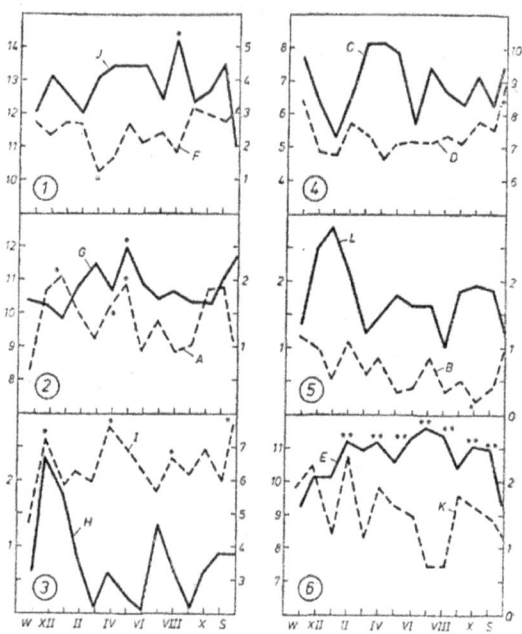

Fig. 35. Change of mood according to Nowlis in winter-over (N = 20) during the annual Antarctic isolation

Source: own elaboration of V. Nowlis, 1965.
Legend: 1 – Activation – inactivation: J – energy, F – fatigue; 2–3 – Positive – negative social orientation: G – social feelings, A – aggression, H – skepticism, I – egotism; 4–5 – Positive – negative sign of experienced emotions: C – cheerfulness, D – excitement, B – anxiety, L – sadness; 6. Level of control: K – nonchalance, E – concentration; W – baseline tests; XII, I, II – monthly tests; S – ship tests; * p <0.05; ** p <0.01 – between the initial and subsequent tests.

The shown Figure 35 characterises changes in mood depending on a number of situational factors. For example, a sharp increase in the J factor (energy) is associated with a periodic improvement in the weather in August (p<CQ,05) and with the installation of a ski lift, which created the conditions for active outdoor recreation.[26] In terms of the F factor (fatigue), two periods of increased fatigue are observed. The first, covering the months of December, January,

---

multiplying the sum of the ratings of the individual adjectives within a specific factor (four-point scale) by a fixed scale factor.
26 A detailed description of this event is given in the diary (cf. Terelak, 2021).

February, involves unloading the vessel and station maintenance work. This period is followed by a significant decrease in fatigue (p<0.05),[27] which is related to the departure of the summer team and the deliberate relaxation ordered by the station manager and the bad weather in April (significant cyclonic activity). A second period of increased fatigue occurs during the final three months of isolation (September, October, November) as a result of the intensification of work associated with the scientific tasks and the inventory and cleaning work prior to the imminent handover of the station to the next team of experts. The G factor (social feelings) reflects two stages in the organisation of the station's social activism: in March there was the initial organisation of social life after the departure of the summer team, and in May (p<0.05), when the station's informal activism spontaneously emerged around the so-called Babushka Club.[28] The dynamics of factor A (aggression) corresponds with the data on the stimulatory role of aggressive behaviour presented earlier in Fig. 28, although it completely fails to explain all the aggressive behaviour from a content perspective. A more detailed analysis of the isolation situation at the Arctowski station suggests that the first peak of aggressive behaviour in January (p<0.05) is associated with the process of separation of the group of winter-over participants (as an autonomous group) from all participants (summer and wintering-over group) of the 3rd Antarctic Expedition of the Polish Academy of Sciences. The next peak in May (p<0.05) is associated with the previously mentioned separation of station assets. Both periods were often accompanied by harsh conflicts. Finally, the last peak in October and November is mainly related to the "stratification" of the hitherto integrated task force in view of the return to the country. Complementary, as it were, to the above considerations are the findings in factor I (egotism). The course of the curve indicates that an increase in the level of egotism is associated with difficulties in the integration of the wintering-over group. E.g. the peak of egotism appearing in December (p<0.05) is associated with the summer team taking over "power" at the station together with the leadership of the entire 3rd Antarctic Expedition of the Polish Academy of Sciences. The withdrawal of the winter-over participants in this situation is their reaction to the discrepancy

---

27 Only statistically significant differences between baseline and follow-up studies are highlighted in Figure 34. A number of differences between the studies were obtained, but we have refrained from highlighting them in the figure for reasons of legibility.
28 The establishment and role of this club in the informal organisational structure of winter-over participants at the Arctowski station is described in detail in the Diary of the Third Antarctic Expedition PAS (cf. Terelak, 2021).

between the declared attitude of the summer group ("you are the hosts of the station") and the actual practice (e.g. the actual leader at that time was the leader of the Third Expedition). The effect of this was an increase in aggression towards members of the summer team, as we mentioned when discussing factor A. The next peak in April ($p<0.05$) and August ($p<0.05$) is associated with weather not allowing people to leave the premises. Finally, the last peak on the ship ($p<0.05$) is the result of a complete disintegration of the group. Withdrawal is a defensive reaction to the weariness of the self caused by having to be "face to face" with oneself for a year. The factors H (scepticism), C (cheerfulness), D (excitement), L (sadness) and B (anxiety) indicate that adaptation to the isolating situation is accompanied by certain emotions (positive and negative), and that one of the important conditions for adaptation is an adequate level of emotional control. We are also persuaded by the results for factor E (concentration), which suggest that a high level of concentration ($p<0.01$) can be observed throughout the Antarctic wintering-over period, which is the result of being aware of the responsibility for the tasks assigned on the one hand, and of the many real interpersonal risks and stresses on the other. However, as indicated by the results in the K factor (nonchalance), one should take into account the occasional tendency to underestimate risks as a result either of inexperience (e.g. in the initial period of stay at the station) or of becoming accustomed to stress factors (especially towards the end of isolation according to the principle "The devil is not so black"). The fatal accidents still recorded at Antarctic stations, precisely in the periods mentioned, confirm the validity of the above thesis (cf. Beltramino, 1966).

The importance of examining mood under the conditions of isolation is pointed out among others by J.E. Boyd et al. (2007), who – building on previous studies of astronauts and cosmonauts on mood dynamics – hypothesised that mood patterns among ISS crew members would be mediated in multinational groups by culture, regardless of the overall prevalence of these states in each cultural group. Studies on the ISS personnel have pointed to cultural differences in crew mood during space flight. Thus, the results of Russian cosmonauts are consistent with the asthenisation model, interpreting asthenia as the state occurring prior to space isolation, while American astronauts treat mood fluctuations as an indicator of stress. Regardless of cultural differences in the theoretical interpretation of mood variability, it is important to standardise its diagnosis in order to develop strategies to deal with such symptoms of isolation stress, regardless of cultural differences, as they may hinder future international space missions, especially long-duration missions to Mars.

However, it has long been argued that patterns of mood swings in the case of Antarctic expedition members are largely homogeneous (Hawkes et al.,

2017). The main objective of the study carried out by C. Hawkes et al. (2019) was to monitor the mood swings of the expedition participants throughout the wintering-over period at the Antarctic station. Results from 423 Antarctic expeditions, the members of which were recruited from all expeditions of the Australian Antarctic Program in 2005, 2006, 2007 and 2008 (304 men and 119 women), were subject to statistical analysis. Respondents filled in the Hopkins Symptom Checklist-21 (HSCL-21), consisting of 21 items for self-description of current health, well-being and distress (quoted from: Green, Walkey, McCormick and Taylor, 1988). Analysis of the research results showed that during the expedition, the participants experienced a significant decrease in their mood levels in two phases: the initial wintering-over period and in the last month of their stay at the station. In the discussion, the authors point out, among other things, the important fact that the expedition's mood swings are largely homogeneous, which may be the result of a rigorous "select in" procedure resulting from the extreme stress experience of Antarctica, as evidenced by individual personality profiles, characterised by traits such as stress resistance, emotional equilibrium, conscientiousness, agreeableness, and extraversion, among others. From a practical point of view, the identification of a homogeneous pattern of mood variation established over such a large heterogeneous sample and irrespective of the station's location in Antarctica and the length of time the crew has been there, can be regarded as an analogue of a future long-term space mission. However, one has to agree with the methodological reservations of the authors, which concern the small number of longitudinal studies with possibly frequent measurements; the ex post interpretation of results; homogeneity in terms of gender, nationality; limitations regarding the direct application of the application in aerospace. It is not without significance that there has been a significant change in the model of classical isolation, without easy communication access with the outside world, and the model of modern isolation with widespread access to satellite communications (Internet, TV, tablets, etc.) and modern station infrastructure with its scientific and technical instrumentation. This includes the existence of new tools for data aggregation and multidimensional statistical analysis of common factors responsible for human performance in ICE environments (Norris, Paton & Ayton, 2010).

To summarise this part of the discussion, it can be said that mood indicators are a good "thermometer" to measure the psychological cost of adaptation to the extreme conditions of Antarctic isolation, which is a less stable system than it appears – when considered only from the formal point of view of its stimulus importance (Palinkas & Houseal, 2000; Hawkes & Norris, 2017).

The studies of mood cited above are subjective, especially in the part concerning the interpretation of situational determinants of mood change, which was done "post hoc," on the basis of narrative analysis or rating scales, which are also characterised by subjectivity. However, they add credibility to the research on overt anxiety. For this reason, among others, C. Hawkes et al. (2019), subjecting common findings of relatively homogeneous patterns of mood swings as the axial syndrome of Antarctic winter-over participants, examined the dynamics of mood swings among members of 423 expeditions from the Australian Antarctic Research Program from 2005–2009. Although the authors used advanced statistics (e.g. latent class growth analysis) in their data analysis, the results supported the common view that the patterns of mood swings of Antarctic expeditions were largely homogeneous.

5.2.3.2. Dynamic of anxiety

State-Trait Anxiety Inventory *(STAI)* by C.D. Spielberger[29], based on the symptom concept of anxiety, distinguishes between two types of anxiety: (1) state – understood as situational anxiety, characterised by subjective, consciously perceived feelings of tension and anxiety as well as increased activity of the autonomic nervous system[30] and (2) trait – as a relatively constant disposition to react to a situation perceived as a threat (Spielberger, 1966). Anxiety was measured once a month throughout the one-year isolation period and on the ship on the way back home. The results are shown in Fig. 36.

---

29   The authors of STAI are Spielberger C.D., Gorsuch R.L., Lushene, R.E. (1970) and with the permission of the authors it was used exclusively for experimental research at the Psychological Laboratory of Jan Strelau, whom I thank for making it available in its Polish version.
30   This definition was, among other things, the basis for the use of anxiety thus understood as an indicator of mood.

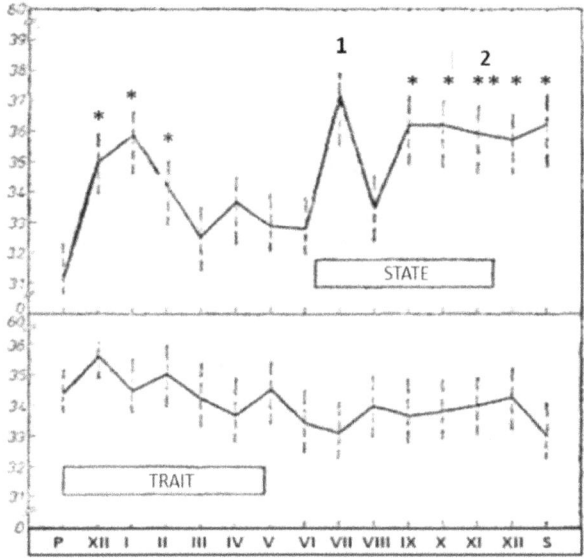

Fig. 36. Dynamics of the anxiety level of winter workers (N = 20) during the annual Antarctic isolation at the Arctowski station

Source: own elaboration.
Legend: P – pretest in the country; I, II, III,… XII – monthly surveys at the Arctowski station; S – on-board research on the way back home; 1 – Middle of Antarctic Winter; 2 – experiment "power plant failure;" * p <0.05; ** p <0.01 – significance of differences between the pre-test and subsequent tests.

Two conclusions can be drawn from the analysis of Fig. 36. Firstly, anxiety as a state reflects reactions to a threatening situation. For the first three months of stay at the Arctowski station, statistically significant differences between baseline and December, January and February (p<0.05) indicate an increase in anxiety levels, which are associated with anticipation of the stress of isolation. The renewed increase in anxiety in July (p<0.01) is also associated with anticipation of the threat posed by the onset of Antarctic winter and actually being "cut-off from the outside world." By the end of the isolation (p<0.05) and on the way back to the country (p<0.05), anxiety is, in turn, associated with anticipation of re-adaptation to normal living conditions and the need to resume many former social, professional and family roles. Similar data on the fear of returning to one's country were obtained by, among others, C.S. Mullin (1969), A.J.W. Taylor (1973), E.K.E.Gunderson (1968).

The results of the research carried out with wintering machines at a medium-sized station in Antarctica correlate with the research carried out at a small Arctic station, which is illustrated in Fig. 37.

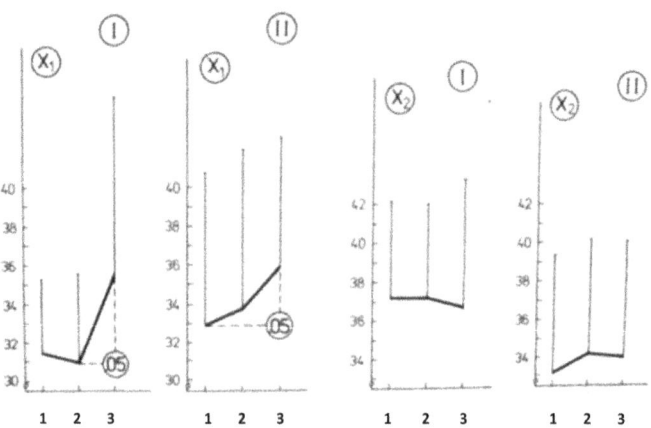

Fig. 37. Level of anxiety: condition (X 1) and trait (X 2) of Arctic (I) and Antarctic (II) winter trees

Source: own elaboration.
Legend: 1. Before winter; 2. The middle of winter; . After winter.

The comparison of the results presented in Fig. 42 in relation to the symptoms of anxiety also shows differences in the patterns of aggressive behavior that are characteristic of small and medium-sized polar stations.

The presented results of the research on the level of anxiety carried out in the conditions of a natural experiment in Antarctica also correlate with the research conducted by D.A. Rockwell et al. (1976) in a 105 day laboratory experiment. The authors tested three groups at 10-day intervals in terms of STAI Spielberger: two experimental, consisting of polar explorers in a 105-day isolation and 21-day isolation, and one control group. Firstly, the pretest yielded significant differences in high levels of anxiety in the control group, which is understandable given the better distribution of scores. Polar researchers, as previously subject to natural selection, were characterised by low levels of anxiety as a trait and condition. Thirdly, in the first phase of the experiment, an increase in the level of anxiety was observed in all the groups studied in relation to the baseline, with the polar explorers' level slightly decreasing after 20 days of the experiment and the control group only after 50 days. Anxiety levels then stabilised in both 105-day isolation groups until day 90 in the control group and day 100 in the polar explorers' group, after which they increased sharply in both groups. Post-test results were relatively lower than in the pretest. Our own research as well as that discussed above suggests the existence of a certain characteristic pattern of response to the stress of social isolation, characterised by a mechanism of anticipation of imagined threat, which is revealed at the beginning and at the end of isolation. This threat may be purely symbolic (reduction of comfort), especially under laboratory experiment conditions, or a real threat (e.g. death) – under natural experiment conditions. This is confirmed by our own research in September (Fig. 36), reflecting concern about the prolonged information deprivation associated with the maintenance of the broadcasting facilities in the home country and the lack of communications for some time.

This situation was experienced as a major psychological discomfort. In addition to the situational nature of anxiety, M. Malkiewicz & J.F. Terelak (2016) point to culturally determined anxiety attitudes associated with the changing civilisational threat, which may be relevant to the assessment of multicultural space habitat crews.

According to O.N. Kuznetsov and V.L. Lebedev (1972) in a situation of isolation in which a small group of polar explorers find themselves, the exchange of information between people is greatly reduced after a while ("everything has long since been said"). Hence, the ability to exchange information with the outside world rises to the level of a basic need. Thus, it can be said that with the duration of Antarctic isolation, the need for information – both specific and non-specific – increases. Specific information under Antarctic isolation conditions is also greatly reduced, as whether it is transmitted by telex or radio, it is characterised by high selectivity. In this situation, there is an increase in the demand for non-specific information, which, regardless of its content, reduces its deficiency. Thus, Arctowski Station saw a custom of gathering in a group after radio communication sessions to "gossip." This way, non-specific information performed the important task of maintaining a general "mental tonus," lowering anxiety levels slightly (cf. Jones, Wilkenson & Braden 1961).

The group reacted with similarly high levels of anxiety ($p<0.01$) during the power plant accident in October ($p<0.01$). When one considers that an efficient power plant is the "heart" of a polar station, heightened anxiety when this comfort is threatened is a justifiable reaction. Also noteworthy is the fact that in fifteen consecutive measurements (see Fig. 34) anxiety as a trait maintains its relatively constant level. This proves, among other things, that the behavioural changes discussed so far are situational in nature and testify to the great plasticity of the human psyche in terms of adaptability.

In summarising the dynamics of anxiety responses in one-year Antarctic isolation, we will highlight the drawing stages of adaptation to stressful conditions: (1) An initial period of high anxiety, correlating positively with the magnitude of the individual's real or anticipated threat; (2) A period of reduced anxiety, accompanying the familiarisation with the physical, psychological and social situation of the Antarctic station; (3) A final period, involving the disintegration of the task group, which in the previous months had been the emotional support and guarantor of security, and the fear of taking on old or new social roles again in the near future (Mocellin et al., 1991). Occasionally, however, polar researchers observe stochastically occurring depressive states (despite prior psychological and psychiatric screening) that require *ad hoc* psychological support (Palinkas, Johnson & Boster, 2004).

5.2.4. Cognitive perspective

Within this perspective, the dynamics of perception of the Antarctic isolation situation and intellectual performance were investigated.

5.2.4.1. Dynamics of perception of the situation.

Situation perception, understood as the perception of the totality of subjective conditions that determine whether the surrounding reality has stimulative, purposeful or organisational meaning for humans, can be an indicator for assessing the degree of annoyance of the situation of annual Antarctic isolation in relation to the psychological costs of individual adaptation discussed so far.

Two Situation Perception Scales constructed by M. Ciosek (1977), containing 20 negative and 20 positive opposing adjectives, ordered randomly on the sheet, were used to study the perception of isolation. For each term, the respondent had a choice of one of five responses: 5 – yes, 4 – rather yes, 3 – yes and no, 2 – rather no, 1 – no, 33.[31] Own elaboration took place on a monthly basis during the one-year isolation period and on the ship on the way back home. Two separate indicators were used: positive and negative (sums of weights separately for 20 positive and 20 negative dimensions), assuming that the situation of Antarctic

---

31 Applying both scales to the situation of a deep-sea fishing voyage, Ciosek found the dynamics of changes in the perception of the situation, which determined the choice of this technique with the author's permission, for which I offer my sincere thanks.

isolation is so complex that one comprehensive indicator cannot be used. The results of the study presented in Figure 38 convince us of this.

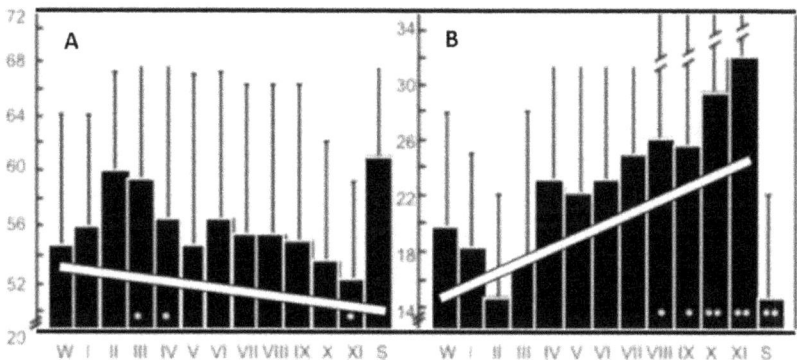

Fig. 38. Changes in the perception of the situation in "winter workers" (N = 20) during the annual Antarctic isolation
Source: own elaboration.
Legend: XII, I, II – monthly surveys; S – ship surveys en route to the country * $p < 0.05$; ** $p < 0.01$ between the initial (XI) and subsequent tests); a) positive perception of the situation, b) negative perception of the situation.

As can be seen from Fig. 38, the subject elements of the isolation situation of winter-over participants at the Arctowski station are perceived more in negative than positive terms. An interesting pattern in the perception of the situation can be seen from about the middle of wintering-over (July), when it shifts towards negative ratings ($p<0.05$ and 0.01) assuming the highest negative value (and the lowest positive) on the ship on the way back home. This is consistent with data on the declining attractiveness of participating in an interesting adventure such as an expedition to Antarctic regions (Gunderson & Nelson, 1962; Lugg, 1973; Natani, Shurley & Joern, 1973).

The findings in terms of perception of the situation are somewhat surprising. Thus, the situation is perceived most positively by "winter-over participants" in two periods of isolation: in February and March ($p<0.05$). Although on the basis of the anxiety state index (cf. Fig. 34) we previously considered this period to be stressful, however, the results presented in Figure 35 do not rule out another interpretation, e.g. that the positive perception of the situation at this time is related to some relief that the winter-over participants felt after the departure of the ship to the country with the summer crew. This is because they became the actual hosts of the station, with the associated feelings of pride, less physical work than during the summer, the beginning of the implementation of individual plans, etc. The perception of the situation from April onwards (i.e. from the moment of realisation of actual isolation from the outside world, involving the departure of the last ship before the Antarctic winter from Admiralty Bay) maintains a certain negative trend. A more detailed analysis of events tends to link elements of negative perceptions of the situation mainly to the difficult Antarctic winter and general fatigue and growing nostalgia (cf. Terelak, 2021). This is supported by the findings of some polar psychologists on the genesis of the "winter-over syndrome" concerning the negative perception of the Antarctic isolation situation (cf. e.g. Gunderson, 1968; Palmai, 1963a; Ilyin et al., 1970). The results of my own research are also partly consistent with those of M. Ciosek (1977), which were conducted during a deep-sea cruise. The author stated that the evaluation of the situation during the initial period of the voyage is definitely positive, which is related to the novelty of the situation, which provides an appropriate dose of excitement. In the middle phase of the cruise, characterised by monotony, the situation was perceived in negative terms. Towards the end of the voyage, perception is again positive, which, according to the author, is linked to the anticipation of a "normal" situation. Similar data is cited by J.H. Earls (1969) from observations of a 140-man crew of the nuclear-powered submarine USS "Nautilus" during a 60-day dive. This means, among other things, that we are dealing with a non-specific phenomenon, primarily related to long-term social isolation.

The studies cited so far on the perception of the isolation situation entitle us, as we believe, to conclude that, as a non-specific aspect of this phenomenon, the perception of the isolation situation changes over time and determines the current relationship with both the physical environment (the stimulus and spatial-temporal aspect) and the psychological environment (the individual's values and abilities), conditioning, among other things, the process of emotional-social adaptation.

An important thread explaining, among other things, the negative perception of the Antarctic isolation situation along with its duration, is the study of attitudes, which, in addition to the emotional thread, contain a cognitive component, which is in line with the definition of attitudes by L. Festinger, L. (2007), creator of the Theory of Cognitive Dissonance.

5.2.4.2. Dynamics of attitude

Since it is extremely difficult to study attitudes through behavioural observations especially in natural situations, the Semantic Differential (DS.) of Ch. E. Osgood, G.J. Suci & P.H. Tannebaum (1967), which is a general scheme for a connotational differentiation method whereby the concepts under study are measured on a scale that is modified depending on the issue under study, was applied in own research. DS. consists of several or more (as appropriate) bipolar adjectival scales. These scales take into account the following elements determined by factor analysis, i.e. the most important dimensions of connotative meaning: value, evaluation (e.g. good–bad), activity (e.g. fast–slow) and strength (e.g. strong–weak). Adjective scales are usually 7- or 9-point scales. The technique described was also used by its authors to study attitudes – the authors relied on the assumption that an attitude is identical to the value dimension of semantic space, i.e. the placement of the attitude object at the appropriate point in that space (Osgood, Suci & Tannebaum, 1967). Thus, an attitude indicator is a set of choices (expressed numerically) on scales, highly burdened with the value factor.

In own research, *the value factor* concerned the following headwords: "my wife," "my parents," "my children," "my supervisor," "my colleagues," "me as a father," "me as a man," "me as a husband," "me as an employee," "me as a colleague." In terms of *the assessment factor*, the following adjectives for social concepts have been used: good–bad, effective–ineffective, tense–relaxed, attractive–repulsive, false–true, competitive–cooperative, safe–threatened, wise–stupid, cruel–friendly. Abstract notions were not used. The *power factor* was determined by three pairs of adjectives: intense–non-intense, masculine–female, colourful–colourless. The activity factor was defined by two pairs of adjectives: passive–active, cold–warm. The higher number of scales in the evaluation factor than in the others is mainly related to the fact that the object of measurement was attitude, and that the reliability of the factor increases with the number of scales used. According to the instructions, the respondent rated on a seven-point scale (from 1 – not at all, through 4 – neutral, to 7 – very much) the extent to which a given adjective is related to the concept being evaluated. Tests were conducted quarterly throughout the one-year isolation period and on the ship on the way back home. The measure of the value of attitudes in our case is the value of the evaluation factor, which is the sum of the obtained values on seven-point adjective scales. The results of the study are shown in Fig. 39.

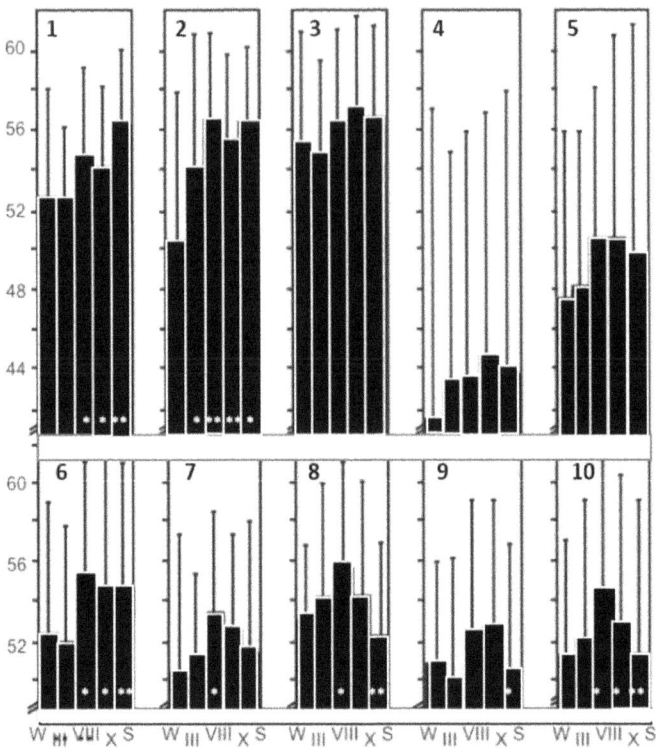

Figure 39. Changes in attitudes (DS Osgood) in relation to "other and the roles played in the country in wintering-over (N = 19) during the annual Antarctic isolation
Source: own elaboration.
Legend: 1 – "my wife," 2 – "my parents," 3 – "my children," 4 – "my superior," 5 – "my friends," 6 – "me as a father," 7 – "me as a man," 8 – "me as husband," 9 – "me as an employee," 10 – "me as a colleague," W – initial examination; III, VIII, X – monthly examination; S – ship survey; p <0.05; p <0.01.

As shown in Fig. 39, statistically significant differences were obtained (t for correlated variables p<0.01 and p<0.05) for categories such as: "my wife," "my parents," "me as a father," "me as a man," "me as a husband," "me as an employee" and "me as a colleague." Attitudes towards loved ones left behind change here along with the duration of isolation towards an idealisation of their image. The direction of the relationship also applies to the category "my children," although no statistically significant differences were obtained. Such idealisation does not occur as clearly in the case of people with whom the respondents had less emotional connection (e.g. "my supervisor," "my colleagues").

Similar tendencies occur in terms of idealising the self. The role of the father ('I, as a father' – statistically significant differences p<0.05 between March and August and October and; p<0.01 between March and February (on the ship survey) is idealised more and more with the duration of isolation. On the other hand, the very intimate role, expressed by the concept of "I as a man," is idealized the most in the middle of wintering-over (p<0.05) between March and August), after which – the closer the return to the country – the closer the attitude is to the initial one. Similar trends are drawn in the equally intimate role defined by the concept of "I as husband" (p<0.01) between August and the ship tests as well as between October and the ship tests. The data above shows that both the emotional and cognitive components of attitude are modified under Antarctic isolation. In addition, the results of the research in terms of self-assessment categorised as: "I as an employee" (p<0.05; between August and the ship tests, and between October and the ship tests) and "I as a colleague" (p<0.05; between the baseline test and August, and between March and August, and p<0.01; between August and the ship tests) confirm these findings, which suggest a fear of taking up again, after a long period of isolation, a number of important family, work and colleague roles. This is also confirmed by other Antarctic researchers such as G. Palmai (1963a).

5.2.4.3. Dynamics of cognitive performance

K. Fuerst and J.P. Zubek (1968) drew attention to changes in mental performance during different phases of sensory and perceptual deprivation in laboratory conditions and J. Paul et al. (2010) – to the dynamics of changes in mental performance during different periods of isolation in natural Antarctic isolation conditions. Following this line of thought, own research was undertaken in this area using the classical Kraepelin Test128,[32] which is part of a set of psychodiagnostic tools determining the level of ability to perform mental work and the rate of fatigue accumulation under conditions of long-term monotony of the environment. The test consists of adding up numbers (printed in columns one below the other) for one hour and evaluating the total number of actions

---

32 The study used the Kraepelin Test Manual (for group research), edited by Z. Dobruszek from the Psychometric Laboratory of the Polish Academy of Sciences with the approval of the Laboratory's head, Mieczysław Choynowski, Ph.D., for which I express my sincere thanks.

performed, the average number of actions per minute, the spread of results in different phases of the test, the error rate. The final results of the measurement are illustrated by the so-called work or fatigue curves, which will be presented in relation to winter-over participants in the following figures 34 to 36.

The study involved the entire wintering-over crew of the Arctovsky Station in 1979–1980, aged 25–48. The test was repeated regularly every two months. The test subject had to add digits continuously, emphasising each three-minute phase at a given signal. These phases of work were used to plot the mental performance curve. Based on a factor analysis of various features of the work curve, the authors of the Kraepelin test identified factors responsible for various attributes of mental work, such as: (a) measures of productivity – indicating the general level of mental performance; (b) measures of energy and endurance – indicating resistance to fatigue; (c) measures of speed of adaptation to work conditions indicating the level of capacity for immediate effort; (d) measures of fluctuating levels of emotional resilience; (e) measures of accuracy indicating the reliability of the subject. Each of these factors has several numerical indices. Ten of these were selected for analysis in this study. Statistical analysis showed that four of them allow a clear interpretation in terms of: (1) *Global sum of operations* (as a measure of the total number of operations performed in one hour), which is an indicator of the level of overall mental fitness; (2) *Percentage of skill gain* – as a measure of the percentage change in the rate of mental work when comparing the initial and final segments of the work curve, which is an indicator of the level of mental fitness (3) *Convexity of the work curve* – as a measure of the difference in the rate of addition between the mean and extreme segments of the one-hour work curve, which is an indicator of the capacity for immediate effort; (4) Fluctuation of the curve – which is a measure of the degree of change in the rate of mental work in its final phases, which is an indicator of the influence of emotion on the quality of mental work. Fig. 40 illustrates measures of the total number of operations performed in one hour, which are indicative of the level of overall mental performance of the winter-over participants studied.

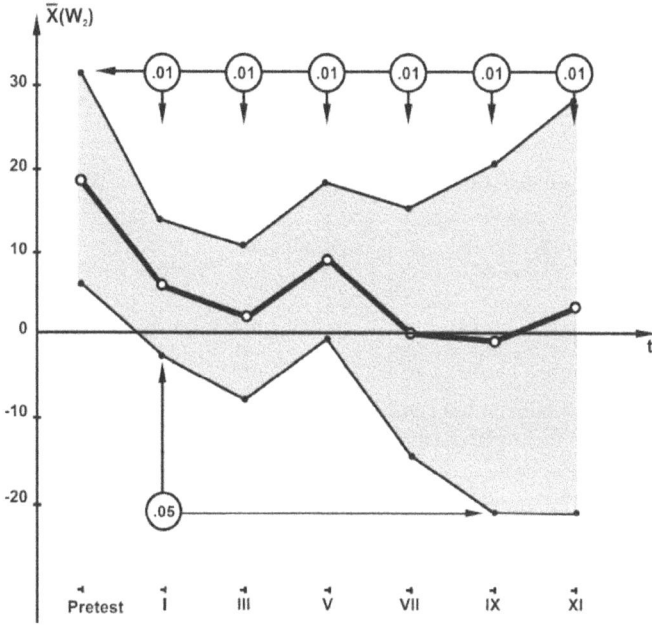

Fig. 40. One-hour Kraepelin test measures of general mental capacity (increment percentage) during one-year Antarctic isolation (N 19)

Source: own elaboration.
Legend: .01 – p <.01; $x(W_2)$ – mean percentage of work tempo increment; t – duration of Antarctic isolation.

As can be seen from Fig. 40, the mean for the pretest (pre-expedition survey) is significantly lower than all averages during wintering-over. The mean for the third month in Antarctica is significantly lower than the final average mean. This indicator is the result of practice and indicates that a so-called learning plateau has already been reached in May. However, the increasing standard deviations from the mean by the end of the year indicate significant inter-individual differences in the overall mental performance of the subjects.

The percentage of gain in proficiency, which is an indicator of the level of mental fitness of winter-over participants, is shown in Fig. 41.

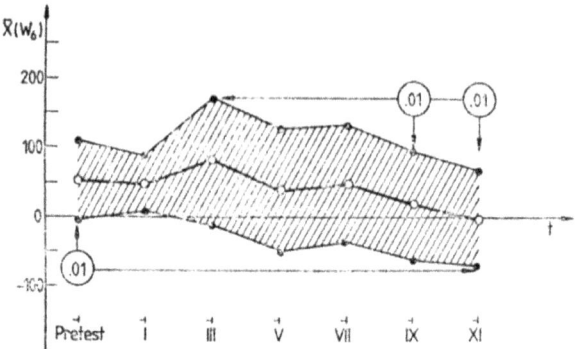

Fig. 41. One-hour Kraepelin test measures of general mental capacity (work curve convexity) during one-year Antarctic isolation (N = 19)
Source: own elaboration.
Legend:. 01 – p <.01; x(W$_3$) – arithmetic mean of the convexity index (absolute eminence of the central segment of the work curve over extreme segments); t – duration of Antarctic isolation.

Fig. 41 depicting levels of mental performance and fatigue resistance, reveals a systematic decline in mental performance and significant differences in mental fatigue resistance, as evidenced by discrepancies in the scatter of standard deviation scores from the group mean, with a slight increase in mental performance (see Fig. 38). However, this makes sense when compared to the analysis of the changes in the convexity of the working curve, which are illustrated in Fig. 39, indicating that the difference in the rate of addition between the mean and extreme sections of the one-hour work curve, indicates a decline in the ability to make immediate effort, which winter-over participants referred to as "intellectual laziness," and which Russian Antarctic researchers refer to as "polar asthenia" (Ilyin & Egorov, 1970). In Fig. 39 it is clear that from the fourth month of Antarctic isolation onwards the convexity of the working curve decreases steadily. Hence, the results of the three previous analyses can be explained by a gradual increase in the rate during the initial isolation period, as a result of practice, and a decrease in the rate during the final isolation phase. Furthermore, it is also worth noting that for the convexity of the mental work curve, the scatter of the results around the annual mean is minimal (standard deviations), which may indicate that the "mental asthenia" is non-specific and may be related to neuromuscular mechanisms of adaptation to information deprivation and long-term monotony of the environment, as characterised especially by the second part of Antarctic isolation. This is partly supported by the curve fluctuation index,

The functional aspect of the Antarctic isolation situation   247

which reflects the influence of emotions on the quality of mental work. This is illustrated in Fig. 42.

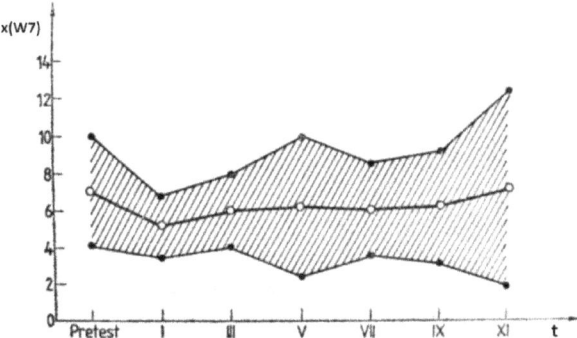

Fig. 42. One-hour Kraepelin test measures of effects of emotions on mental work quality (fluctuation index) during one-year Antarctic isolation (N = 19)
Source: own elaboration.
Legend: $x(W_4)$ – mean value of fluctuation index; t – duration of Antarctic isolation.

Fig. 42 suggests that when analysing the dynamics of performance during one-year Antarctic isolation, we cannot detach ourselves from the context of emotions. In the course of isolation, the whole person changes and the atomistic approach of conducting an analysis of the results from a single perspective, beyond the narrative itself, does not bring us any closer to a scientific explanation of the phenomenon described. An interactive approach, on the other hand, increases our chances. Perhaps the emotional changes discussed in many polar expedition narratives are not just changes in emotional expression, but also have an impact of a motivational nature (motivation as directed emotion), on qualitative and quantitative changes in mental capacity.[33] Therefore, it can be assumed that the changes observed in Figures 34 i 35 are due to factors rooted in the deeper structures of the nervous system. This hypothesis is partially supported by the results obtained by H. Reuning (1957) in his study of the relationship

---

33 "My thoughts started to blur. I can't write more than a few lines at a time. I pick up the books I brought with me, /… / but I can read only one paragraph "(Jean McNeil, 2016, pp. 339–340).

between the curve convexity index of the Kraepelin test and the frequency of alpha waves in EEG studies.

Other cognitive functions were investigated by A.J.W. Taylor and K. Duncum (1987), who conducted a series of visual attention and memory tests at the beginning and end of a year's wintering-over at Scott Base, Antarctica. The study series was repeated over five consecutive years. The results did not reveal any deterioration in this type of cognitive performance, which was contrary to the subjective feelings of the study participants themselves. A similar study of cognitive function during the Antarctic wintering-over of the 24th and 25th Chinese National Antarctic Research Expedition was conducted by G. Yan et al., (2012). Attributes of cognitive processes such as pattern recognition, short-term memory, visual attention, and spatial imagination were the subjects of the study. Tests were conducted four times: in January, March, April and June 2010. It was found, among other things, that short-term memory and pattern recognition abilities were stable, while 82% of the subjects had improved results in spatial orientation and attention tests, which may have been the result of learning. In turn F.U. John Paul, et al. (2010) surveyed scientific and logistical staff (N=23) on mental performance such as comprehension, symbol recognition, visual attention and memory in order to find out if there was a significant change in the cognitive functions of polar explorers staying at the "Maitri" Hindu Antarctic research station for fourteen months. Tests were conducted three times – at the beginning (second month), in the middle (seventh month) and at the end (twelfth month) of the Antarctic stay. Statistical MANOVA analysis showed an increase in task comprehension function indices and delayed recognition of digit symbols as well as lability of visual attention processes requiring short-term memory.

Discussions of previously obtained results from similar studies, were carried out by, among others: G.J. White, A.J.W. Taylor and I.A. McCormick (1983); A.F. Barabasz, R.A.M. Gregson and Ch. M. Mullin (1984); L.A. Palinkas et al., (2005) and John Paul et al., (2010). The discussion focused on the predictive value of mental performance tests in selecting candidates for Antarctic expeditions. It was concluded that the conclusions from previous studies, regarding firstly an increase in mental work rate in terms of a reduction in information processing time under Antarctic conditions and secondly an increase in the average decision time of completing a task, are not entirely reliable and are probably related to the effects of learning over multiple trials with the same test. The authors therefore question the research procedure and the relevance of the data analysis. It is necessary to agree with the authors regarding the difference in study procedures, but also in selected aspects of cognitive processes and how the number of consecutive studies (e.g. only two studies: at the beginning and at the end of wintering-over, which omit the most difficult period of wintering-over), which show the dynamics of adaptation processes at the cognitive level. Hence the unequivocal conclusion drawn from the above studies by A.F. Barabasz et al. (1984) that so far cognitive retardation has not been confirmed as a specific phenomenon of Antarctic isolation is controversial to say the least, as adaptation to social isolation is primarily non-specific and requires a different medley of research and analysis of results. Thus, for example, I. Barkaszi et al. (2016), taking into account the three networks of attention that have emerged in recent years, distinguishing aspects in the process: alerting, orientation and executive attention, conducted a longitudinal study with the auditory attention network test (ANT) among wintering-over participants at the Concordia Antarctic Research Station[34] with recording of brain auditory potentials. Mediating factors between behavioural variables and isolation duration were reduced oxygen partial pressure (due to station's location) caused by moderate hypoxia, and variations in sunshine duration. Six measurements of test attention and auditory brain potentials were made. Behavioural and potential responses were analysed and found, among other things: a decrease in reaction time in response to repeated tests; however, contrary to the authors' expectations, the components of N2, P3, RON potentials related to attentional functions did not show significant changes. The decrease in reaction time and decrease in N1 amplitude in the attention test (ANT) can be attributed to a sustained learning effect. Thus, Antarctic isolation conditions had no negative effect on cognitive activity in the range studied.[35]

This problem requires further laboratory studies with a deeper psychological interpretation based on psychophysiological theories of activation (cf. Hebb, 1965) or on a more detailed analysis of psychological variables, such as genetically determined temperamental traits (cf. Strelau, Angleitner & Newberry,1999), responsible for interindividual differences (cf. the large standard deviations in Fig. 34 and 35).

The results obtained in our study showing the effect of Antarctic isolation on mental performance deterioration (see Fig. 39) are not only interesting in

---

34 The Concordia Antarctic Station is located at 3220 m above sea level on the ice dome of Dome C, which is one of the coldest places on Earth. Winter temperatures can drop below -80°C (the last record was -84.6°C in 2010) (Tafforin, 2002).
35 J. Fan et al. (2002) believe that the network ANT test provides reliable estimates of a single subject's alerting, orienting and executive functions, and further suggests that the performances of these three networks are not correlated. This procedure can be used in addition to psychoneurological examinations to assess the effects of stroke and attention deficit disorders.

themselves, but also have practical implications for the psychoprophylaxis of long-term human-crewed spaceflight.

5.3. The social aspect of Antarctic isolation

Polar psychologists agree that the high professional diversity of members of modern polar expeditions makes these groups heterogeneous in terms of gender, culture, psychological needs, value hierarchies, personality traits, etc. (Gunderson and Mahan, 1966; Strange and Klein, 1973; Palinkas et al, (2004). Other authors point out that professional diversity may not only be a source of interpersonal stress, but also a source of satisfaction associated with a sense of usefulness within the overall goals of the expedition, as well as from belonging to a group of "chosen by destiny" (Gunderson & Nelson, 1963; Natani, Shurley & Joern 1973; Taylor, Wheeler & Altman, 1968).

The basis for the own research was the interactive model of social adaptation to the extreme living conditions of T. M. Newcomb, which includes the following structural elements: (1) *Values*, i.e. extremely important goals around which many specific attitudes can be built); (2) *Attitudes* as generalized states of readiness for motivated behavior concerning the achievement of a planned goal); (3) *Motives* defined as learned, goal-oriented drive states of the organism); (4) *Drives*, i.e. emotional states of the organism that initiate its activity) (Newcomb, Turner & Converse, 1966, p. 45).

A number of studies on social adaptation draw attention to various factors modifying the social adaptation of a small task force under the conditions of a polar station. The main factor, however, is the duration of social isolation, which affects the dynamics of group processes, the subject of research in social psychology proposing many theoretical models, useful for designing own research. We will draw your attention to two models. The first of them is known in the literature on the subject as *The Five-Stage Model of Group Development*, comprising 5 stages: (1). Forming (group members get to know each other); (2) Storming (group members search for mission goals and social norms); (3) Norming (group members establish close social relationships); (4) Performing (group functioning according to goals and designated social positions); (5) Adjourning (disintegration of the task group after accomplishing group goals, requiring restructuring) does not apply to extreme situations and is therefore of little use in explaining the dynamics of group processes of Antarctic winter-over participants (Tuckman & Jensen (1977). It seems that *The Integrative Model of Group Development* by Susan Wheelan (1994), which makes group processes mainly dependent on situational variables and the cooperation of group members, is more useful for studying the dynamics of small task groups. This model includes the following stages: (1) Dependency and Inclusion (significant member dependency on designated leader, safety concerns and inclusion issues); (2) Counterdependency and fight (conflicts over group goals and procedures and development of a unified set of goals, values and operating procedures); (3) Trust and structure (increase in members' trust, commitment to cooperation, openness to tasks and the establishment of an organisational structure and the building of positive working relationships with each other); (4) Work and productivity (time of intense productivity and team effectiveness); (5) Final – termination stage (task-related break-up of the group, causing disruptions and conflicts in some factions of the group or leading to separation, despite expressions of appreciation for each other). In turn, the holistic model of social adaptation, known as the *Social penetration model* by I. Altman & D.A. Taylor (1973), based on the so-called onion models, whose core is the person (self) surrounded in turn by four layers such as: (1) Self-image, encompassing values, needs and feelings, i.e. the most intimate aspects of the self, as well as subjective perception, i.e. our interpretation of the situation; (2) Creating an image, responsible for creating the environment and arranging actual social interactions at the level of verbal, facial and logistical behaviour (e.g. using objects and establishing the territory of interaction); (3) Interface with the interpersonal environment, dealing with such social areas as: self-markers (static), interpersonal distance, anticipation of relations, selection of certain group members within the formed social structure; (4) Interaction with others including, among other things: self-markers (dynamic), verbal behaviour, facial expression and reactivity to the social environment. Many Antarctic researchers believe that the Social penetration model of I. Altman and D.A. Taylor (1973) is best suited to the conditions of long-term social isolation in Antarctic conditions, especially in terms of maintaining the need for intimacy and privacy, which according to R.D. Caplan (1987) are identified in the winter-over syndrome as a result of a mismatch in the person-environment (P-E) relationship that generates discomfort, which may be modified by an anticipated reward after the completion of the mission (e.g., a financial reward or the illusion of becoming a hero or obtaining an academic promotion, etc.). This model considers three perspectives: personal relationships with others, the structure of the social environment and the structure of the physical environment.

Other mediating factors of social adaptation are pointed out, among others, by C.S. Mullin & H.J. Connery (1959), who, on the basis of Antarctic research conducted on the occasion of the IGY, expose: personality, leadership styles, group cohesiveness, expectation, attitude, etc. The authors found, for example, that difficulties in social adaptation were mainly manifested by people characterised by intolerance of others, lack of discretion, professional incompetence, lack of interest, quarrelsomeness, authoritarian personality, aggressiveness, lack of sense of humour, etc. Similar data is cited by S. Smith & W. Haythorn (1972), who observed social behaviour mainly in experimental isolation situations. Our findings using nomination techniques and rating scales support the above suggestions.

5.3.1. Dynamics of interpersonal attraction

The nomination technique, also known as the reputation scale, involved selecting the highest and lowest ranked individuals from among the group members, according to the following instructions: "Within the frames of the categories below, you are to make a selection of four names with the most positive and negative characteristics. If possible, give four names everywhere. However, it is essential that you list at least one name or number that you use in your research in each category. The data obtained will not be shared with anyone and is only used to develop psychological selection methods for future Antarctic expeditions. That is why I am asking for frank statements." The fifteen attributes assessed related to two categories of behaviour, covering social and professional functioning at the station, under the following types of assessment: "I like most – I dislike most," "most impresses me – least impresses me," "is the greatest authority for me – is no authority at all," "least afraid – most afraid," "most trusting – least trusting," "most like to cooperate with… – least like to cooperate," "most like to spend my free time with… – least like to spend my free time." The same categories of wintering-over participants' behaviour (each individually) were assessed by the Arctowski station manager.[36]

The nomination technique was used twice, at the beginning and at the end of the isolation. We treated the high and low sociometric items determined from it as correlates of social adjustment. In addition, categories such as knowledge of the station's profession, diligence, self-control and resistance to frustration, popularity in the group, friendliness, ability to lead others, responsibility, and willingness to winter-over with the same person again were assessed.

The ratio of positive to negative choices was an indicator of social adjustment. We contrasted the scores obtained by the group with those of the station manager (leader), using the behavioural categories listed above. Each member of the

---

36 These techniques were based on the idea of P.D. Nelson & E.K.E. Ganderson (1963b), used in studies at small US Antarctic stations and, after cultural adaptation, applied to studies at Arctowski Station.

The social aspect of Antarctic isolation 253

group was rated on a 9-point scale (1 being the lowest rating and 9 being the highest). The assessment by the manager was a one-off at the end of the annual isolation. The ranks obtained by each group member (1 --- best social adjustment and 19 – worst) in terms of social adjustment are shown in Fig. 43.

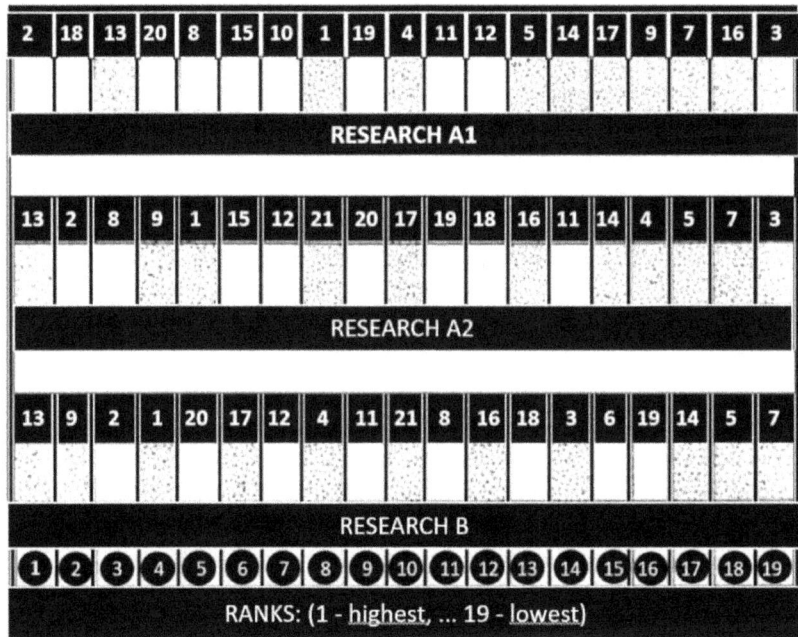

Fig. 43. Dynamics of interpersonal attractiveness in terms of competences and sympathy as assessed by group members and the station manager
Source: own elaboration.
Legend: A1 – the January 1979 survey and A2 – the December 1979 survey) from the perspective of the winter-over participants; B – the December 1979 assessment from the perspective of the station manager; rectangles: white – scientists; dotted rectangles – technicians; ranks: 1 – highest mark – 19 – lowest mark.

Fig. 43 shows there is a relative correspondence between group and leader ratings of the best and worst socially adjusted people. First of all, the electrician responsible for the generators at the station is very prominent (No. 13), doctor-surgeon (No. 2), cook (No. 1) and a meteorologist (No. 9). These were people on whom a sense of security depended in the first place, hence they were subject to a kind of "protection" from negative judgements. This is a good illustration of the fact, commonly stated by polar psychologists, that in a small isolated task group, regardless of education and age, professional skills, high task motivation and pro-social behaviour count the most. This is confirmed by the research of P.D. Nelson & E.K.E. Gunderson (1964) who, on the basis of a number of sociometric techniques, found, among other things, that the sociometric structure of a group is most determined by the functions held at a station rather than by occupational positions in the country. An example of this can be found in the observations of E.K.E. Gunderson (1966b), testifying that on small American Antarctic stations the cook holds an important position in the group, as he works sequentially with all group members who are on cyclical duty throughout the period of isolation.

Further analysis of the results shown in Fig. 44 shows that the persons: No. 3 – technical manager, No. 5 – carpenter, No. 7 – technician-mechanic and No. 14 – radio-telegraph operator, isolating themselves from the group, occupy the lowest positions on the scale of informal group structure of adaptation. At the same time, as can be seen from the comparison of the survey from January with the survey after 11 months in the evaluation of the group, it is primarily the actual skills and qualities that count, not the declarative ones, as confirmed by the change in many items (e.g. microbiologist – survey No. 18, and fisherman – survey No. 4). Noteworthy, contrary to expectations, is the high and relatively stable position of psychologist in the rankings (study No. 15), omitted from the station manager's assessment at all.

The results presented here support suggestions, especially by American psychologists (e.g. Gunderson, 1966b), that a small isolated task group is significantly differentiated in terms of its informal structure. We will return to this problem in further considerations.

A separate, important issue in social psychology is the measurement of *interpersonal attractiveness,* reflecting the process of formation and persistence of *interpersonal networks on the basis of importance ratings or emotional criteria: liking – disliking* (Newcomb, 1968). Interpersonal attractiveness assessed according to the criterion of emotional attitudes towards others has been the subject of research using a scale originally called "The plebiscite of kindness and aversion" by Janusz Korczak.[37] The study consisted in the respondents defining their attitude towards the other members of the group (indicated by name) using

---

37 This technique developed and used in orphanages by J. Korczak, who was murdered with his pupils in the fascist camp at Majdanek, was adapted for quantitative evaluation by the psychologist M. Pilkiewicz (1973), whom I thank for permission to use it.

the following labels: "++" I like very much; "+" I like; "0" – indifferent attitude; "-" I dislike; "– I dislike very much. The study was conducted on a 2-week cycle for the one-year isolation period and on board the ship on the way back home.

Although this technique, like all nomination techniques, is burdened significantly on the social approval variable, due to its "play" form,[38] it allowed their sociometric status to be determined on a 5-point scale on the two dimensions of sympathy vs. antipathy. The positions on the different scales were calculated from the distribution of the scores obtained using the formulas: (1) For the sympathy scale: SS = (100% ++) + (50% +); (2) For the antipathy scale: SS = (100% – -) + (50%). Obtaining positions on the scales allowed each respondent to be classified into five basic groups of the "sociometric scale of acceptance:" (1) *Acceptance* (A_0 – outstanding, A_1 – strong, A_2 – weak); (2) *Averageness* (X); (3) *Polarisation* (Po – outstanding, P_1 – strong); (4) *Isolation* (I_0 – outstanding, I_1 – strong, I_2 – weak), (5) *Rejection* (O_0 – outstanding, O_1 – strong, O_2 – weak). The results of the 26 measurements are presented in Table 5.

---

38 Verbal assessment of others, as our observations have shown, poses a greater social threat to a small isolated group than assessment by graphic signs, which has been shown to be a practical tool for obtaining information about the interpersonal attractiveness of individual group members.

Mon	XII	I	I	II	II	III	III	IV	IV	V	V	VI	VI
Nr	1	2	3	4	5	6	7	8	9	10	11	12	13
1	Ao	Ao	Ao	Ao	Ao	Ao	Ao	Ao	Ao	Ao	Ao	Ao	-
2	Ao	Ao	Ao	Ao	Ao	Ao	Ao	Ao	Ao	Ao	Ao	Ao	Ao
3	X	A₂	J₂	J₂	A₂	J₂	X	Ao	A₂	X	A₂	A₂	A₂
4	Ao	Ao	Ao	Ao	Ao	Ao	A₀	A₂	Ao	Ao	Ao	Ao	Ao
5	A₂	Ao	A₂	A₂	Ao	Ao	Ao	Ao	Ao	Ao	Ao	Ao	Ao
6	-	-	-	-	-	-	-	-	-	A₂	A₂	A₂	A₂
7	A₂	A₂	A₂	A₂	A₂	A₂	A₂	X	X	X	X	A₂	X
8	Ao	Ao	Ao	Ao	Ao	Ao	Ao	Ao	Ao	Ao	Ao	Ao	Ao
9	A₂	Ao,,	Ao	A₂	Ao	Ao	Ao	Ao	Ao	Ao	Ao	Ao	Ao»
10	Ao	A₀	Ao	Ao	Ao	Ao	A₀	A₀	A₀	Ao	Ao	Ao	A₀
11	Ao	Ao	Ao	A₂	A₂	Ao	A₁	A₂	J₂	J₂	A₂	X	X
12	A₂	A₂	A₂	A₂	Ao	Ao	Ao	A₂	Ao	Ao	Ao	Ao	Ao
13	A₂	Ao	Ao	Ao	Ao	Ao	Ao	Ao	Ao	Ao	Ao	Ao	Ao
14	Ao	Ao	A₂	A₂	A₂	Ao	Ao	A₂	A₂	A₂	A₂	Ao	Ao
15	Ao	Ao	Ao	Ao	Ao	Ao	Ao	Ao	Ao	Ao	Ao	Ao	Ao
16	A₂	A₂	Ao	J₂	A₂	Ao	Ao	Ao	Ao	Ao	Ao	Ao	Ao
17	A₂	Ao	Ao	Ao	Ao	Ao	Ao	Ao	Ao	Ao	Ao	Ao	Ao
18	Ao	Ao	Ao	A₂	Ao	Ao	Ao	Ao	Ao	Ao	Ao	Ao	Ao
19	Ao	Ao	Ao	Ao	Ao	Ao	Ao	Ao	Ao	Ao	Ao	Ao	Ao
20	Ao	Ao	Ao	Ao	Ao	Ao	Ao	Ao	Ao	Ao	Ao	A₀	Ao
21	-	-	-	-	Ao	Ao	Ao	Ao	Ao	A₀	A₀	Ao	Ao

Mon	VII	VII	VII	VIII	VIII	IX	IX	X	X	XI	XI	XII	I
Nr	14	15	16	17	18	19	20	21	22	23	24	25	26
1	Ao	Ao	Ao	Ao	Ao	Ao	Ao	Ao	Ao	Ao	Ao	Ao	Ao
2	Ao	Ao	Ao	Ao	Ao	Ao	Ao	A₀	Ao	Ao	Ao	Ao	Ao
3	A₂	X	X	X	X	X	X	X	X	X	X	X	X
4	Ao	Ao	Ao	Ao	Ao	Ao	A₁	A₁	Ao	A₂	A₂	X	A₂
5	A₀	A₁	A₁	A₁	X	X	X	X	A₂	X	X	X	A₂
6	Ao	Ao	Ao	Ao	Ao	Ao	Ao	Ao	Ao	Ao	Ao	Ao	Ao
7	X	X	X	X	X	J₁	X	X	X	J₂	X	X	A₂
8	Ao	Ao	A₀	A₀	Ao	Ao	A₀	Ao	Ao	Ao	Ao	Ao	Ao
9	Ao	Ao	Ao	Ao	A	Ao	Ao	Ao	Ao	Ao	Ao	Ao	Ao
10	Ao	Ao	Ao	Ao	Ao	Ao	Ao	Ao	Ao	Ao	A₁	Ao	A₁
11	X	A₂	A₂	P₁	X	A₂	A₁	X	A₂	A₁	A₁	A₁	A₁
12	Ao	Ao	Ao	Ao	Ao	Ao	Ao	Ao	Ao	Ao	Ao	Ao	Ao
13	Ao	Ao	A₀	A₀	Ao	A₀	Ao	Ao	Ao	Ao	Ao	Ao	Ao
14	Ao	A₂	Ao	A₀	Ao	Ao	Ao	Ao	Ao	Ao	Ao	Ao	Ao
15	Ao	A₁	A₁	A₁	Ao	Ao	A₀	A₁	Ao	A₁	A₁	A₁	Ao
16	A₀	Ao	Ao	Ao	A₀	A₀	Ao	Ao	Ao	Ao	Ao	Ao	Ao
17	Ao	A₀	A₀	A₀	Ao	Ao	A₀	Ao	A₀	Ao	A₀	Ao	Ao
18	A₂	A₂	A₁	A₂	X	A₂	X	A₂	A₂	X	A₂	A₁	A₁
19	A₀	Ao	Ao	Ao	Ao	Ao	Ao	Ao	Ao	Ao	Ao	Ao	Ao
20	Ao	Ao	Ao	Ao	Ao	Ao	Ao	A₀	Ao	Ao	Ao	Ao	Ao
21	Ao	Ao	Ao	Ao	Ao	Ao	Ao	Ao	A₀	Ao	Ao	Ao	Ao

Table 5. Results of 26 polling measurements of friendliness and aversion of winter-over participants during the annual isolation at the Arctowski station

Source: own elaboration.

As can be seen from Table 5, most winter-over participants gave mainly the highest ratings of "A_o – outstanding acceptance," which is characteristic of isolated small task groups (Terelak, 1985). This is among other things due to the fact that in the situation of the so-called "commonality of fate" a tendency not to violate the intimacy (self-disclosure) of other members of the small task group emerges, as pointed out by W. Derlega and A.L. Chaikin (1975) in their monograph that in informal relationships, which are quite a difficult psychological process, people encounter resistance to revealing themselves (self-disclosure) to other people, albeit sometimes a necessary one, because in order to get information from someone one must first give them information about oneself, reciprocally expecting a return. The authors distinguished two dimensions of intimacy: (a) depth – when someone reveals those sides of themselves that relate to the central "Self" and are most at risk of judgement or rejection; (b) breadth – covering the number of areas of the person's life that are revealed. Lack of self-disclosure is an often overlooked aspect of communication in small task groups that hinders the formation of informal interpersonal relationships. However, as social psychologists have shown, there are various ways of disclosing oneself to others that are personality-driven, e.g. a low level of self-knowledge (Skarżyńska, 1979). The matter becomes more complicated as discussed in the research conducted by L. Davidová (2019), which focused not only on the psychosocial dynamics of intra-group processes, the patterns of which are generally known from studies of Antarctic crews, but on the communication of the spacecraft crew with the management of the MCC (Mission Control Centre). The author's own research involved observing crew cohesion and communication with the MCC in two terrestrial lunar habitats as mission analogues: "Lunar Expedition-0" and "Lunar Expedition-1." Based on a standardised interview and individual new DOTI visualisation method (Dotty Overview of Team Interactions), the patterns of group dynamics and limited trust in MCC communication exposed in the literature, limiting the autonomy of the Martian habitat crew, were confirmed.

Returning to own research, in Fig. 46 we present the sociometric positions of the six selected winter-over participants rated rather controversially on the acceptance vs. disliking scale in 25 subsequent study sessions.

The position of the six selected winter-over participants rated rather controversially on the acceptance vs. disliking scale in the following surveys is illustrated in Fig. 44.

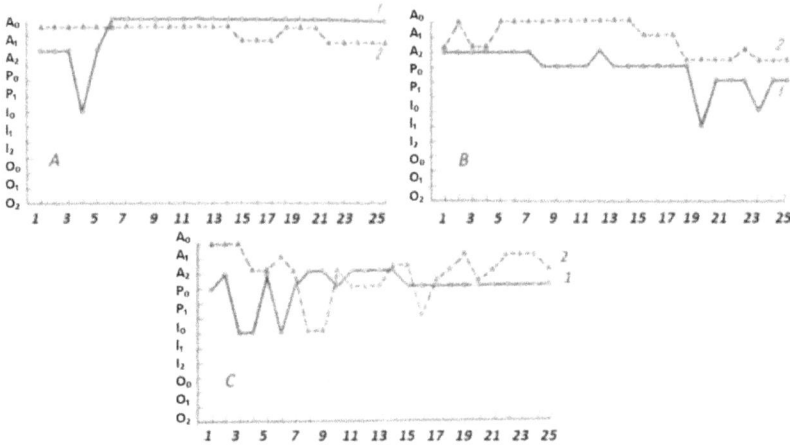

Fig. 44. Examples of the dynamics of interpersonal attraction of winter-over participants during one-year Antarctic isolation

Source: own elaboration.

Legend: Figure A – respondents: (1) radio operator vs. (2) – psychologist; Figure B – respondents: (1) carpenter vs. (2) seismologist; Figure C – respondents: (1) amphibious vehicle driver vs. (2) locksmith; Sociometric scale of acceptance vs. dislike: (1) Acceptance: A_0 – excellent, A_1 – strong, A_2 – weak; (2) Polarity: P_0 – excellent, P_1 – strong; (3) Isolation: I_0 – excellent, I_1 – strong, I_2 – weak; (4) Rejection: O_0 – excellent, O_1 – strong, O_2 – weak.

Analysing the results of the survey presented in Table 5 and Figure 44 together, it should be noted that apart from three cases where categories such as "mediocrity" and "isolation" were used in assessing attractiveness, in most cases only the "acceptance scale" in its three variations was used: A_0 – outstanding, A_1 – strong, A_2 – weak, omitting other scales expressing negative emotions in interpersonal relationships, which is not consistent with my observations presented in the diary (Terelak, 2021). However, based on the results obtained from the "acceptance scale," a certain typology of interpersonal attractiveness of winter-over participants can be distinguished. First of all, a "conservative model" is observed in most cases, characterized by maintaining a relatively stable, good position on the acceptance scale throughout the whole period of isolation, and this applies to the "radio operator" (position 1) and the "psychologist" (position 2) (cf. Fig. 45A). Secondly – a "disadaptation model" is also revealed, characterised by changes towards a decline in attractiveness in the case of such respondents as: "carpenter" (position 1) vs. "seismologist" (position 2) (cf. Fig. 45B). It is also possible to distinguish a "fluctuation model," characterised by a certain cyclicality of position changes on the acceptance scale in both directions and this applies to the following respondents. "amphibious vehicle driver" (position 1) and "locksmith" (position 2) (cf. Fig. 45C). The importance of the informal group structure based on the subjective criterion of friendliness vs. aversion is emphasised, among others, by G.K. Sandal, & H.H. Bye (2015), who investigated the consistency among crew members (n = 6) participating in a 520-day simulated mission to Mars with the use of a different method (behavioural monitoring and evaluation by so-called competent judges). Among other things, they found that in-group tensions are generally associated with a decrease in levels of benevolence and cohesion with the duration of social isolation. This is also consistent with the results of our own research into interpersonal attractiveness using psychometric methods, as illustrated in Fig. 451.

Summarising the results of the study on the dynamics of the informal structure of the wintering-over group during annual Antarctic isolation, it should be noted that the used nominative methods have their advantages and disadvantages. The advantage of the method is its simple graphical form (as opposed to descriptive), making it easy to assess the person using +/- signs. This method can be used, for example, to study the informal structures of spacecraft crews confined in a small space. The disadvantage of this method (as with all nomination methods) can be considered its audience, which is a discomforting situation that undoubtedly has the effect of reducing the reliability of the results (the public approval variable). However, the content analysis of the individual protocols, confronted with the ongoing observation (e.g. taking into account the number of recorded conflicts between individual group members), showed, among other things, that the subjects were afraid to express not only dislike or hostility but also indifference towards their colleagues. Therefore, in such situations they chose to declare only weak acceptance (cf. Fig. 45 scale A_2). Thus, the lack of differential interpersonal attraction cannot be explained by unfamiliarity with partners but by high social threat pressure. This, of course, does not only apply to the research technique discussed above, as it is typical of the defence mechanisms of a small isolated group, as experimental research in social psychology, among others, convinces us. E.g. D.E. Flinn et al. (1961) conducted a series of 30- and 17-day social isolation experiments using a space simulator (2-seated SAM simulator). Among other things, the authors found significant individual differences in social behaviour, depending on the individual personality traits of the pairs of subjects studied. Continuous observation of the individuals' behaviour revealed, among other things, that the reticent person became irritated when engaged in conversation with a talkative partner, whereas previously (i.e. outside the situation of social isolation) such features as "talkativeness" were not apparent. It was also noted that a meticulous and scrupulous person became irritated with a disorganised partner. It has been shown, for example, that during such long and forced contacts with another person, seemingly harmless habits and manners lead to irritation and serious conflicts. However, the highly motivated task situation itself meant that any "hostile" reactions were so controlled that they did not materially reflect on the level of task performance, nor were they recorded objectively by the technique used by R.F. Bales (1951), evaluating 12 categories of analysis of group interaction processes, such as: A. Social and emotional area: (a) positive: (1) showing solidarity, (2) relieving tension, (3) expressing agreement; (b) negative: (10) expressing disagreement, (11) showing tension, (13) showing antagonism; B. Task-neutral area: (4) giving suggestions, (5) giving opinions, (6) indicating orientation, (7) asking for orientation, (8) asking for opinion, (9) asking for suggestion.

Within the above categories, the subjects in the Flinn et al. (1961) experiment, assessed the social-emotional and task behaviours of all pairs of subjects in an isolation situation simulating spaceflight 45.

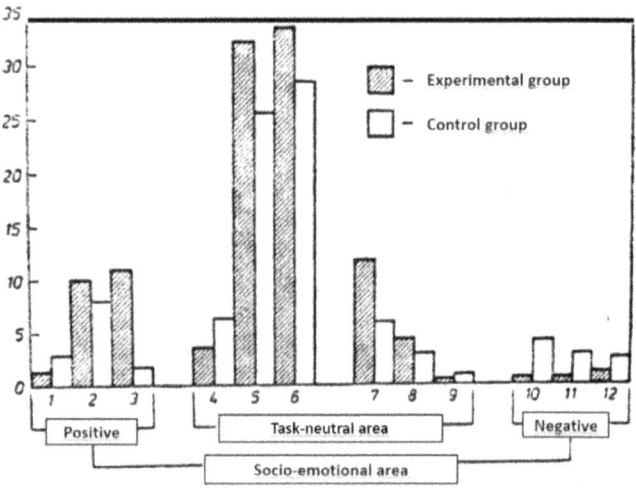

Fig. 45. Evaluation of social interaction using Bales' categories during 30-day isolation in a space simulator

Source: own elaboration of Flinn et al., 1961.

Fig. 45 shows the percentage capture of group interaction within Bales' 12 categories during the 30-day task isolation compared to the control group. As can be seen from the analysis of the figure above, it is possible to speak of certain interactions that are characteristic of situations of isolation. E.g. the social-emotional sphere, especially the negative one, is well controlled and manifests itself in various forms of behaviour: "relieving tension" (e.g. jokes, laughter, showing satisfaction) or "expressing agreement" (e.g. passive acceptance, showing understanding, unanimity and obedience) while at the same time there are latent antagonisms that are usually suspended until common tasks are completed. These results are consistent with the subjective experiences of those who reported after the study that the interactions taking place between members of the "crew" were entirely formal and "polite." Many negative emotional reactions were displaced and directed at experimenters outside the simulator.

It seems that the phenomenon described above is specific to a small isolated task group, which was revealed in the case of the previously presented research using J. Korczak's "Plebiscite of kindness and version." This is also confirmed by my observations of hidden aggression towards the psychologist, which was expressed anonymously through verbal threats, placed on cards under the "psychiatric" door, and only revealed itself in person on the ship on the way back to the country, where several people refused outright to participate in the last study. And yet, during their one-year isolation, they did not reveal an aversive attitude towards the psychologist[39] (cf. Terelak, 2021).

5.3.2. Sociometric structure of group dynamics

Group dynamics at both macro- and micro-group scales are increasingly being studied by social psychologists (Bussmann, U. & Schweighofer, S. (2014). The dynamics of informal task group structure during the various phases of Antarctic isolation were the subject of our own research using the *subjective sociometry*.[40]

---

39 An example of how a psychologist in a small isolated group interacted uncomfortably with the other winter-over participants is evidenced by the fact that during meals with an allocation of alcohol, loud suggestions were made "to get the psychologist drunk, otherwise he will do research and spoil the fun." I describe such situations in detail in my Antarctic Diary (cf. Terelak, 2021).

40 This technique, with the approval of prof. V. Derlegi (USA), its author, was made available for Antarctic research carried out by prof. J. Grzelak, to whom I would like to express my sincere gratitude (Derlega & Grzelak, ed., 1982).

This method involves graphically expressing the relationship between the subject and other group members, as well as between individuals.[41] The manual stresses that this technique only expresses the psychological closeness between people and says nothing about the type of feelings (liking or hostility). This means that even the person who is placed furthest away from the respondent can be liked a lot, just less compared to others placed closer to the respondent. We should add that this research technique was applied on a bi-weekly cycle throughout the Antarctic isolation period and on the ship on the way back home. A detailed cluster analysis of the results obtained from the 26 matrix arrays included the following steps. Summary statistics were made of the data received, which relate to a person's subjective view of the relationship between a small group of people living in difficult conditions for many months. They converted data from information about the coordinates of points marked on graphs to distances, resulting in the loss of reliable patterns that people used to mark their opinions. By using distances ordered by people according to the 26 sessions, one has described how these have changed over time, as illustrated by example graphs of peer relationships that can be viewed as measures of proximity. The central points of mean proximity across all sessions were also plotted. Only descriptive findings will be presented, as more detailed and extensive analyses will be the subject of a separate study and publication. An example of the results of subjective sociometry from the position of one respondent, reflecting the subjective perception of the structure of the wintering-over group in the four stages of Antarctic isolation is illustrated in Fig. 46.

---

41 This is an 18 X 18 cm grid. The subject's task is to arrange the points (i.e. themselves and other people) on a grid so that the distance between them reflects their daily interactions with each other. Whenever the subject has no idea what the actual relations between two or three colleagues are, he or she "places them" according to his or her own attitude towards them and then goes round all these points in a circle. The subject describes his or her place on the grid as "I," while the other points are described by the initials of the names (in our case the numerical index used in all tests) (Derlega & Chaikin, 1975).

The social aspect of Antarctic isolation 263

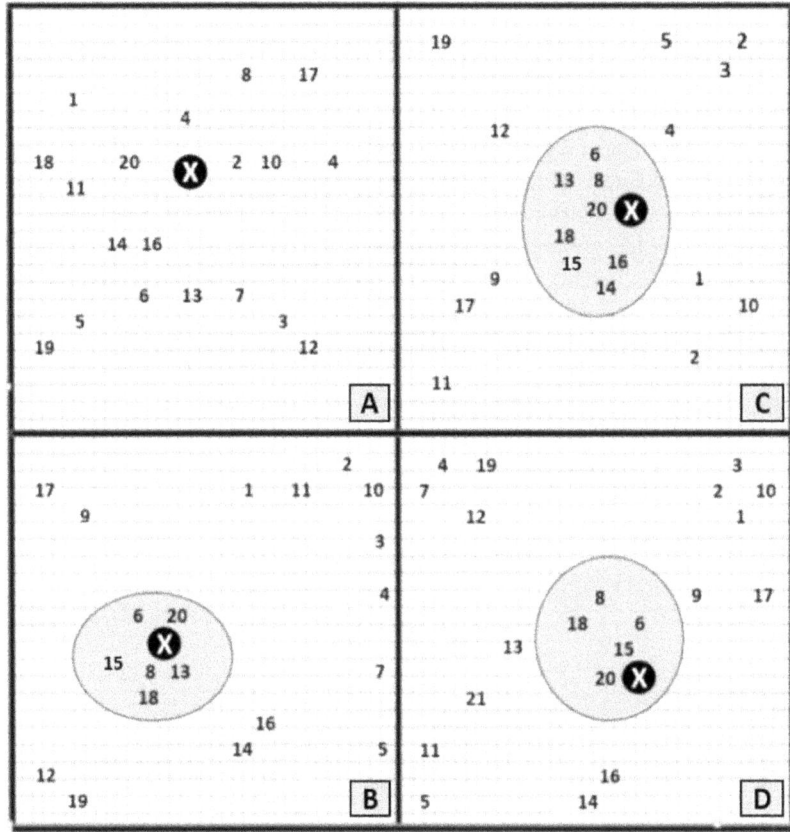

Fig. 46. Dynamics of the informal wintering-over group structure during the period of annual Antarctic isolation as assessed by respondent 15

Source: own survey.
Legend: a) December – 1978; b) July – 1979; c) December – 1979; d) January – 1980;
X – respondent No.

As can be seen from Fig. 46 the data cited is only meant to illustrate certain trends of group integration in the four selected stages of the research from the perspective of the studied No. 15 (psychologist), however, a preliminary analysis of the material obtained from the study using subjective sociometry confirms the thesis that over the course of 26 measurements, four stages of group integration can be distinguished as an indicator of social adaptation.

Thus, in the first stage, i.e. during the one-month boat cruise to the Antarctic and the first month at the Arctowski station, the group is open, (Fig. 46A) because each member of the group is interested in each individual and the group as a whole. First social contacts and first impression friendships begin to emerge. This gives rise to the formation of fractions, mainly in the nature of similarity of occupational or leisure interests, which are characteristic of the second stage of adjustment, lasting approximately 4–5 months. This is consistent with the graph analysis[42] of the results of the averaged group structure based on the distance index between members during the first week of stay at the Arctowski Station, as illustrated in Figure 47A.

---

42 Graph analysis was written by the co-authors of this book chapter: Marcin Dahlen and Lukasz Subramanian, in the Python programming language; for the purpose of calculations, native libraries for the language and the NumPy library were used. The few algorithms reported in publications were not included in available libraries and they required implementation. The NetworkX library was used to develop the graphs. I used Matplotlib to plot and illustrate the graphs. [https://github.com/marcindahlen/Antarctica_research_2020].

The social aspect of Antarctic isolation 265

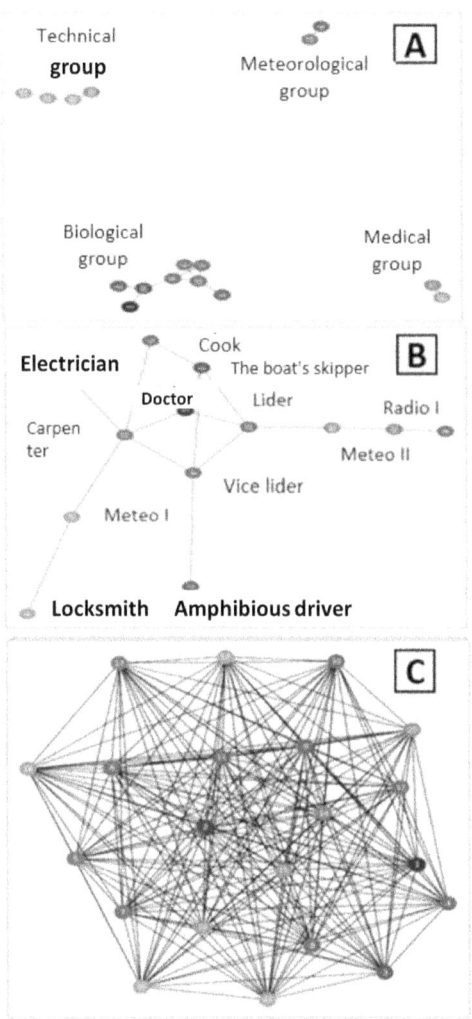

Fig. 47. Central points of average proximity determined by graph analysis from 1, 3 and 13 sessions of 26 two-week longitudinal measurements: A, B, C, reflecting the dynamics of the wintering-over group structure under Antarctic isolation conditions

Source: own elaboration.

Legend: A – survey on a ship on its way to Antarctica (November 1978); B – survey at the Arctowski Station (December 1979); C – averaged sociometric structure of the group of winter-over participants in the middle of wintering-over (July 1979).

As can be seen from Figure 47, the focal points of average proximity in session 1 (Fig. 47A – pre-test on the ship en route to Antarctica) do not reflect the existence of a task force, as the graph nodes are selectively clustered in a small sociogram space and illustrate close relationships prior to participation in the 3rd Antarctic Expedition of PAN. It is clear that the initial social relationships that are established are mainly due to shared accommodation in multi-person cabins. In turn, as can be seen from Fig. 47A, two weeks after landing at Arctowski Station, the task force had not yet been formulated, and the structure of the wintering-over group was divided into three professional subgroups: technical, biological, medical one. Initial social relationships are established, resulting mainly from working together to unload the ship and taking on new responsibilities. In session 3 (Fig. 47B – the preliminary period of Antarctic winter) we can see that some graph nodes reflect the existence of certain fractions and show that not everyone met the requirement of a sufficiently close relationship with at least one person. The formal structure of the group is based primarily on a clearly defined managerial and technical cluster. Other group members' accounts were scattered throughout the sociometric space. It reflects well the averaged sociometric group structure on the basis of the distance index[43] between members in this phase of group structure formation (Figure 47B as well as the earlier presented Figure 46B), which indicate that in this phase of building the informal sociometric group structure, occupational clusters clearly stand out: "biologists" (respondents No. 6, 12, 18, 20), "radio people" (14, 16), "common mountaineering interests" (respondents No. 12 and 19). The other members of the group were "observers," characterised by a wait-and-see attitude or independence from the group. This stage had been transitional in nature, until the "social core" (asset) separated – the beginning of the third stage of group adaptation, shown in Figures 47C and 46C, confirmed by the spatial representation of the group structure in Session 13 (July 1979) after eight months at the Arctowski Station. As it results from the comparative analysis of the structure of the wintering group at the Polish Arctowski Station (Fig. 48 A) and the American Amundsen-Scott South Pole Station (Fig. 48B) (Johnson, Boster & Palinkas, 2003), clusters of scientists groups clearly stand out in the middle phase of wintering. and technicians.

---

43 Distance – the positions marked by the respondent on the sociogram, reflecting their two-dimensional coordinates, which in graph analysis are interpreted as positions of points on the plane and defined as Euclidean distance – for each person and for each session. The individual sociograms illustrate the perception of others in the wintering-over group.

The social aspect of Antarctic isolation 267

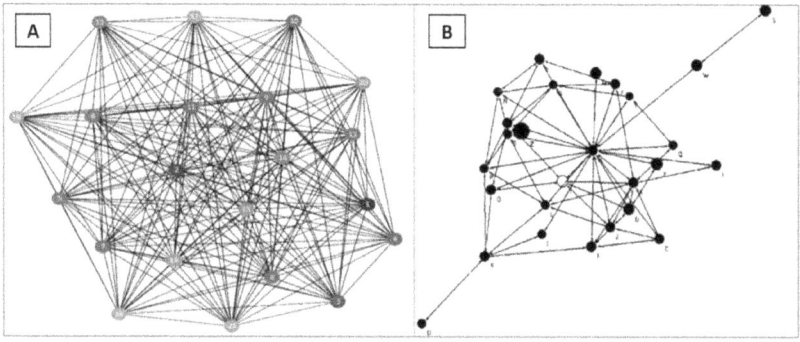

Fig. 48. Cohesion of the wintering group at the Polish Arctowski Station (Fig. A) and the American Amundsen-Scott South Pole Station (Fig. B) in the middle wintering phase.
Source: Fig. A – own elaboration; Fig. B – J.C. Johnson, J.S., Boster & L.A. Palinkas, 2003).

As can be seen from the above figure 48, in the middle phase of wintering, when the station is completely isolated, winter worms are characterized by a group consensus regarding potential external threats, in contrast to the remaining phases of wintering, characterized by the separation of peripheral factions and cliques outside the station's assets.

Finally, in session 21 (Fig 46D – a month before coming back to the country) and after the end of winter in Antarctica, one can see that it clearly reflects the important phenomenon of the disintegration of the structure of the task group, consisting, among others, in the depletion of the cadre cluster and the emergence of new peripheral sub-clusters as e.g. the "management" group: respondent No. 2 (doctor), respondent No. 3 (technical manager), respondent No. 21 (fisherman-skipper), respondent No. 10 (head of the wintering group); the "technical" group: respondent No. 5 (carpenter), respondent No. 7 (mechanical technician), respondent No. 16 (radio operator-electrotechnician), respondent 14 (radio operator); and so-called "satellites" – isolating themselves from the group, such as sub-clusters: survey no 9 (magnetologist) survey No. 11 (seismologist), survey No. 19 (parasitologist). Characteristic is the phenomenon of the emergence of typical cluster conflicts at this stage: scientific vs. technical one, reminiscent of the state before wintering-over, and the so-called satellites are unchanged throughout the Antarctic isolation period. Content interpretation of the results obtained requiring additional information on personality.[44]

Analysis of mean proximity scores across all 26 sessions of the sociometric structure study of Antarctic winter-over participants presents evidence that mean perceived proximity to the other man and interactions between other people is an intrinsic characteristic of every human being that does not satisfy the need for intimacy within the task group *under the conditions of isolation*.

This is also confirmed by researchers of task group dynamics at other Antarctic stations, such as G. Palmai (1963a) and R.E. Strange & W.J. Klein (1973), who found that regardless of situational factors, small isolated task groups are characterised by similar specific adaptation dynamics: stage one – openness to all members; stage two – polarisation of the group into cliques and factions (e.g. occupational, recreational, age factions, etc.); stage three – amalgamation of members of the "social core" group (asset), with the exception of the so-called satellites, isolating themselves from the group, and peripheral cliques. The accompanying group conflicts are also characteristic of different stages of social adaptation, especially as they are not generally stable (Derlega & Chaikin, 2010). This is confirmed by further analysis of Fig. 47D, which illustrates that the study group which has already reached a certain stage of integration (Fig. 48C, has regressed again to the "intimate communities" stage.[45] In summary, the dynamics of changes in the structure of the group operating in the annual Antarctic isolation are shown in Fig. 49.

---

44 Characteristics of the influence of personality on group dynamics at Arctowski Station are presented in the diary (Terelak, 2021). Moreover, the very concepts of community and social distance, are not rigorously defined and require a certain degree of arbitrariness, as Santo Fortunato writes in "Community detection in graphs" of 25.01.2010 (https://arxiv.org/abs/0906.0612.).

45 The "Intimate Community" is a non-normal microstructure of people who judge each other well, trust each other and enjoy working and spending time with each other.

The social aspect of Antarctic isolation 269

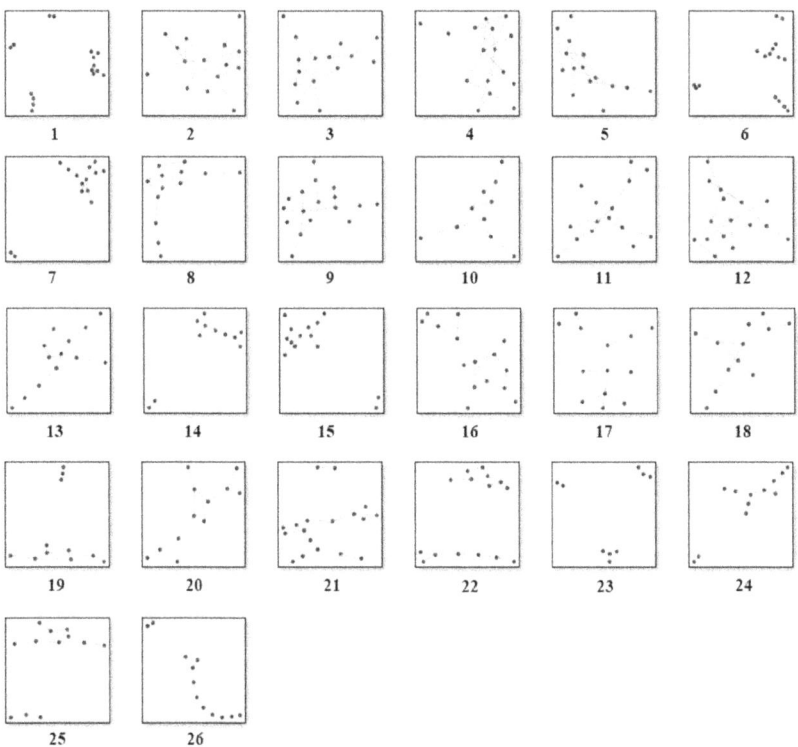

Fig. 49. The dynamics of changes in the structure of the group operating in the annual Antarctic isolation at Arctowski Station
Source: own elaboration.

Although the dynamics of the group structure over time, taking into account all 26 sessions, allows for a quantitative graphical representation of the changes occurring between individuals in terms of their place in the different nodes of the graphs, it is not always comprehensible without a qualitative assessment, which does not allow for a statistical check of the changes occurring in the group.

The dynamics of changes in group structure during the period of annual Antarctic isolation at the Arctowski Station presented in Fig. 47 reflects a close reciprocal relationship from both sides i.e. the perceiver and the perceived. At the same time, it is clear, in practically all charts, that the technical group was more situated in the central sociometric space as one integrated in terms of proximity (respondents No. 1, 3, 4, 5, 7, 13, 14) compared to the researchers' group, whose relationships were more dispersed in terms of distance. Noteworthy is the neutral position of the station manager (10), whose relations with the other group members are situated at the centre of the sociometric structure, which is related to a rather liberal management style that minimised sources of interpersonal conflict (cf. Terelak, 2021). This is pointed out among others by L.L. Schmidt, J.A., Wood & D.J. Lugg (2004), who based their analysis of the performance of groups of polar explorers participating in the Australian Antarctic Expeditions between 1996 and 2000, found that the role of the leader in creating team climate was perceived in terms of leadership effectiveness at 77% of the variance, which accounted for 14% of the overall variability in team climate.

The direction of the dynamics trend of the change in structure over time is evidenced, among other things, by the decreasing number of edges,[46] which indicates a decreasing number of mutual relations fulfilling the criterion of group formation, which is consistent with previous observations of individual distance measures (cf. Fig. 48). The compactness of the graph is directly proportional to the number of edges and inversely proportional to the square of the number of nodes. For each regression line, a measure of the slope is given. This is consistent with the descriptions of winter-over participants at various Antarctic stations, exemplified by the narratives of winter-over participant Jean McNeil (2016, p. 309): "There was less and less going on at the base. We no longer had evening lectures and group film screenings. Our only entertainment was sitting in the dining room, chatting and playing cards and flicking through mountain magazines and old issues of *National Geographic*, which we had already read thirty times." These observations are confirmed by the values presented in Figure 50 for the number of edges in the group graphs for successive sessions, and the compactness measures of these graphs.

---

46 For graphs in individual sessions, all nodes are connected to each other by edges that have different weights due to different distances (Chakraborty, at. al., 2018).

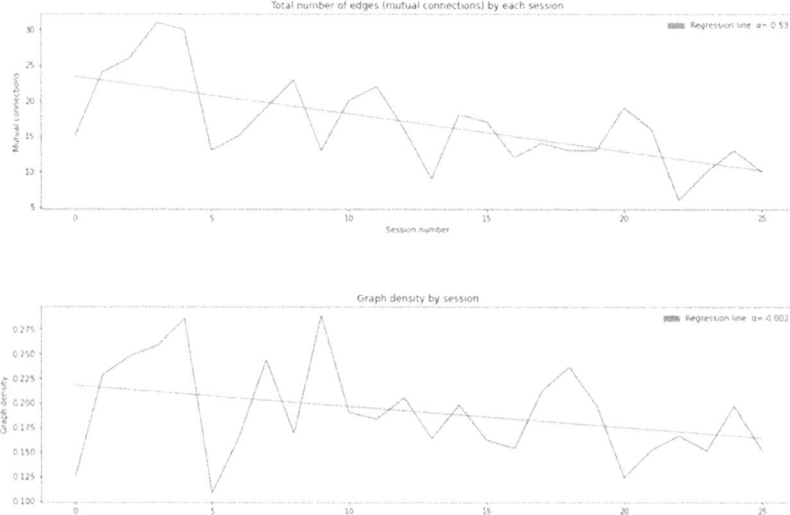

Fig. 50. Trends in the dynamics of the annual structure of the wintering-over group assessed on the basis of changes for (a) the total number of edges (mutual connection) by each session (y = -0.5303 + 23.966; R^2 = 0.4196; R = 0.6477) and the density of the graph by session (y = -0.009 + 0.2837; R^2= 0.3729; R = 0.6106)
Source: own elaboration.

As the main research problem were the changes in the group structure, the above indicators were calculated on the basis of the obtained group graphs, which allowed to present the changes taking place with the successive sessions. Fig. 50 presents the values of the number of edges in the group graphs for subsequent sessions, which document the growth of the distance between winter-over participants over time and suggest an occurrence of significant situations between sessions 4 (11 February) and 5 (27 February). On the basis of the documentation collected during the stay at the Arctowski station, published in the form of the "Antarctic Diary" (Terelak, 2021), two stressful situations were identified, which have to do with the preparation of the group of "summer workers" to return to the country (date 11.02.1979) and with the departure of the MS Garnuszewski ship to the country and only the wintering-over group staying at the station (date 27.02.1979)."[47]

---

47 Excerpt from Antarctic Diary: "27/II (1979)18) – "At 5.30 a.m. the alarm bell sounded in the "airplane." It meant that "Garnuszewski" had raised the anchor. We dressed

Notwithstanding such detailed explanations, it can be generally concluded that the structure of the wintering-over group formed and evolved over time, correlated with the situational context at Arctowski Station.

These conclusions correspond with the studies conducted by J.C. Johnson, J.S. Boster, & L.A. Palinkas (2003) at the Amundsen-Scott South Pole station in three consecutive Antarctic wintering-overs. In the course of analysing the dynamics of formal and informal social networks under conditions of isolation, he found, among other things, that globally coherent networks among winter-over participants in the middle phases of wintering-over were associated with wintering-over participants' consensus about potential external threats, in contrast to the other two phases characterised by separation of station asset and peripheral factions and cliques. When comparing the three wintering-overs, the authors highlighted differences in group cohesion. In the first wintering-over, the highest group cohesion was found at the end of this period, which was characterised by its continuous increase. During the second wintering-over season, the pattern of informal group structure showed a slow decline in cohesion, coinciding with the end of wintering-over, and this pattern is comparable with the results of the network analysis of winter-overs participants at the Arctowski Station. Finally, the third wintering-over at the Amundsen-Scott South Pole Station is characterised by a transition from the highly cohesive structure at the beginning of winter to a more fractional structure at the end of winter. The authors of the research justify these differences by varying personal composition of individual wintering-over crews and by the human resource management styles (e.g. "expressive leadership"). One has to agree with the comment of the research authors that the specificity of a group of winter-over participants operating in extreme isolation is that selection of a clear informal leader by the crew has a clear advantage over the appointment of a formal leader. However, if a clear

---

warmly because of the negative temperatures and went out in front of the "airplane." It was a beautiful sunny morning. As I hoisted the flag on the mast, "Garnuszewski" appeared at the exit of Ezcurra Fjord and began a half-hour parade in front of our station.

I fire the first rocket and get a response from the ship. From now on, coloured fires go up from both sides again and again. It may not be a symbol of joy as in greeting, but it is certainly an expression of mutual understanding of the turn of events. As I fire the last rocket, the ship is already at the entrance to Bransfield Strait. The "majestic silhouette of Garnuszewski" slowly disappears behind the horizon. We will longingly await its return when it arrives for us in 10 months' time." (Terelak, 2021, p. 96).

formal and informal leader meet at the same time, conflicts are to be expected (cf. Davidová, 2017).

In conclusion, it should be assumed that both the author's own research and the research by J.C. Johnson, J.S. Boster, and L.A. Palinkas (2003) prove that the leadership of a formal leader alone is not enough for a coherent social network structure to automatically form and maintain itself. The redirection of professional and social competences of different individuals, from competitive to consolidating attitudes, may have a pro-development trend of group dynamics, guaranteeing the efficiency of group functioning even in extreme situations of Antarctic isolation, providing an analogue for future missions to Mars. Further research on variables mediating group dynamics, such as gender or culture seems justified (Johnson & Finney, 1986).

In 2011, the recommendation of the National Academy of Sciences Decadal Survey was released, entitled: "Recapturing a Future for Space Exploration: Life and Physical Sciences for a New Era," which highlighted the need for a full understanding of genders and the differences between them, which prompted the development of research on the subject within the NASA community as well. An example of such research is the paper by N. Goel et al. (2014), which summarises the state of research (published and unpublished) related to gender differences in adaptation to space flights, inter alias, in terms of: (1) personality, interpersonal relationships, as well as performance and job satisfaction; (2) resilience to stress. Differences between men and women in these areas were found to significantly affect adaptation and differentiated medical care during space flight. Furthermore, the authors emphasise that during long-duration space missions, a mutual understanding of the impact of gender on crew interpersonal relationships is essential.

In contrast, J. Blackadder-Weinstein et al. (2019), aware that women's participation in Antarctic wintering-over is still underrepresented, studied women participating, in the male group, in an expedition that traversed the Antarctic continent. The findings indicated clear individual differences in personality and personal values, along with the duration of the expedition they were an increasingly frequent point of tension due to two strongly dominant individuals who had a negative impact on the team dynamics. This was partially confirmed by the research conducted on the impact of heterogeneity (in terms of occupation and gender) of the International Space Station (ISS) crew composition by A. Vinokhodova & V. Gushin (2014) from the perspective of crew cohesion dynamics and interaction with the Mission Control Centre (MCC). The purpose of the on-board experiment were the behavioural patterns of small groups (space crew) on the ISS 19–30 mission with twelve Russian crew members. The data obtained showed that the system of values and personal attitudes of most of the participating cosmonauts remained generally stable during extended space flights, but the mediating role of such personality traits (e.g. motivation, intellectual level, self-knowledge, self-discipline and sociability), in the effective performance of professional activities and in maintaining good social relationships, was indicated. This is confirmed by A. Sarris & N. Kirby (2007), who highlight the relationship between cross-cultural behaviour and organisational culture in terms of behavioural norms and expectations. A study of the culture of participants (N = 116) in wintering-over at Australian Antarctic stations showed that although the overall culture profile at Antarctic stations reflects a culture focused on satisfaction, friendship and participation, gender differences in perceived norms and expectations also emerged. Thus, men described an open and group action-oriented culture, while women were oriented towards procedures, hierarchical group structure and individualism.

J.E. Boyd, et al. (2009), assuming that cultural differences between crew members and mission control personnel can affect long-duration space missions, analysed three cultural variants: national (American versus Russian); professional (crew members versus mission control personnel); organisational (ISS). The study was conducted on a sample of crews from the Mir space station and the ISS, as well as US and Russian control centre employees. Respondents answered questions about mood and climate in the group on a weekly basis. The ISS sample also completed a questionnaire on culture and language. The results showed that cultural differences were more likely to affect social climate than mood. The analyses revealed differences between the Mir and the ISS in the social climate primarily among the spacecraft crew and addressed the high level of work pressure among the American crew. M. Basner et al. (2014), in turn, studied adaptive responses in an international group of six healthy men closed in a 550 m^3 Martian habitat for 520 days. Once a week throughout the mission, crew members completed the Beck Depression Inventory-II (BDI-II), the short form Profile of Mood State (POMS), the conflict questionnaire, the Psychomotor Vigilance Test (PVT-B) and a series of visual analogue scales on stress and fatigue. Significant interindividual differences were observed for mood disorder (POMS) and depression (BDI-II) in 93% of the mission weeks. In addition, interpersonal relationships were studied, which involved identifying the two crew members with whom they most frequently interact during the mission. It was found that the level of closeness between individual members of the international group studied varied quite considerably and that conflicts increased with the duration of the mission.

Coinciding with the views presented above are the proposals of small-group sociologists, who also draw attention to the fourth stage of group dynamics, characterised by a breakdown into numerous subgroups which, however, do not have a factional nature (Boriskin & Slewicz, 1968). One of the reasons for this, based on experience at US Antarctic stations, was pointed out by R.E. Strange & W.J. Klein (1973), who put the blame for the final process of group disintegration squarely on the unfavourable style of station management responsible for scrupulous adherence to group norms. However, this seems to be an overly narrow explanation that does not take into account many other factors, although in our case there is something to it, as the wintering-over manager preferred the liberal management style[48] of the Arctowski station, which is generally disliked by small groups under stress (Mroziewski, 2005). In further interpretation, attention should be drawn to the interesting findings of H.R. Schaeffer (2009), who found, among other things, that an increase in threat causes an increase in affiliative motivation. It is therefore not out of the question that a decrease in the threat after the Antarctic winter causes a reverse reaction. This suggestion seems to be confirmed by the research of E.K.E. Gunderson and P.D. Nelson (1963), who found, among other things, that under stress (difficult weather conditions, failures at polar stations) an increase in behaviour expressing positive emotions and a decrease in behaviour tinged with negative emotions was observed in group interactions. It is important in this case to locate the sources of danger. If it was located outside the group, an increase in positive attitudes and group consolidation was observed as a result of considering the group as a source of safety. In contrast, if the source of the threat was located within the group, disintegration is observed. The authors further found that towards the end of an isolation characterised by a decline in threats, a deterioration in interpersonal relationships emerges, resulting, among other things, from self-fatigue and the loss of the task-based nature of the group. We are convinced of this by observations of the behaviour of former winter-over participants on the ship on the way back home,[49] which is caused by the loss of the task-based nature of the group and a

---

48 The consequences of this are pointed out by P.D. Nelson (1962), stressing that a factor that can facilitate or hinder social adaptation is group leadership style and identification with one's formal leader, as the emergence of an informal leader makes group integration much more difficult.

49 "You can make friends here for a moment with a pilot, a plumber, a marine biologist, a radio technician, a writer, someone who is ten or twenty years older or younger. Freed from the bonds of demographics, peer groups, shared interests and activities, we can get to know anyone. It is such a social experiment, an ice ark. After leaving Antarctica, however, this capsule of casual acquaintanceship will disintegrate" (Jean McNeil, 2016, p. 214).

return to the social roles performed before the Antarctic expedition (Johnson, Boster & Palinkas, 2003). This suggestion arises if only from an analysis of the essence of tasking, as argued by P.A. Hare (1962), who emphasises that small task groups, in addition to the interactions that take place between their members, are also characterised by features such as: (1) the existence of common motives or group goals that determine the direction of group action; (2) the existence of a set of group norms that determine the rules of group action; (3) the stabilisation of group roles; (4) the formation of a network of interpersonal ties based on like–dislike relationships.

An analysis of the situation in which the group we studied found themselves on the ship on the way back home shows, among other things, that the characteristics listed by Hare have radically changed in importance compared to the period of stay at the Arctowski station. The common group goal of responsibility for the station and the group's survival in extreme Antarctic conditions was fulfilled when the station was handed over to the next polar expedition (the moment of handing over the Polish flag, singing the national anthem and handing over the symbolic keys to the Arctowski station). Consequently, certain norms of group action have necessarily been modified. Finally – with the change of the task situation and the anticipation of the "old" social roles performed in the country, the previous arrangement of group roles and the system of group values were also destabilised, which lost the meaning developed as a pattern of adaptation to conditions of isolation, which is also typical of the situation related to the end of the space mission (Suedfeld, 2006).

A separate issue is the individual perception patterns of individual members of the wintering-over group, shown in Fig. 48, which correspond to the previously presented results in Fig. 51 obtained through the nomination method and the participant observation method by the station manager.

The social aspect of Antarctic isolation   277

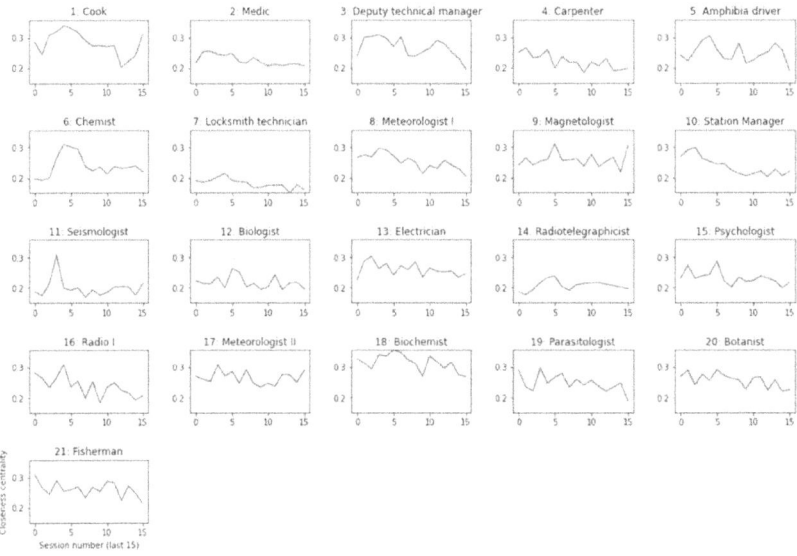

Fig. 51. Individual patterns of perception of others by individual winter-over participants during 15 consecutive sessions
Source: own elaboration.

As can be seen from Figure 51 above, measures of central proximity were calculated from the defined graphs and observed to change over time.[50] As can be seen from the fifteen graphs shown, each person's change over time was different, so averages were calculated over the whole study period to identify those with the highest or lowest scores. Thus, analysis of the individual charts suggests that the highest consistent proximity was for the biochemist (No. 18 – nicknamed "Small Manager"), and for the cook (No. 1) and the electrician (No. 13), and the station manager (No. 10) in the first four sessions.

When analysing individual relationships, it is important to note that they have changed over time in the form of the creation of new fractions and changes in closeness ratings towards individuals. These are not radical changes, as the respondent may still have considered someone to be close, although the assessment of the other person may have changed significantly. This is characteristic

---

50 The graph refers only to the last 15 sessions, as this was the only time all the respondents were present.

of the situation of social isolation of small task groups in the harsh winter conditions of Antarctica, characterised by periodically lost contact with the outside world and a lack of interpersonal alternatives.[51]

However, the above summary shows the electrician as the person unequivocally closest to all the respondents[52] for almost the entire winter-over period. The others revealed their closeness only incidentally (for 2 or 1 session) or even distanced themselves from the group (0 sessions), especially in the second half of wintering-over. This is in line with the previously presented Plebiscite of kindness results (cf. Fig. 47). The simultaneous occurrence of several individuals results from an equal number of relations fulfilling the ex aequo requirement. Considering only the highest number of relationships may not be sufficient, as one of the subjects could consistently come second in many sessions, which would also be significant, as evidenced by participant observation of the chemist who served as an informal station manager for a long period of wintering-over, or the cook who was in the top three positions almost throughout the year in plebiscite surveys. (cf. Terelak, 2021).

Summing up the discussion on social aspects of Antarctic isolation, it should be stipulated that we have signalled only a part of the problems related to the functioning of small gender-homogeneous groups.[53] The generalisability of the

---

51 Contemporary studies of isolated groups, especially in laboratory experimental settings, may be interrupted at the request of participants due to discomfort or accidents, which is a very different situation from the study group in the Antarctic isolation situation. The participants in this study had a constant sense of danger and knew that if a threat was made they could not simply walk out of the experiment safe and sound. It is also worth noting that conditions in Antarctica are very different even from those in the Arctic, as Antarctica has never been inhabited by humans and has no civilising infrastructure. (cf. e.g.: https://www.nationalgeographic.com/science/2019/01/see-how-astronauts-simulate-mars-mission-on-earth/).

52 Under the date of 04.12.1979 in the Diary I wrote the following characterisation of the Electrician: "In an informal plebiscite we chose the "best winter-over participant" – 1979." Jasio Swołek, the main power engineer, mechanic and a "golden handyman" as well as s a man of many positive features, showing a lot of initiative at work, conscientious, friendly, modest and unusually reticent, got the title. Indeed, he is the most worthy of the designation of the "best of the best." The result of the plebiscite coincides with the marks he received in many psychological tests. So this is no coincidence. (Terelak, 2021, p. 259).

53 We realise that heterogeneous groups would have very different group dynamics in annual Antarctic isolation, as suggested by some literature data (Rosnet et al., 2004; Schmidt, Wood & Lugg, (2005).

conclusions of the studies cited above encounters certain difficulties related to the low external validity of studies carried out under natural conditions. Social psychology data shows, among other things, that experimental groups differ from naturally functioning groups in many respects, e.g. interests, common or divergent goals, motivation, and mainly emotional bonds. Especially representatives of experimental social psychology (e.g. E. Aronson, 2009) show a higher external validity of laboratory studies. It seems, however, that if the methodological arguments (ensuring sufficiently good conditions for the control of many important variables intervening in the social phenomena under study) are valid for groups staying together for a relatively short period of time, the dynamics of the internal group structure (especially informal group structure), mainly in the case of long-term social isolation, is beyond the possibilities of unit experimental research, since, according to Th.M. Newcomb, R.H. Turner & Ph.E. Converse (1966), interaction processes are the resultant of the simultaneous action of three variables, namely: (1) the personalities of individual group members, (2) characteristics of group dynamics, (3) a wide range of situation attributes. Thus, experimental research is unable to capture the most salient group phenomena that, under the conditions of natural Antarctic isolation, interact with real danger and a sense of the "inevitability of fate."

Being aware of certain methodological limitations associated with psychological research conducted in the natural conditions of Antarctic isolation, where any psychological research is a major threat to the sphere of privacy, it should be believed that it nevertheless allows for a fairly comprehensive description of the dynamics of individual and group adaptation to extreme situations. This can have, on the one hand, practical significance (e.g. selection of members of any small task force forced to work in isolation) as well as theoretical significance (attempting to explain the many complex mechanisms of adaptation to an extreme situation and inspiring many detailed experimental studies).

We have already discussed a number of these problems by drawing attention, on the one hand, to the specific psychological costs associated with adaptive behaviour and, on the other hand, to the fact that an individual's adaptive behaviour is controlled to a greater extent by situational variables than by individual characteristics (cf. Argyle & Little, 1972; Grzelak, 1982).

At the conclusion of social adaptation to social isolation, we will still touch upon the atavistic phenomenon related to the territoriality of behaviour and the dynamics of the perception of the group as a reference point for the stimulating role of the social situation under conditions of long-term isolation.

5.3.3. Dynamics of territoriality of behaviour

Territoriality of behaviour[54] was assessed using a participant observation method and a situational experiment involving the use of the station manager's order to change seats at the table every month.[55]

Regarding the results from the observation of the territoriality problem, it was found, among other things, that the limitation of living space, characteristic of a small polar station, causes an increase in "territorial behaviour."[56] It was observed, for example, that the fractions distinguished in Fig. 44 quite strictly followed the rule of exclusive use of certain rooms outside the "aircraft" (e.g. biological laboratories, meteorological laboratories, radio station, power plant). Outside working hours, these rooms were intended for the organization of social life, thereby expanding the "psychological space." Their use was conditioned by membership of a particular "professional" or "recreational" fraction and this principle was usually strictly observed.

Another manifestation of territorial behaviour was the reservation of regular seats at a common table or chairs during film screenings. Fig. 52 illustrates the phenomenon of territoriality with the example of the "struggle" for fixed seats at the common table.

---

54 Territoriality is an atavistic need characteristic of all animals including Homo Sapiens. Restriction of territoriality in a situation of prolonged isolation is stressful, as can be illustrated by the example of industrial chicken farming as opposed to "free-range" hens, whose eggs are advertised as more valuable for consumption.
55 This method has been used for centuries in many male and female religious orders to increase group integration and prevent the formation of informal cliques and factions. Counteracting the appropriation of territoriality are the monastic rules, which are a set of principles governing all manifestations of life in a given community, recorded in an uninterrupted oral or written relay. One such method is the transfer of monks to another monastery every few years (cf. Dekert, Gronowski, Hrżycki, eds., 2013).
56 T he concept of "behavioural territoriality" is defined by us as the activity of a person, focused on using space for the purpose of creating for themselves acceptable living conditions and optimal functioning.

The social aspect of Antarctic isolation 281

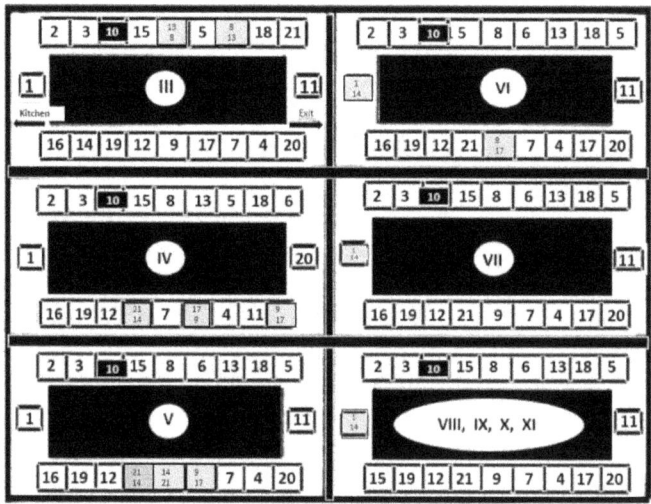

Fig. 52. The phenomenon of territoriality in a small isolated task group (N = 21) on the example of the analysis of the distribution of seats at the table
Source: own elaboration.

The data presented in Fig. 52 are consistent with the results of the experimental studies by L. Altman and W.H. Haythorn (1965), who, on the one hand, stated that the situation of isolation causes an increase in territorial behaviour in everyone and, on the other hand, that some personality traits (e.g. dominance or affiliation) considerably modify this type of behaviour. The analysis of personality traits of people who fought for the exclusivity of their seat at the table for quite a long time (see Fig. 45) confirms the Altman and Haythorn's observations. For instance, those characterized by dominance (respondents No. 14 – radio-telegraphist and No. 17 – meteorologist II) have been fighting for the exclusivity of the seat at the table for the longest time. In another system, in which one person was characterized by a high tendency to affiliate behaviour (e.g. respondent No. 9 – meteorologist I and 5 – amphibian driver) and another one by a tendency to dominate (respondents No. 5 – chemist and 13 – electrician), the rivalry lasted shorter and was generally not accompanied, opposite to the first system, by conflicts, although "territorial behaviour" could also be felt (cf. Haythorn, 1970, Matsuda, 1964).

However, in a stressful situation, territorial behaviour went by the wayside and the whole group was guaranteed emotional support. For example, when hurricanes periodically raged outside, almost everyone, regardless of the time of day, gathered in the main dining room and silently assessed the threatening force of the wind.[57] Pro-social group behaviours were highlighted in rescue situations when a winter-over participant needed help outside the station (at sea or on a glacier). Detailed descriptions of such situations can be found in my Antarctic Diary (Terelak, 2021).

Similar to my research on group dynamics using subjective sociometry, C. Tafforin (2004) conducted research in which he referred to Konrad Lorenz and Erich von Holts' ethological theory of territorial behaviour. In his own research, C. Tafforin focuses on describing the spatial behaviour of the crew of the the French Dumont d'Urville Antarctic station, as an analogue of a space mission. The method used consisted of the subjects (n = 13) drawing their position (on a special topographical map of the dining room) during midday and evening meals every day throughout the Antarctic summer and once a week during the winter. The quantitative data was based on a geometric approach, involving the analysis of coordinates of occupied positions and distances between crew members, and also included an egocentric orientation, reflecting positions and distances from the assessor's perspective. The data analysed covered three adaptation periods: the first three months, the intermediate two months and the last three months of isolation. The results showed that in the first week after arrival at the station and in the last week before departure, the team members occupied different positions at the station, while during the winter period the team members gathered in a common space, but the distances between the winter-over participants were increasing with the duration of isolation. A similar effect was observed in an experiment by the ethologist Erich von Holst, who found that under threat conditions, grazing animals of different species moved freely throughout the pasture, while in a threat situation (a thunderstorm with lightning) they concentrated in a species group in one place. Whether the results obtained by C. Tafforin can be useful in case of herding behaviour as an analogue, as he suggests, of future space flights, is highly controversial due to the small usable space. In another study conducted at the French-Italian "Concordia" Antarctic station, C. Tafforin (2009) described from a perspective the

---

57 Similar atavistic animal behaviour is narrated by the ethnologist Schachter, who observed, in a natural grazing experiment, that in calm situations the various species of animals in different parts of the grazing area mixed together, whereas when storm signals were approaching they clustered together in species-homogeneous groups (e.g. horses, cows, sheep, etc.).

dynamics of a multicultural and gender-mixed group under the conditions of one-year isolation and seclusion, as an analogue of a long-term space mission.

These were based on an atavistic tendency for territorial behaviour, which manifested itself in behaviours such as, among others, social orientations, station space division, place preferences (e.g. due to sex and nationality during meals), social cohesion (e.g. large in a threatening situation). The author suggests that ethological observations have the theoretical and methodological nature of a transdisciplinary approach, connected with anthropological perspectives on the micro-societies of future Martian habitats. Not entirely in agreement with such a directly derived conclusion is P. Suedfeld (2018) who, while comparing research conducted on Antarctic stations from the psychosocial perspective of the analogue of space missions, such as Skylab, Mir and ISS in terms of attributes such as social isolation, entrapment, novelty, discomfort, danger and distance from civilisation, nevertheless points out that analogues should not only look similar, they should produce similar adaptive effects. The author gives the answer to this postulate with a question mark, submitting it to discussion.

5.3.4. Dynamics of perception of group

The dynamics of social adaptation during Antarctic isolation can be summarised by examining the perceptions of the wintering-over group, using the method previously discussed when examining perceptions of the situation in positive and negative terms (cf. Fig. 35).

The perception of the group under conditions of several months of isolation in the conditions of a deep-sea voyage on a fishing vessel was the subject of M. Ciosek's (1977) previously discussed research on the perception of the situation, which showed that as the duration of the voyage went on, the sailors surveyed tended to evaluate their colleagues more and more negatively. In our study of winter-over participants under the conditions of annual Antarctic isolation, we obtained similar results, which are illustrated in Fig. 53.

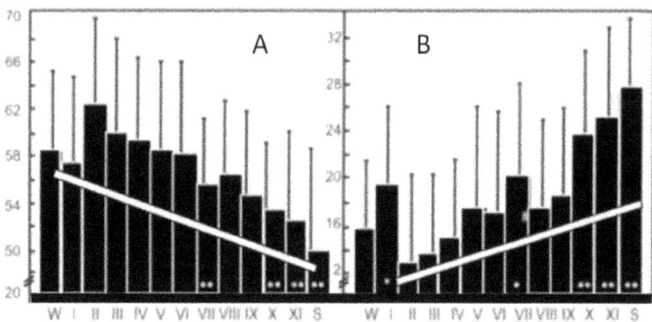

Fig. 53. Changes in the perception of the group wintering-over (N = 20) during the annual Antarctic isolation

Source: own elaboration.
Legend: XII, I, II – monthly surveys; S – ship surveys en route to the country * p <0.05; ** p < 0.01 between the initial (XI) and subsequent tests.

As can be seen from Fig. 53, the personal and behavioural elements of a group in a situation of long-term isolation are perceived more in negative than positive terms. An interesting regularity in the perception of the group is drawn approximately from mid-winter (July), when it changes in a negative direction (p<0.05 and p<0.01) assuming the highest negative value (and the lowest positive value) on the ship on the way back home. This is consistent with the data on the previously presented in Fig. 42 on the decrease in interpersonal attractiveness, caused by a kind of "burnout in the role of an Antarctic polar explorer" associated with the need to stay for such a long period of time "face to face" with the same people (Lugg, 1973; Natani, Shurley, & Joern, 1973). In our case, the group is most positively (and least negatively) perceived by winter-over participants during two periods of isolation: in February and March (p<0.05). This is due to the fact that, during this period, the winter-over participants felt some relief after the ship left for the country with the summer team and the joy that they were finally the only managers of the station. This is confirmed by the results of the first two months, which show a negative perception of the group, caused by congestion in a small station, significant professional, gender and cultural differences, which often led to conflicts between the summer and the winter group in relation to some of the privileges associated with staying at the station (e.g. restrictions for the summer group in the access to toilets and hot water in the main building and restrictions on the use of mineral water, etc.). The data quoted above only indicates the problem of a change in the perception of oneself and others in isolation over time. Being tired of each other manifests itself in a drastic way on the ship on its way back to the country, as evidenced, among other, by breaking all emotional relations, and communication which is not about talking but rather about issuing short messages.[58] Research on communication in the task force in the small isolated Antarctic station can be treated as an analogy of communicative behaviour in a spacecraft isolated from outer social space (Gushin et al., 2016).

Nor can another aspect of the Antarctic isolation situation, rarely considered in psychological research, be overlooked, namely the problem of the effect of the constant change of time zones on human functioning. This problem is dealt with by a separate scientific discipline, namely chronobiology.

5.4. The chronobiological aspect of Antarctic isolation

In living organisms, rhythmic changes are observed in almost all life activities.[59] A separate discipline, chronobiology, is interested in the analysis and description of cyclical recurring life phenomena (Halberg, 1969, 1980). The basic synchronisers[60] of functional rhythms in biology are all the cyclical changes occurring in the natural environment, such as the alternation of day and night. From the point of view of the duration of biorhythms, the following rhythms are distinguished: ultradian – i.e. rhythms shorter than 20 hours (e.g. peristaltic rhythms, heart rate, respiration, electroencephalographic rhythms, etc.); circadian[61] – i.e.

---

58 This is confirmed by the reactions of external observers, who were members of the crew of the ship m/s Antoni Garnuszewski, who commented succinctly on the prevailing silence during shared meals: "These are, after all, winter-over participants or the "psychotic ones."
59 "The rhythmic structure of biological systems can be considered a principle complementary to that of homeostasis" (Jurgen Aschoff, 1979, p. 225).
60 "A self-sustained oscillator with an inherent frequency underlies human 24-hour periodicity" (Jurgen Aschoff, 1965, p. 1427). These rhythms, which have a significant influence on the course of all life processes, are called synchronizers or "time-givers" (or Zeitgebers in German). It can be imagined as a "servo-mechanism," operating on the principle of negative feedback and serving, in relation to living organisms, to bring two or more phenomena, processes, etc., into a state of relative temporal congruence (e.g. congruence of periods, phases, etc.) (Oatley, Goodwin, 1971).
61 We are talking about circadian rhythm and not diurnal rhythm, because it is not perfectly adapted to the astronomical rhythms of the Earth (24-hour rotation around its own axis), as the biological "clock" of the organism is influenced by various factors regulating it to the current environmental conditions. In relation to humans, the role of synchronisers of biological rhythms is played not only by photo-ecological factors, but also by civilisational, psychological factors, called social synchronisers.

circadian, lasting from 20 to 28 hours (e.g. sleep – wakefulness, metabolic processes, etc.); infradian – i.e. lasting more than 28 hours (e.g. several days, monthly, seasonal, annual, several years; e.g. menstrual cycle, hormonal cycles, etc.). Rhythms can be determined: (1) endogenously, when they are conditioned by biological oscillations within the organism (e.g. rhythmic activity of nerve cells, organs, physiological processes, etc.) and reflect the natural frequency of spontaneously acting bio-oscillators;[62] (2) exogenously, when they are conditioned by the variability of external phenomena not manifested in constant environmental conditions.

5.4.1. Photoecological aspects of Antarctic adaptation

Most important in the adaptation of living organisms to the conditions of the outside world are circadian rhythms, in which the primary astronomical synchroniser in plants and animals is the social synchroniser that regulates alternating times of activity and rest (Czeisler, CA, et al., 1999). The beginning of the study of biological rhythms was the demonstration of a sinusoidal diurnal variation in body temperature and the association of this observation with changes in life activity, showing its maximum in the afternoon and minimum at night, with fairly large individual variations.

---

[62] Thus, for example, evidence confirming the endogeneity of circadian rhythms is provided by the occurrence of these rhythms among cells transferred outside of the organism (in tissue cultures), which still have a 24-hour rhythm of cell divisions (+/- 1 hour) (Von Mayersbach, ed., 1967; Halberg, Nelson, Runge & Reynolds, 1971; Kwarecki, 1985). This is due to the existence of a master biological clock, (called the circadian rhythm pacemaker), which is located in the hypothalamus in the nucleus suprachiasmaticus, which regulates the subordinate clocks located in individual cells and organs and, at the same time, has the capacity to respond representatively to changes in environmental conditions for the whole organism (Klein, Moore & Reppert, Eds., 1991).

The chronobiological aspect of Antarctic isolation 287

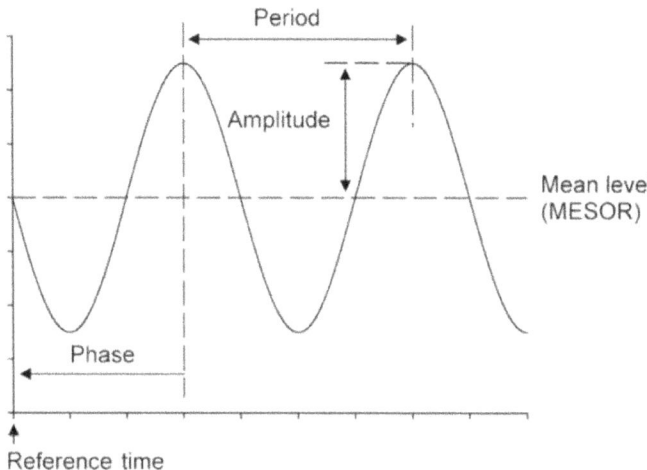

Fig. 54. Diagram of oscylatory and four parameters: mesor, period, amplitude, phase
Source: own elaboration.

As shown in Figure 54, each circadian rhythm can be described by these four parameters: (1) Phase – the span between the maximum (acrophase) and minimum (nadir) of the trait being evaluated in a cycle; (2) Period – the duration of one cycle; (3) Amplitude – half the span between acrophase or nadir and mesor; (4) Mesor – the mean circadian value. A fifth parameter can also be specified: the frequency, i.e. the number of cycles per time unit.

The issue of circadian rhythms is of great importance for understanding the adaptive mechanisms of humans living in modern civilisation, in which many so-called social synchronisers disrupt the atavistic circadian rhythm originally in line with the astronomical synchroniser.

Mismatch between rhythm phases and synchronisers leads to "external" desynchronisation, while being in such a situation (such as when staying in a new time zone) leads to "internal" desynchronisation. When external synchronisers are switched off (e.g. in a darkened isolation chamber, underground confinement in mines and caves), free-running rhythms occur in humans, in which, on the one hand, a lack of correspondence is observed between biological time (counted by the Endogenous Self-Sustaining Oscillatory Clock) and local astronomical time, which leads to an inner inconsistency between the basic rhythms of the body's activities for life (Menaker, 1961).

The occurrence of free-running circadian rhythms may be indicative of genetic, endogenous determinants of biological rhythms. The body's response to a change in the phase of the environmental synchroniser is adaptation to the new conditions, known as the resynchronisation process. The resynchronisation mechanism can be explained by the action of the circadian rhythm phase regulator, shown schematically in Fig. 55.

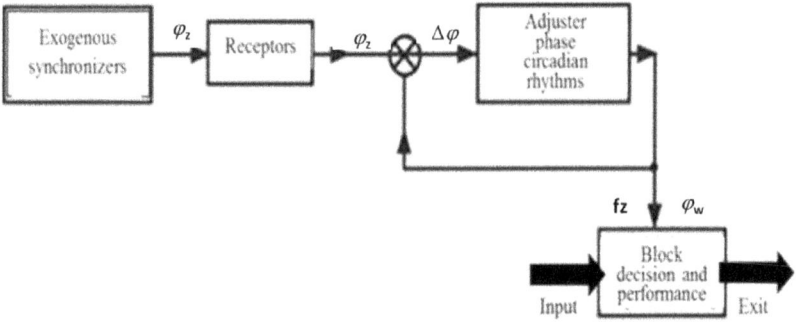

Fig. 55. Negative feedback loop in the circadian rhythm phase regulation system
Source: own elaboration.

In the stationary state, there is an equality between the phase of the external synchronisers φ_z (the "set" signal) and the phase of the circadian rhythm φ_w (the "measured" signal). The frequency of the external synchronisers f_z determines the frequency of the circadian (internal) rhythm f_w in this model. In the case of a fast change of the φ_z phase of external synchronisers at the input of the circadian rhythm phase regulator block, a difference $\triangle \varphi$ of incompatible phases of the set φ_z and the measured φ_w signal will appear. After comparing the signals, the system changes the output value φ_w so as to reduce the value of the "error signal" $\triangle \varphi = \varphi_z - \varphi_w$. In order to characterise such a system, the concepts of speed and nature of response, among others, are introduced in the control theory. A good example of how negative feedback works as a regulator of circadian rhythms was *homo sapiens'* experience related to the introduction of jet-powered passenger aircraft with transcontinental range in the mid-20th century, which revealed desynchronisation and its biological mechanism. This involves rapidly crossing time zones (every 15° longitude there is an hour's difference in astronomical time between neighbouring time zones), which, in addition to its benefits, has its own negative effects of a physiological and behavioural nature that occur, starting with a change between two time zones i.e. a difference of 30° longitude. Flight attendants and air passengers on ranscontinental lines maintain their own endogenous biorhythm, incompatible with the local astronomical time at the place of arrival. The lack of correspondence between biological time (deduced by the CNS, organs and cells) and local astronomical time causes the so-called *jet-pilot synd rome*, which has various names (*jet-lag desynchronosis, transmeridian dyschronism*, etc.) (Wittmann et al., 2006).

The mechanism of desynchronisation of biological rhythms and re-synchronisation to local conditions, or resynchronisation, is based on the degree of divergence between the indications of the three clocks: astronomical, physiological and civilisational. The symptoms of desynchronisation are described as, among others: sleep disturbances, daytime sleepiness and fatigue, situational disorientation, loss of attention, memory problems, irritability, irrational behaviour, gastrointestinal disorders, dehydration, peripheral fluid redistribution (e.g. swelling of the lower limbs), etc. (Klossowski, 2002). These symptoms resolve after a few days of adaptation (resynchronisation). Failure to adhere to the resynchronisation period risks ill health (Chai, & Flaherty, 2019). A number of studies have shown that the rate of resynchronisation varies according to the direction of time zone changes (e.g. westbound around 90 min/day and eastbound 50 min/day).

The results of studies of the resynchronisation process, carried out with changes in time zones, allow us to conclude that the circadian rhythm phase control system is similar to an integrating regulator, while the speed of response is about 1 hour per day. As a first approximation, the resynchronisation process can therefore be represented as a function of time t with the use of the following equation: $\triangle \varphi = \varphi_z - \varphi_w = \triangle \varphi_o$; where r is the resynchronisation rate and $\triangle \varphi_o$ is the size of the phase change of the external synchronisers at time t_o. The resynchronisation process ends at t_k, when the error signal $\triangle \varphi$ reaches zero. The individual aspects of the resynchronisation process are shown in Figure 56.

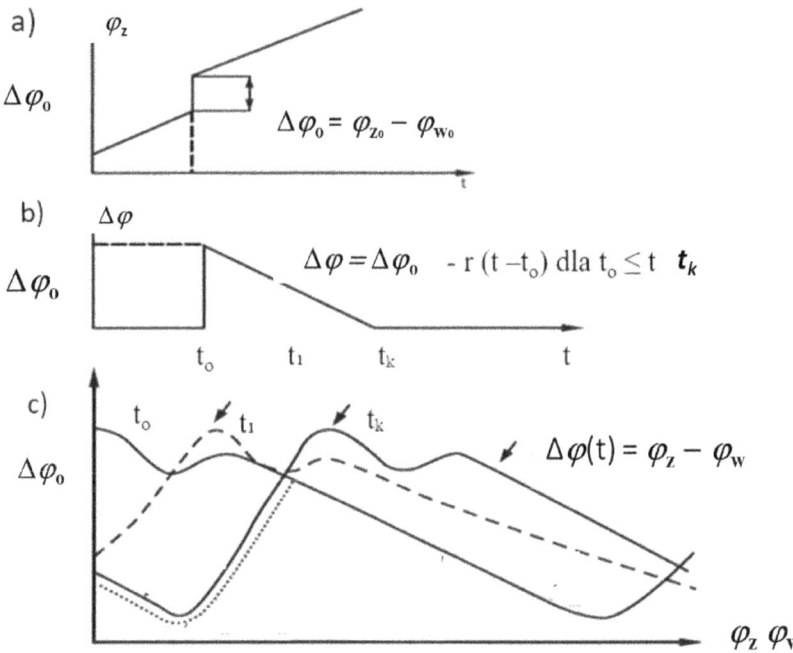

Fig. 56. The course of the resynchronization process: a) change of the external synchroniser phase; b) change of phase differences as a function of time; c) circadian rhythm in subsequent stages of resynchronisation

Source: own elaboration.

(Legend: to — the moment of external synchronisers phase change; t_k — the moment of completion of the synchronisation process; $\triangle \varphi_o = \varphi_{zo} - \varphi_{wo}$ — the initial value of the phase difference (equal to the size of the external synchroniser phase change); $\triangle \varphi(t) = \varphi_z - \varphi_w$ — the value of layer difference at the moment t.

The simple exogenous model of the phase regulation system presented in Fig. 56 is sufficient to explain phase dependencies in the resynchronisation process, although it does not allow the explanation of human functioning in an environment devoid of natural external synchronisers (e.g. bunker isolation, during mining disasters or in caves without artificial lighting, but also during long polar nights), in which the so-called drifting of circadian rhythms (free-running) occurs (Kennaway, Van Dorp, 1991).

The situation changed dramatically when man invented artificial lighting, which, referred to in chronobiological literature as the "civilian dusk," significantly and arbitrarily lengthens the period of activity, at the expense of shortening the resting phase. It turns out that adaptation to the artificial cycles determined by social synchronisers is possible, but entails specific behavioural and health costs (Aschoff, 1979). An example of revealing atavistic circadian rhythms is, on the one hand, the *"post-lunch phenomenon"* described by W.P. Colquhoun (1971), known as the afterlunch siesta phenomenon (drowsiness, reduced mental performance, etc.) or the decrease in all activity between 3 and 5 a.m., as evidenced by higher accident rates compared to daylight hours (O'Donnell et al., 2005). The lack of light during the night causes increased melatonin production (Zeitzer et al., 2000). It was found that there is a negative correlation between melatonin levels and mental performance. This is confirmed by own research, which shows, among other things, that cognitive performance decreases during winter periods, which is linked to the fact that melatonin levels are significantly higher in Antarctica in winter than in summer. However, it is worth noting that the negative effect of a prolonged period of darkness also lowers mood, which has been termed "seasonal affectivity." In addition, the production of the T3 thyroid hormone decreases at night (Palinkas et al., 2001), which was the basis for the separation of a distinct "polar T3 syndrome," characterised by increased fatigue, depression and anger and lower cognitive and psychomotor performance in terms of reaction time (Pääkkönen etal., 2008).[63] Hence, there have been many studies of diurnal fluctuations of various human mental performance, both under natural conditions (e.g. Klein, Wehmann & Hunt, 1972) and laboratory experiments, among which we will highlight our own elaboration, which aimed to determine the effect of circadian rhythm (studied every 4 hours around the clock) on simple mental functions (divided and concentrated attention, mental performance, simple reaction time and eye-hand coordination). The results of the study

---

63 The secretion mechanism of the two primary hormones of wakefulness (cortisol) and sleep (melatonin), reflects the original photo-ecological situation before the invention of artificial lighting (Claustrat, & Leston, 2015). Imagine a situation at the beginning of the origin of the species Homo Sepiens, in which an hour before waking up, 80% of cortisol, belonging to the class of glucocorticosteroids (the so-called stress hormone), is delivered to the body, which secures energy reserves more or less until about 2 p.m., at which time a drop in its level causes the "post lunch phenomenon." The replenished remaining 20% of cortisol should last until astronomical twilight, which is the signal for the secretion of melatonin, the sleep hormone. Melatonin secretion slowly ceases around 3–5 a.m. and cortisol production slowly begins to provide the energy to wake up and act after about 2 hours (Dijk et al., 2012). Therefore, in these critical two early morning hours, the body, if not asleep, is not yet able to function efficiently, and can fall back asleep without signs of drowsiness.

confirmed that all the curves of the mental faculties studied are sinusoidal in nature with a maximum at around noon and a minimum at night, with a drastic fall between the hours of 3.00–5.00 a.m.

5.4.2. Dynamics of circadian rhythm of cognitive functioning

Psychologists' interest in the issue of biorhythms in the context of Antarctic isolation is primarily related to the fact of the specific geographical location of the polar station in relation to the geographical location of one's own country. The consequence of this fact in our case is a specific relation for the polar station: day–night and a permanent shift of the time zone (about 4 hours backwards in relation to the Polish winter time and 5 hours – in relation to the summer time). In addition, the absence of many important social[64] (different rhythm of the day than the one we have been used to so far) and civilisational synchronisers (e.g. work, family, social life schedule of the week) is also not insignificant (Arendt, 2012).

Thus, as suggested by research conducted in polar regions to date, this chronobiological aspect of isolation is not without its effects on human functioning, especially during the period of adaptation to new photo-ecological and social conditions (cf. e.g. Boriskin, 1967; Shurley, 1973; Steel et al., 1995)..

This is confirmed by our own research on circadian variation in cognitive and psychomotor performance, carried out quarterly throughout the year at the Arctowski station. The selection of research dates was guided by the duration of illumination in the area of Arctowski station in particular quarters. The research involved: (1) mental performance in visual attention (five-minute alternation of adding and subtracting columns of digits:[65] (2) visual-motor coordination at an imposed work rate of 107 and 125 light signals per minute (Piórkowski coordinometer and cross support); (3) subjective assessment of elapsed time (using a stopwatch and measuring 10- and 20-second periods). All the aforementioned tests were performed six times a day during the following hours: 6.00 a.m. – 10.00 a.m.– 2.00 p.m.– 6.00 p.m. – 10.00 p.m. – 2.00 a.m. The same sequence of studies and subjects was always maintained.

---

64 The role of the individual synchronisers is not equivalent, hence the distinction between: (1) dominant synchronisers, whose action leads to clear, strong effects (e.g. light, diurnal variations, ambient temperature, etc.); (2) weak synchronisers, which are meant to replace the dominant ones
or interact with them (e.g. gravity, atmospheric pressure); (3) subtle synchronisers, whose operation is not yet precisely understood (e.g. magnetic field, electromagnetic radiation, etc.).
65 6 parallel versions of the test were used to minimise the learning effects of successive attempts.

Coffee and alcohol were prohibited a day before and during the study. An adequate amount of sleep the night before the study was observed. No medium or high intensity work was performed on the day of the study. Before going on to discuss the results obtained, we will draw attention to the specificity of the mathematical-statistical analysis method used, called cosinor mapping or "cosinor wheel,"[66] was based on the method proposed by F. Halberg (1980). This is the sinusoidal model, shown in Fig. 57, which in our case was used to assess the variability of mental functioning over time.

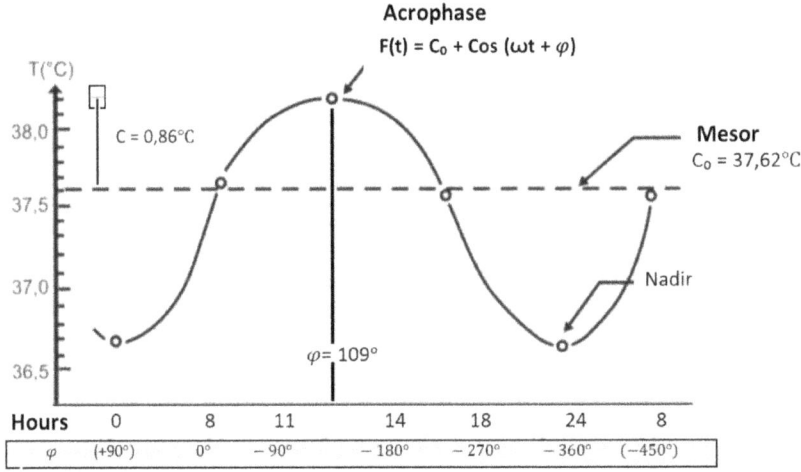

Fig. 57. The circadian rhythm of human body temperature
Source: own elaboration of F. Halberg, Y.L. Tong and E.A. Johnson, 1967.
Legend: C_o – mean value (mesor); C – amplitude; ω --- angular frequency; t – time; – acrophase in degrees from 0; Frequency – number of cycles per unit time; Period – Duration of one cycle; Amplitude – Half the span between acrophase and nadir.

---

66 We will skip discussing the different ways of analysing biorhythms, as this is the subject of many studies (e.g. Harris, 1974), while a detailed description of the assumptions of the cosinor mapping method and the method of analysing the results is presented, among others, by K. Zużewicz et al. (1977), W. Nelson et al. (1979), C. Bingham et al. (1982).
R. Refinetti, G.C. Lissen & F. Halberg, (2007) present various contemporary numerical methods for the analysis of circadian rhythms, such as: (1) Fourier analysis; (2) Lomb – Scargle periodogram; (3) Enright periodogram; (4) Linear cosinor; (5).ARIMA.MLE in SPLUSb; (6) Recursive Lomb – Scargle periodogramc; (7) Nonlinear cosinor; (8) Simulated annealing.

According to the principle of least squares, the cyclical curve "best fit" to the empirically obtained results was determined according to the equation presented below:

$$F(t) = C_0 + C\cos(\omega t +)$$

where: C_0 – daily average for a given time series, C – amplitude (degree of deviation of a given process from the average state, i.e. from the daily average), – acrophase (phase corresponding to the occurrence of a maximum value in relation to midnight). In determining the aligned curve, a pair of values is assigned to the time series, i.e. amplitude and acrophase. Having a minimum of five time series describing the same phenomenon, it is possible to perform their statistical analysis in the so-called cosinor wheel, shown in Fig. 59.

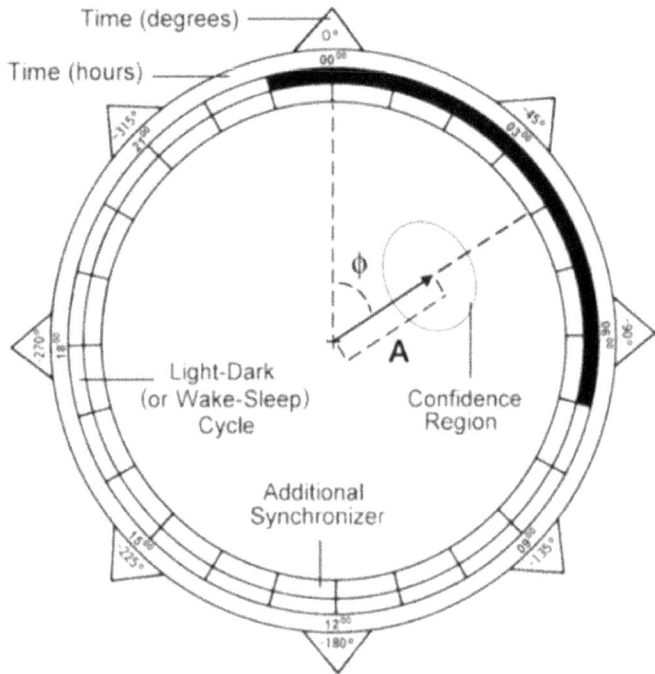

Fig. 58. Statistical methods for estimating and comparing cosinor parameters
Source: own elaboration of C. Bingham et l., 1982.

As can be seen from Figure 58, the model consists of cylindrically placed circles whose rotation of the pointers moves by an angle of $\{-360°\}$,[67] corresponding to 24 hours, and thus corresponds to the period of the time function. The point of distinction is midnight, i.e. 00.00 = 24.00, from which the angle – the acrophase of the rhythm – is measured. For the radius of a circle, the unit to be used is the unit of the characteristic under examination. The amplitude is represented as a vector whose origin is at the centre of a circle, of length equal to Co of the direction determined by the value of the oriented angle (acrophase). Thus each point in the cosinor wheel is determined uniquely by a pair of values (C,). With the pairs determined for individual time series, the mean amplitude and acrophase can be determined. Assuming that these pairs represent dependent variables with a two-dimensional normal distribution, a confidence interval is determined for the mean vector (C, φ), with a pre-established confidence coefficient, e.g. p<0.95%.

Thus, the basis of the statistical method for estimating and comparing cosinor parameters, concerns the polar representation of the amplitude and acrophase of the rhythm. Whereby the acrophase (φ) is defined by the angle of the vector whose length corresponds to the amplitude (A). The ellipsis indicates the 95% confidence area for the combined A and φ cores. An estimate of the 95% confidence intervals for A and φ can be obtained by drawing concentric circles tangent to the error ellipse of A, thus determining its intersection with the length of the vector (A), or by drawing tangents to the ellipse for φ. In the case of polar coordinates, an error ellipse is obtained, which is shown in Fig. 59, illustrating self-reported results characterising the impact of changes in circadian rhythm of winter-over participants in terms of mental performance, recorded in four quarters of the year during isolation at the Arctowski station.

---

67 Note that all phases are negative angles for reasons of consistency with the mathematical designation of directional angles (in mathematics, the direction of designation of positive angles is counterclockwise).

296 Transactional Characteristics of the Winter-Over Syndrome

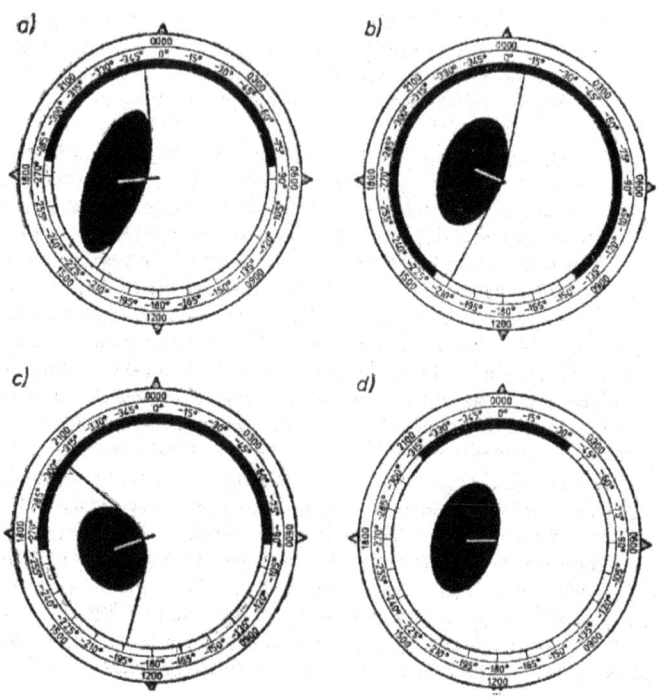

Fig. 59. Dynbamics of circadian rhythm of cognitive functioning of the winter-over of the 2nd Antarctic Expedition PAS
Source: own elaboration.
Legend: a) March, b) June, c) September, d) December.

Figure 59 shows the circadian cyclical changes in the number of solved arithmetic tasks in four consecutive periods corresponding to the seasons in the Arctowski area. By plotting the error ellipse on the cosinor wheel, it was possible to establish whether we were dealing with a biological phenomenon subject to cyclical changes during the day, i.e. the circadian rhythm. The error area covers the centre of the cosinor wheel, which confirms that the phenomenon analysed is a biological rhythm. This means that the phenomenon can take all the values of the acrophase 0–360°, i.e. it is not possible to indicate such a time interval during the day in which there would be an increase in the value of the examined variable. It is different when the ellipse does not cover the centre of the cosine wheel. Both situations can be traced in Fig. 56, in which the darkened semicircles correspond to night time, the light ones to day time. The black ellipses indicate the error characterising the amplitudes and acrophases and the black lines indicate their confidence interval. The white line in the centre of the ellipse illustrates the timing of the acrophase. Thus, for example, the circadian rhythm is marked in two months: in March(Fig. 56a) and in September (Fig. 56c). In both cases, the acrophase falls at around 5.00 p.m. local time. Characteristically, in both of these months the ratio of shade to light is similar to national conditions. This applies more to September than March, as indicated by the greater spread of results in March. In the case of the month of June (Fig. 56b) the circadian rhythm is marked very weakly, while in December (Fig. 56d) – it does not occur at all. Acrophase occurs in June at around 7: 20 p.m. and in December at around 5:20 p.m. It is characteristic that in both cases the rhythm disturbances occur under conditions radically altered from the domestic ones in terms of the light-dark cycle. In June, conditions on King George Island are similar to the "polar night," while in December there are "white nights." Fig. 56b and 56d indeed reflect the change in the activity-sleep relationship observed during these periods. More complaints of sleep disturbance were recorded in both periods compared to the other periods. For example, many people, especially in the scientific group, shifted their mental work in June and December to night hours (i.e. between 10 p.m. and 3 a.m.) at the expense of getting up in the morning for breakfast. Such cases are also described by other polar psychologists (Mullin, 1960; Palmai, 1963b; Shurley, 1973; Boriskin, 1973), although due to the different geographical location and photo-ecological conditions of Antarctic stations, these changes fall into different phases of the day, while maintaining their cyclic pattern.

If the significance of differences for a given variable was not obtained in the cosinor analysis, the mean values of the examined indicator were presented in the x, y coordinate system (where: x – indicator value, y – time) with the standard error and significance of differences. The lowest value of a given indicator per day was compared with the others. This way of analysing circadian variability in psychomotor performance and subjective time estimation is shown in Figure 60.

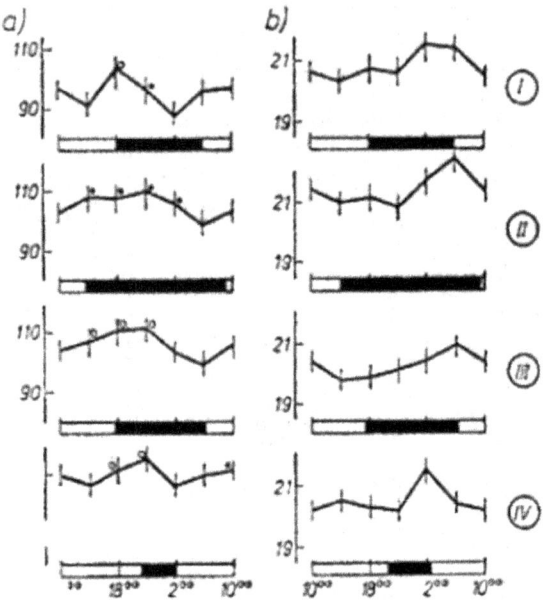

Fig. 60. Daily variability of the correctness of the: a) Visual-motor attention test and b) Subjective time assessment wintering during his stay at the Arctowski station
Source: own elaboration.
Legend: mean and ± mean error; I – March, II – June, III – September, IV – December; darkened area – night time; 10.00,18.00, 02 – hours; white dots o – $p < 0,01$, black dots – $p < 0.05$.

It can be seen from Fig.60 above, among other things, that the ability to perform a test examining eye-hand coordination (coordinometer) and estimation of elapsed time does not show circadian rhythmicity. There is some synchronisation of rhythm in the results concerning the level of eye-hand coordination (especially in test III), but not statistically significant. The highest values differ from the lowest at the $p<0.01$ level.

In attempting to discuss the results presented, some methodological difficulties, that need to be pointed out, arise. During the stay of the 2nd Arctic Expedition of the Polish Academy of Sciences at the Arctowaki station (wintering-over group), the obligatory hours of waking up and main meals (8.00 a.m., 2.00 p.m., 7.00 p.m.) were observed. This was undoubtedly a strong social factor synchronising the circadian rhythm of mental performance. On the one hand, this may have reduced the magnitude of acrophase drift during the changing photo-ecological conditions while on station. On the other hand, it prevented the differentiation of individual members of the wintering-over group into those preferring "nightlife" and "daylife" observed in such a situation.

5.4.2. Sleep disturbance

In Fig. 61 shows the average durations of smu and daytime naps for winter worms against the photoecological conditions prevailing at the latitude of the King George island, where the Polish Antarctic station named Henryk Arctowski (In Fig. 62, the light and dark spots reflect daylight and night time).

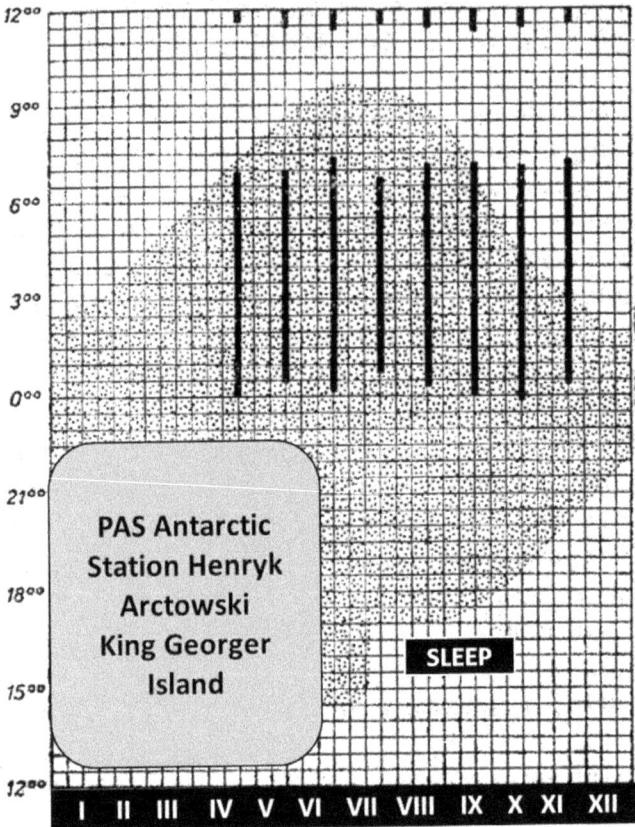

Fig. 61. Duration and time of sleep of the winter-over of the 2nd Antarctic Expedition PAS
Source: own elaboration of W. Kowalski, 1981.
Legend: the dotted field – time of day between sunset and sunrise; Roman numerals and numbers – months; vertical lines – bedtime duration for winter-over participants; short vertical lines at the top of the figure – average time of "naps" during the day.

Although Fig. 61 suggests a more or less equal duration and time of sleep for all subjects, however, a tendency to prefer "nightlife" did occur in some subjects, especially those who, out of necessity, performed a variety of late night activities (e.g. meteorologists, radio engineers, power engineers). They actually had altered sleep patterns (e.g. multiple naps during the day). In some people these changes were very severe, including insomnia throughout the day. Examples of significantly different individual sleep patterns are shown in Figure 62.

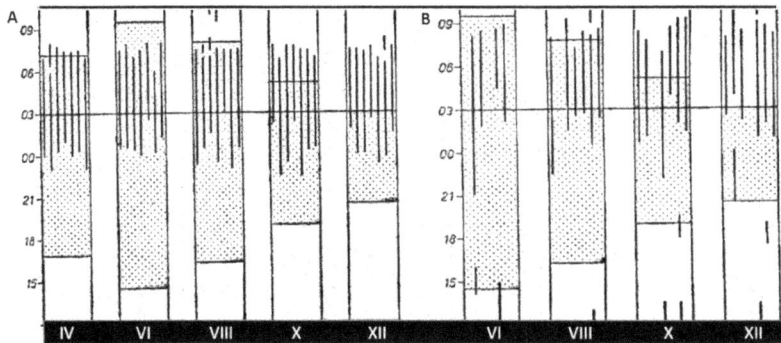

Fig. 62. The course of sleep in selected scientists during a one-year stay at the Arctowski station: A – geophysicist, not on night duty; B – meteorologist, performing night shifts every 3 hours in each first week of selected months

Source: own elaboration.[68]

Legend: the horizontal line represents the usual mid-sleep time; the dotted field represents the part of the day between sunrise and sunset.

As can be seen from the above figures 62A and 62B, it is characteristic to go to bed quite late during the whole stay at the station and to have different sleep patterns (night sleep and day naps).

In summary, the interest of Antarctic researchers in the problems of sleep patterns and wakeful activity of temperate zone humans incidentally inhabiting the polar regions stems from four main motives which, following J.T. Shurley et al. (1970), can be formulated as follows: (1) Empirical verification of previous anecdotal reports of frequent and uncomfortable insomnia symptoms (so-called "big eye") complained of by North American and European polar explorers; (2) Estimation of the effects of changing sleep-wake cycles under novel photo-ecological conditions; (3) Description of the change in the circadian sleep-wake cycle in the absence of multiple social synchronisers, associated, e.g., with free choice of times to get up, eat, and work; (4) Assessment of the effects of other polar

---

68 The co-author of the study was a physician-researcher participating in the Second Antarctic Expedition of the Polish Academy of Sciences, Dr W. Kowalski (1982).

zone-specific environmental stressors on sleep and activity patterns, reflecting the adaptive capacities of polar explorers, which can be considered as analogues for the space habitat environment (Dijk DJ, et al, 2001). To answer some of the above-mentioned research questions in our study, we obtained empirical arguments regarding the circadian variability of mental performance, which can be characterised by the following statements: (1) There is desynchronisation of rhythms during Antarctic winter and full summer relative to resynchronisation during the following agreed-upon periods: "spring" (Q3) and "autumn" (Q1); (2) Particular consistency of rhythms is observed in Q3, i.e. in the period of prolongation of the bright phase; (3) The observed changes in the time of occurrence of acrophases and amplitudes of individual rhythms are not so significant as to cause disturbances threatening their complete desynchronisation with a significant decrease of mental efficiency in the studied features and the possibility of adaptation to stay in this zone; (4) The existence of a social synchroniser (established and respected rules of the day) and the absence of a complete polar night seem to be decisive here (Buguet et al., 1987). Therefore, it can be said that during the stay at the Arctowski Polish Antarctic Station no major disturbances of sleep and bedtime occur, if the synchronising social factor (fixed daily schedule) is maintained. However, in a multi-shift work situation, some individuals experience a shift in activity hours to late at night, often leading to disorders in the form of insomnia, which is known from the chronobiological literature, which emphasises unequivocally that one of the most important modifiers of work capacity in multi-shift work is nighttime sleep and daytime sleepiness (Bhattacharyya et al, 2008), as this indicates a shift in the rhythm phase, when the rhythm curve and amplitude do not change but the acrophase (i.e. the time of occurrence of the highest value of the observed trait), shifts from its usual position. When the phase shift reaches a maximum value of 180 degrees (half a period) we speak of rhythm reversal, which according to A. Torbjörna, K. Görana & G. Mats (2007) is characteristic of people relatively adapted to night work. Thus, factors such as night sleep and daytime naps are responsible for intraindividual differences in work ability (cf. Boivin & James, 2002; Zeger, Irizarry & Peng, 2006). These conclusions, in their general form, were also confirmed at the level of physiological indices (temperature of the mouth and skin, heart rate and blood pressure, diuresis and excretion of sodium and potassium in a portioned daily urine collection, etc.), which, in relation to the studies of the Arctowski landers, was presented in detail in the work of W. Kowalski (1982). The use of results from Antarctic databases on the circadian variability of human activity under altered photo-ecological conditions as a criterion for the reliability of research conducted during a long stay on the ISS in space was reflected in a study conducted by NASA researchers M. Basner, et al. (2013) in the MARS-520d laboratory ground experiment, simulating a future mission to Mars. Sleep and wakefulness dynamics were observed in a six-person multinational crew during 520 days of isolation from terrestrial photo-ecological conditions. Measurements included light exposure (4.396 million min) and weekly computerised neurobehavioural assessments (n= 888) to identify changes in crew activity levels, quantity and quality of sleep, during the 17-month experiment. Among other things, it was found that with reduced light exposure, most crew members experienced impaired sleep quality and changes in the amplitude and phase of their circadian rhythm. Thus, the results of both natural and laboratory studies indicate the need to maintain terrestrial primary photo-ecological synchronisers (e.g. light exposure) and civilisational synchronisers (e.g. time of food intake and work performance) to maintain optimal activity levels during long-duration space missions.

Far-reaching caution in interpreting the presented research results is primarily related to certain methodological difficulties arising in the case of psychological research conducted in natural conditions. This is pointed out among others by K.E. Klein, H.M. Wehmann & B.I. Hunt (1972), who point to factors that modify the course of circadian rhythms. These are the following factors: the present psychophysical state of the organism, experience, motivation, sleep deprivation, etc., mainly change the magnitude of the amplitude of the rhythms and thus the magnitude of the performance level. Other variables, such as temperamental characteristics, day-night cycle changes, primarily cause phase shifts in rhythm. Other factors modifying photoperiodicity, such as e.g. seasonality (summer vs. winter), geographical location of the Antarctic station and therapy, are pointed out by J. Arendt (2012). The photoperiodic disturbances of various physiological processes described in the literature as: SAD – true winter depression, decreases in work productivity, trouble sleeping and sleep quality, according to the author, can be modified with hormone therapy (e.g. the hormone melatonin or its metabolite 6-sulfatoxymelatonin) or phototherapy (adjustment of artificial lighting e.g. 2 hours of bright white light, 1 hour in the morning and evening to restore summer time). Among other things, the experimental data shows that the severity of maladaptive symptoms varies according to the latitude of the station's location. Thus, for example, at 75°S, Antarctic station staff adapt the diurnal system to night work during the week, in contrast to temperate zones where full adaptation rarely occurs. Temporary light treatment helps to restore the normal phase of the sleep-wake rhythm in a small number of people, suggesting that suboptimal lighting conditions are detrimental to health and require further research from a space analogue perspective.

Therefore, one has to take into account the influence of the so-called biological or psychological noise on the results obtained, as well as certain individual differences which were not taken into account by us due to the small size of the study group. However, numerous studies confirm that such structural modification does not alter the sinusoidal pattern of circadian biorhythm (Natani & Shurley, 1975; Borland et al., 1986; Vgontaz, Pejovic & Karataraki, 2007). Figure 63 schematically illustrates the above results from the study.

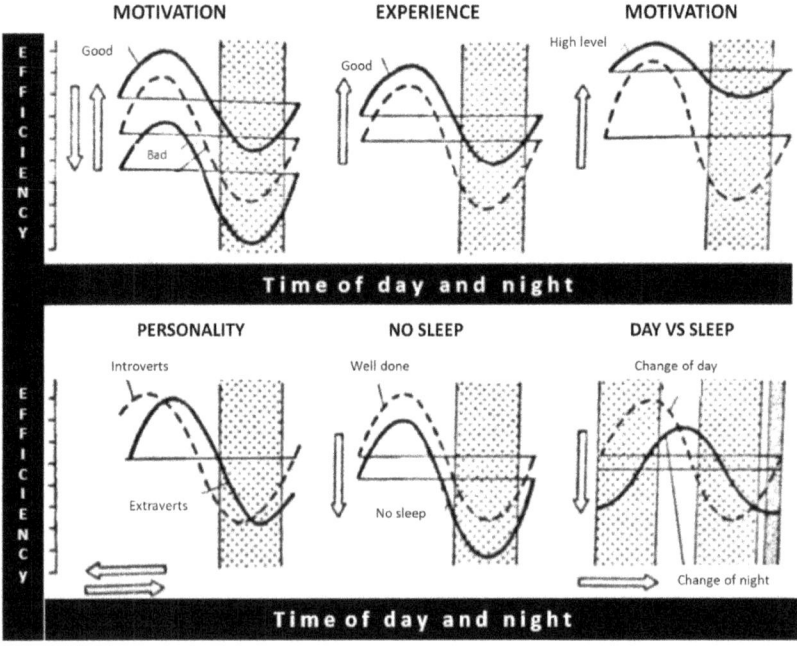

Fig. 63. Some variables modifying the characteristics of circadian rhythms (cognitive and motor skills)
Source: own elaboration of Klein et al., 1972.

As can be seen from the analysis of Fig. 64, none of these modifiers affect the sinusoidal course of the circadian curve but modify only the amplitude height and its spread (acrophase and nadir), which demonstrates the high plasticity of the biological circadian rhythm system (Scheer, et al., 2007). One of the most important factors modifying the ability to work multiple shifts is nocturnal sleep and daytime sleepiness (Santy et al., 1988; Torbjörn, Göran, Mats, 2007). One of the most difficult symptoms of the impact of sleep disorders on performance is the subjective element of sleepiness assessment. This was confirmed, inter alia, by the experimental studies of M.O. Griof et al., (2011) conducted while staying for 37 days in the terrestrial Mars habitat Flashline Mars Arctic Research Station (FMARS) on Devon Island, Canada. The study involved a crew of seven (four men, three women) who were monitored by ECG for signs of sleep disturbance and subjective fatigue and completed mental tests to identify the impact of sleep disturbance on their cognitive skills. Subjective data showed a change in sleep quality, while objective data (ECG) showed no significant change in sleep patterns. There was also no cognitive decline during the mission.

All the factors discussed so far are responsible for intra-individual differences (e.g. leisure, motivation) as well as inter-individual differences (work experience, temperament, health, etc.) in work ability. One of the most important factors responsible for interindividual differences in the masking of circadian rhythms recorded at the level of work ability is personality, which was pointed out as one of the first researchers by W.P. Colquhoun (1971), who found a high correlation between diurnal temperature patterns and personality traits such as extraversion and introversion, which was the basis for dividing people into "morning" and "evening" types. The "morning" people reach a peak in body temperature in the afternoon (12.00 a.m. – 2.00 p.m.) and a minimum just after midnight. These people achieve high performance in the morning and a marked reduction in the evening. The "evening" people are characterised by a later peak in body temperature (4.00 p.m. – 6.00 p.m.) and a higher efficiency falling in the evening hours. Similar results were obtained by M.J. Blake (1967), who found, among other things, that introverts reach higher values of body temperature more quickly in the morning and peak earlier compared to the course of temperatures in extroverts. He further observed that the overall increase in activation levels in both types occurs during daylight hours and that activation levels increase more during the day in extroverts than in introverts.[69]

The aspects (stimulatory, functional, social and chronobiological) of the Antarctic isolation situation discussed so far provide arguments for treating it as an extreme situation; as a highly complex difficult situation that may be an analogue

---

69 Numerous self-assessment questionnaires are based on the above assumptions, designed to isolate the behavioural syndromes characteristic of the "morning" and "evening" chronotype. The first questionnaire of preferred times of activity and rest in English was published by J.A. Horne & O. Östberg (1976).

of long-term space flight and existence on other extraterrestrial planets (cf. Harrison, Clearwater & McKay, 1989; Lugg, 1991).

5.5. A medical perspective on the cost of adaptation to Antarctic isolation

Medical examinations were conducted using a structured observation sheet[70] filled in by the station doctor every fortnight during the one-year isolation. The assessed adaptation cost categories and the results are provided in Fig. 64.

Figure 65 shows, among other things, that adaptation to the situation of Antarctic isolation is a process characterised by specific dynamics. Two periods of adaptation can be distinguished. The first stage involves a two to three months-long stay at the station, during which people become accustomed to their new environment. This is followed by a certain stabilisation of adaptation to the physical, psychological and social conditions of the summer period. Then again, a certain regression of adaptation can be observed, related to becoming accustomed to Antarctic winter conditions. In late June and early July, on the other hand, we see a re-stabilisation of adaptation to the Antarctic winter. The last two months of Antarctic isolation are again characterised by a decrease in adaptation levels, which may be related to the fatigue of the isolation situation itself, people fatigue and impaired emotional control. This line of interpretation of the results is supported by data from other Antarctic psychologists who observed a decline in satisfaction with performed tasks and a general decline in group morale during recent periods of isolation (Gunderson & Nelson, 1963; Crocą, Rivolier & Cazes, 1973; Nicolas, et al., 2015), which is consistent with own research presented above on the dynamics of affective, social, cognitive and group adaptive processes to an ICE (Isolated, Confined, Extreme) environment, captured by the classic "winter-over syndrome" term, which reflects the typical pattern of stress perception and coping in long-term Antarctic isolation.

There is an interesting phenomenon in the area of alcohol consumption. Well, in the most stressful situations associated with the Antarctic winter period, an increase in alcohol consumption is noted. This periodically raises the level of adaptation and, as the experience of many doctors and psychologists has shown,

---

70 This technique is an adaptation of the Biweekly Checklist – L. Crocą, J. Rivolier & G. Cazes (1973), which the authors used at the French "Durvill" station. At the Arctowski station, the assessment was based on group averages of biweekly surveys compared to annual averages (group average of 26 measurements over the year).

A medical perspective on the cost of adaptation to Antarctic 307

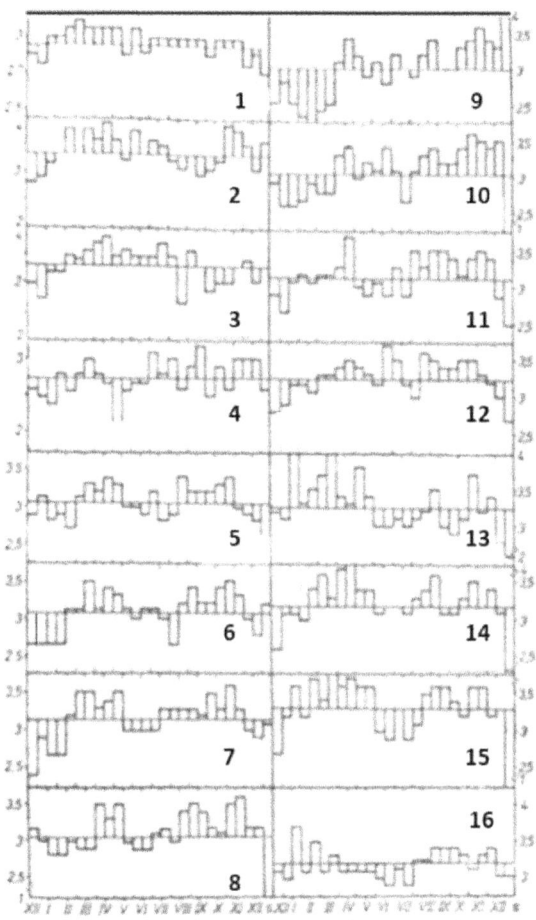

Fig. 64. Dynamics of wintering adaptation to the situation of annual Antarctic isolation as assessed by the expedition physician (N = 20)
Source: own elaboration of Croca, J. Rivolier and G. Cazes, 1973.
Legend: Scale – the higher the index on the 5-point scale, the better the adaptation (conversely, in the case of alcohol consumption); continuous horizontal line – annual average; 1 – mood; 2 – sleep; 3 – appetite; 4 – alcohol consumption; 5 – social talks; 6 – general activity; 7 – individual classes; 8 – group activities; 9 – free time; 10 – household work; 11 – kindness; 12 – sociability 13 – attitude to the station manager; 14 – roles in the group; 15 – attitude to the group; 16 – attitude to the doctor.

incidentally helps to resolve adaptation difficulties, if only by organising a social life (Gunderson, 1968, Hachisuka, 1971). This is due, among other things, to the very nature of alcohol, which in small doses is a stimulant, while in excess it is a depressant (Terelak, 2019). For this reason, alcohol was under the control of the doctor and the station manager.

An interesting opinion on the usefulness of medical-psychological research, based on the evaluation of the effects of the work of a doctor and psychologist of the 3rd Antarctic Expedition of the Polish Academy of Sciences to the Henryk Arctowski was presented by the outstanding Polish Antarctic polar explorer Stanisław Rakusa-Suszczewski (1980, p. 139), who stated, inter alia, that apart from scientific research conducted by a doctor in the field of acclimatization to cold and disturbances in biological rhythms in the circadian cycle and logitudinal studies conducted by psychologist on the impact of social isolation on basic psychophysical functions, more than 80 medical and 20 psychological consultations were provided, and 4 appendicitis surgeries were performed on employees of deep-sea fishing vessels, which proves that Antarctic stations can be treated as natural scientific laboratories, and the research conducted in them they allow to extrapolate the results in terms of the selection of cosmonauts for long-term space flights and use them as analogues of future Martian habitats.

C.C. Flynn, C.F. (2005) states that although NASA participation has considerable experience in dealing with space crew medical problems, the prospect of long-term space missions (LDM) is a difficult challenge of pre-flight, in-flight and post-flight medical support. The behavioural group at NASA-Johnson Space Centre (JSC-BHPG) focused on key factors that can improve astronaut health and performance, and these are: sleep quality; minimisation of circadian sleep-work cycle disruption; preventive health care; psychological adaptive factors; and the human-system interface (the relations between the astronaut and the mission role). In orbit, the crew medical officer becomes a valuable extension of the doctor's work on Earth. However, this requires training of the crew within the scope of these factors, and the crew medical officer needs tools to help predict, prevent, monitor health problems, for which intelligent autonomous devices will be needed, in order to implement the planned autonomous interplanetary missions. Some of these problems are being tested on Antarctic stations. Thus, for example, J. Wood, et al. (2005) describe the cooperation already begun in the late 1980s in this area between NASA and the Australian National Antarctic Research Expeditions (ANARE). The authors presented a system for collecting and analysing data on the health of Australian winter-over participants from the Antarctic station from 1994–2003, and their syntheses conducted from the perspective of astronautics by the Johnson Space Centre and their conclusions on health adaptation to extreme environments as space mission analogues. A similar study of the health of Antarctic polar explorers from the perspective of the space habitat analogue was presented by D.J. Lugg (2005), who believes that despite the absence of such an important space flight feature as microgravity in Antarctica, the changing photoperiodicity and social isolation may provide important data for health and therapeutic procedures for space crews. The importance of the analysis of quasi-experimental data collected over the last half-century by NASA-Johnson Space Centre, Houston, TX for the creation of new mission profiles with very long time scales (months, years) for the ISS and planned exploration missions to the Moon and Mars, is highlighted by B.S. Caldwell (2005). The conclusions of these analyses concern the increased public and technical awareness of the effectiveness of space crew operations as well as support from the mission control. However, M. Shepanek (2005) believes that the use of the classical clinical model of health diagnosis is not sufficient to address behavioural health problems in planned long-term space missions.[71] In turn, D. Gössmann-Lang et al. (2019) evaluated pharmacological support for alleviating extreme isolation stress from the perspective of intersex differences in winter-over participants (10 women and 16 men) participating in three annual expeditions to the German Neumayer III Antarctic Research Station. The study was physiological in nature and involved the analysis of blood, saliva and urine samples taken one to two months before the expedition and then monthly during the expedition and three to four months after returning from Antarctica. Based on the analysis of the so-called stress hormones: cortisol, catecholamine and endocannabinoids, among others, it was found that social isolation and the extreme environment of Antarctica induce pronounced stress documented at the immunological level, with the exception of cortisol and blood count, which show no significant gender-specific differences. These results from the perspective of long-term space missions require further detailed research.

---

71 An example may be the four appendicitis surgeries described in my diary of the 3rd Antarctic Expedition of the Polish Academy of Sciences by a team in which obo chirurha, an untrained psychologist, played the role of an anesthesiologist or instrumentalist (Terelak, 2021, pp. 52–53) and the amazing story of a Soviet surgeon Leonid Rogozov, who, as a participant of the 6th Soviet Antarctic Expedition to Novolazarevskaya Station, in less than two hours, appendectomy of his own appendix in a recumbent position and under local anesthesia with the help of unqualified crew members: a meteorologist (holding retractors) and a driver (holding mirrors) and other scientists (handing over surgical tools) during which he lost consciousness several times.

Another factor, conditioning social adaptation in a situation of isolation, is pointed out, among others, by R.E. Doli and E.K.E. Gunderson (1971), who believe that group size should be considered alongside the small task force attributes discussed earlier. In their study, they found, among other things, that the number of polar winter-over participants at the station (8–10 vs. 20–30) significantly affected social perception and group coexistence. E.g. irritability in social interactions is more common in small stations (around 10 people) than in larger stations (around 30 people). The authors explain this with less living space and more limited social interaction. In contrast, there is no difference according to the above-mentioned authors in the rates of aggression, which probably depends on the composition of the group. We decided to follow this trail further and, in our own research, assess the psychological cost of individuals wintering-over at a small, medium or large research station. This is an interesting strand of research looking for analogues between the conditions of polar stations and space vehicles, which are characterised by low crew numbers. "Small" and "medium" stations are thought to be particularly stressful due to social isolation, characterised by limited interpersonal attraction (Blackburn, Shurley and Natani, 1973; Gunderson and Mahan, 1966; Terelak, 2021).

Conclusion and Findings

T.M. Mills (1967) lists four cognitive reasons for studying small task groups: (1) *Pragmatic* – resulting from the fact that every person in everyday life belongs to a small group (e.g. family, professional, social, etc.).(2) *Psychological* – related to learning the regularities of group influence on an individual and vice versa; (3) *Sociological* – a small group is a good model for testing theoretical concepts explaining intra- and intergroup relations; (4) *Ambivalent* – a small group is a micro-system, being a reflection of large communities, in terms of such features as: division of labour, ethical codes, management, social prestige, etc. Mills' postulates can be successfully applied to Antarctic research on small group functioning in situations of prolonged isolation, the application of which goes far beyond the immediate interests of polar psychology. The second and third reasons for the presented research reflect their own motivation to continue from the perspective of their application to Mars flight, especially since the specificity of Antarctic conditions (e.g. compared to the Arctic) as more extreme, can be considered as an adequate analogue of the Martian habitat.

Summing up the presentation of the results of our own research, it should be stipulated that they do not exhaust all the problems related to human functioning in a situation of social isolation, with particular emphasis on Antarctic isolation. As one of the few longitudinal studies conducted with a high frequency (even at biweekly intervals), it has the potential to complement many empirical data, allowing for more adequate conclusions based on hard empirical data and not only on extrapolation of subjective observations. The reference to the classic Antarctic research of the 1970s and 1980s is a value in itself from the perspective of treating Antarctic research as an analogue of interplanetary missions (Palinkas, et al., 2000), as research in the last two decades is drastically changing the pattern of social isolation due to satellite connectivity abolishing the information barriers of polar explorers practically to zero. Also, the density of Antarctic station locations in the Antarctic Peninsula region, creates a large, colony of stations facilitating personal contacts between "neighbours," which is used for cross-cultural research that complements work on the analogue of orbital, but not necessarily interplanetary flight.

The analysis of the results adopted by me was dictated not so much by the desire to capture all the possible formal relationships, but mainly those that, at this stage of the research, can be explained by both the theoretical and situational context. Hence, the diary of the Antarctic winter-over participant that was elaborated (cf. Terelak, 2021 – second edition), containing narratives that I frequently referred to, as it is an integral part of the monograph presented. Adopting, following T. Tomaszewski (1978), a model of situation analysis that takes into account its multifaceted nature (stimulus, functional, social and chronobiological aspects) (cf. Fig. 1), in addition to its function of ordering the results according to a specific criterion, is to draw attention to the fact that the situation of Antarctic isolation should be considered as a complex extreme situation. This broad understanding of the situation, which is derived from Tomaszewski's (1998) environmental psychology, provides a common basis for unifying terminology and does not preclude the application of a range of temperamental, personality, social, chronobiological concepts to explain many of the details, nor does it preclude the design of new studies, especially experimental research, on human functioning in ICE situations.

From a slightly different perspective, one can find attempts to analyse existing studies of people functioning for long periods of time in small task groups from an ICE perspective, such as with authors like N. Smith et al., (2019) who did a matanalysis of the literature on explaining the role of personality in explaining individual differences in adaptation to Antarctic isolation. Among other things, they found that out of 284 scientific articles, only 23 made a significant contribution to science. These differences in personality traits sometimes have a negative impact on both individual and group functioning of winter-over participants. The authors suggested an attempt concerning the theoretical integration of personality concepts within a more holistic gestalt metamodel, which may have greater explanatory power for individual specific ICE adaptation behaviours but is not necessarily suitable for explaining the non-specific mechanisms of these behaviours.

There are also attempts to construct psychometric tools for monitoring the winte-over syndrome, which have been undertaken in recent years by, among others, M. Nicolas, et al. (2019). The authors developed and validated a short (19 items) Isolated and Confined Environments Questionnaire (ICE-Q) on several groups of French Antarctic winter breeders (n = 140; 25 women and 115 men; means 34 years old) psychological adaptation in four key area (Confirmatory Factor Analysis): (a) social (e.g. relationships, social support), (b) emotional (e.g. emotional changes, boredom, depression), (c) professional (e.g. commitment to work and spending free time)) and (d)) physical (e.g. fatigue, well-being. Although the psychometric value of the questionnaire is satisfactory, it requires further validation due to the fact that the polar winters and the composition of the respondents are an unrepresentative sample of all ICE environments (e.g. ground or space habitats).

The new model of polar and space psychology and the construction of new psychometric tools to monitor the states of adaptation to ICE are called for by, among others, K.Norris, K., D. Paton, & J. Ayton (2010) among others. A similar view is taken by M. Zimmer et al., (2013), who also conducted a systematic literature review by mapping studies on the psychological mechanisms responsible for adaptation to prolonged residence in extreme polar conditions. Literature data were assessed on two thematic axes:

(1) *Negative* psychophysiological and behavioural *effects* due to long-term exposure to ICE. It was found that 79.5% of the scientific articles analysed, especially those from the first few decades, dealt with the symptoms of winter-over syndrome, such as: cognitive disorders (63.6%), depression and lowered mood (56.8%), anxiety (47.7%) and irritability (45.4%), sleep disorders (40.9%) (including sleep problems and insomnia). Social isolation and imprisonment as a source of discomfort for winter-over participants, were cited in 84.0% of the publications identified as seasonal patterns of health and psychological cost symptoms under: Winter-over syndrome (Palinkas, 1991), Polar T3 syndrome (Reed et al., 2001) and subsyndromal affective disorders (Palinkas, Houseal & Rosenthal, 1996). Despite research into symptoms and their relationship to changes in the adaptation to environmental conditions, the scientific status of this research is still *in statu nacendi*.

(2) *Positive (salutogenic) effects* associated with effective adaptation were reported in 65.9% of the publications, which related to manifestations such as improved mood (18.1%), effective coping with situational stress (52.2%), cultural and gender heterogeneity (15.9%), leadership management (6.8%), high levels of motivation (22.7%), low levels of neuroticism (9.0%), and low levels of assertiveness (9.0%). Salutogenic effects are made possible by psychological and logistical support, as although humans adapt to the widest variety of environmental conditions, living and working in the Antarctic environment, despite being a one-off episode for some polar explorers, has a prolonged salutogenic value, enhancing the sense of self-efficacy associated with coping with stress, as shown by retrospective research (Moult et al., 2015). Nevertheless, the constantly acquired new knowledge resulting from Antarctic studies of individuals and task forces, especially longitudinal ones (Decamps & Rosnet, 2005), is of applied importance for space exploration, of which the ICE environment is currently only partially recognised (Wood et al., 2000; Lug, 2005; Binsted et al., 2010). Although we already know a little about the attributes of artificial space habitats from recent research

conducted in terrestrial space habitats and especially on the orbital space stations Skyllab, MIR or ISS, such as: narrow spaces, exposure to unforeseen situations, social isolation, limitations in communication with the outside world, heavy workload and limited possibilities to evacuate the environment, life-killing cosmic radiation, etc. (Bishop, 2011), it is still too modest a knowledge to be anticipated literally. However, what we already know on the basis of the analysis of the literature on the subject and our own research seems to support the verified thesis that the process of emotional and social adaptation to the situation of Antarctic isolation is characterised by specific dynamics and that the empirically established "winter-over syndrome" should be considered in the category of adaptive behaviour, involving psychological costs characteristic of different stages of isolation, and not clinical disorders of a psychopathological nature. This has certain theoretical implications as well as practical implications. The theoretical implications relate to the need to re-evaluate concepts such as adaptation vs. maladaptation, which are too narrow in radically changed operating conditions. Thus, for example, the symptoms included in the "winter-over syndrome" can be classified by various authors either as maladaptive effects (e.g. Mullin, 1960) or as adaptive behaviour related to coping with stress, as argued among others by stress researchers such as M. Barbarito, S. Baldanza & A. Peri, A. (2001), who investigated the coping strategies of winter-over participants on nine Antarctic expeditions with the use of COPE. The survey was conducted three times during the year. The results from the first study showed that the most common coping styles used during this period included a focus on problem solving, while in the middle of wintering-over there was an increase in distancing from problems and the social environment, which the authors of the study called the "freezing" symptom, which included symptoms such as lethargy and apathy. It is this strategy of coping with Antarctic stress, that is an important element of the winter-over syndrome, which in prolonged social isolation well reflects the adaptive behaviour to the stressful situation, which some researchers interpret in salutogenic terms (cf. Suedfeld & Mocellin, 1989).

The optimistic conclusion from the decades of research conducted in extreme Antarctic conditions so far can be reduced to the statement that the described significant cognitive and behavioural changes of polar expedition members are not clinical disorders but are situational in nature, with salutogenic effects, which, being associated with a positive sense of self-efficacy, not only influence recovery and psychological comfort but, above all, reliable performance during

ICE. This makes it reasonable to assume that in this respect they are analogues for long-term space missions. The practical implications of the salutogenic approach are related to the fact that considering the process of adaptation on a continuum: active environment – active person, significantly expands the repertoire of human adaptive behaviour, especially to situations other than "normal" ones. This is of great importance, especially in modern times, characterised by the need for many task forces to operate in situations of social isolation (e.g. radar station operators, submarine sailors, scientific missions, etc.). Moreover, the dream of mankind connected with the "colonisation" of near and distant space depends, among other things, also on the knowledge of the "possibilities" of human adaptation to extreme situations, considered not only on the ontogenetic level, but also on the phylogenetic level (cf. Kwarecki & Terelak, 1981).

Own research cited in the empirical part is an exemplification of a broader problem related to human functioning in social isolation in general. However, research conducted in the natural conditions of Antarctic isolation, involving either an authentic or anticipated threat, provides a basis for treating the Antarctic isolation situation as an extreme one. The relatively high frequency of psychological research and its wide range allow, on the one hand, to characterise adaptive behaviour to difficult situations as fully as possible and, on the other hand, to set further directions for research. As far as the general conclusions of the presented research are concerned, they can be formulated as follows:

(1) Annual Antarctic isolation is a complex situation that is difficult from both a stimulus, functional, social and chronobiological perspective;
(2) Based on the dynamics of adaptive behaviour, two periods can be distinguished: summer and winter, which differ in their stressful nature;
(3) Analysis of individual adaptation suggests that it is associated with specific psychological costs and that it is characterised by quite significant intra- and interindividual differences;
(4) The dynamics of social adaptation of a small task group is characterised by typical stages of polarisation and integration, which, when the task nature of the group is lost, results in its complete disintegration;
(5) Due to the rather large professional and cultural diversity of the study group and the burdening of the study results with the variable of social approval, extrapolation of conclusions with respect to other task forces in social isolation (e.g. due to the Covid-19 pandemic in 2020) is advisable only in terms of general conclusions.

Evaluating previous research directions on the basis of the literature (cf. Suedfeld, 1991), as well as in the presented own research, too few empirical studies of the dynamics of the informal structure of a small task group depending, for example, on the leadership style, the working style of individual group members, are conspicuous. There is also no convincing evidence of a periodic decline in mental performance during prolonged social isolation. Various fields of psychology, especially applied psychology, are interested in explaining the above phenomena, and among them mainly occupational (industrial and management), military and especially astronautical psychology. However, the study of the above-mentioned phenomena in a situation of natural isolation encounters certain difficulties related to the social risk posed by psychological research, especially those using nominative techniques based on mutual evaluations of individual group members. Minimising the impact of assessing the variable of social approval is difficult from a methodological point of view, as almost all personality tests are based on self-assessment. It also raises hitherto unresolved ethical and professional issues related, for example, to the question to what an extent the psychologist as an expert, who is perceived as a threat to the private sphere of the examinee, has the right to impede the already difficult social situation.[1]

The knowledge gathered so far from psychological Antarctic research seems to support the validity of the view of the veteran in this field – J.T. Shurley (1973) – that a small Antarctic station could be considered a "natural laboratory" in which all human sciences could conduct research on adaptation to the most extreme Earth and space habitat (lunar, Martian) situations. Astronautics has been using the results of this type of research for years, especially in terms of learning about the psychological mechanisms of individual and group functioning in situations of space isolation (cf. Bechtel, & Berning, 1991) from the perspective of preparing astronauts for long-term interplanetary flights and stays on non-terrestrial planets. An analysis of the literature for the past 30 years indicates a shift in research paradigms in the selection of candidates for Antarctic expeditions. This applies, first of all, to the paradigm of individual differences in personality (Gavalas, 2011), e.g. in H. Eysenck's view or the so-called Big Five (Barrick & Mount, 1991), which at the level of psychometric tools, show little predictive validity in extreme situations (Rosnet, Scanff, Sagal, 2000). This is confirmed by our own research on extroverts' prediction of adaptation to Antarctic isolation, which shows that their process of adaptation to the extreme situation involves the activation of the introversion mechanism (cf. Fig.27) at

---

1 The following question is still open: Does mere consent given in person before the test reduce the subject's discomfort?

the end of social isolation (the so-called third quarter syndrome) (Bechtel & Berning, 1991).

However, some of the research findings on group cohesiveness dynamics are somewhat controversial, asthere is little empirical data conducted from the perspective of long-term social isolation. This is pointed out among others by J.E. Mathieu, et al. (2015), who carried out a meta-analysis of the results of 17 studies on this subject, using longitudinal data regarding 205 members, comprising 57 student teams, who competed in a complex business simulation for 10 weeks. The thesis that consistency and performance are positively correlated was confirmed, however the relationship of consistency vs. performance was significantly higher than the inverse relationship. Additionally, it was noted that performance vs. consistency increased over the duration of the experiment in contrast to the performance vs. consistency relationship, which was relatively constant. Mediating variables, among which group leadership is prominent, may be responsible for this outcome, as partially confirmed by experiments known under the codename Biosphere-2 described by M. Nelson et al. (2015). The authors, assuming that terrestrial CELSS (Controlled Environmental Life Support System) habitats with human participants could be adequate analogues of space habitats, conducted a two-year study to identify stressors affecting the functioning of a small task group under conditions of long-term social isolation. It was found that the dynamics among the participants in the Biosphere-2 experiment were similar to those evident in other isolated teams, manifested in the formation of factions reflecting personal relationships and conflicts regarding applicable mission procedures. Conflicts were mainly associated with power struggles, although in general the crew was able to overcome them and carry out both operational and scientific tasks related to sustaining life in an autonomous habitat.

In turn, yet another mediating variable, namely the adopted value system, is noted by A. Vinokhodova & V. Gushin (2021), who conducted a study concerning self-development and changes in the value system of crew members participating in two simulated space flights of various durations: SIRIUS-17 (17 days) and SIRIUS-19 (120 days). Among other things, it was found that an increase in self-development was associated with social behaviour regardless of the duration of the experiment, gender or nation, and was primarily concerned with attributes such as universalism, benevolence, self-control, and attitudes towards tradition, which the authors believe has application value for the selection of future Mars mission crews. A similar pilot study was previously conducted by P. Suedfeld (2006), who undertook an analysis of the autobiographies of two high-level NASA administrators, three female astronauts and seven male astronauts, in terms of four value system indicators: achievement, pleasure, kindness and transcendence. There was no significant change across the study groups in the first three indicators of the value hierarchy in all subjects, however, among the astronauts of both sexes, transcendence, defined as the combination of spirituality and universality, clearly increased in the hierarchy after space flight.

Another mediating variable between group cohesion and space crew performance was pointed out by G.M. Sandal, H.H. Bye and F. J. R. van de Vijver (2011), who conducted interesting research, stemming from a social psychology paradigm, on the possibility of the emergence of a "Groupthink" syndrome during a mission to Mars that correlates with high levels of autonomy and cohesion. They assume that long-term social isolation and lack of contact with the outside world can be a mediator. These issues were investigated in an experiment lasting 105 days with an isolated multinational crew (N=6), simulating a mission to Mars. The "Values Portrait Questionnaire" was used, and interviews were conducted at the beginning and end of the experiment. Multiple regression analysis showed no evidence of crew "groupthink" at the level of homogeneity of value attitudes, with a parallel counter-trend, as evidenced, among other things, by the fact that 35 days before the end of the mission significant variation in the three sub-groups' value hierarchies of hedonism, kindness and attachment to traditional values appeared to occur. The results suggest that personal value preferences should be taken into account in the selection of crews for long-duration missions.[2]

Due to the internationalisation of current space research, new theoretical (Pierce, 1991) and methodological (Miller, 1991) challenges arise in relation to the study of multicultural space crews, faced by the world-dominant Anglo-Saxon psychology, which in the cultural part is derived from a monotheistic Christian culture. Taking into account other cultures that influence the formation of personality, especially in terms of value systems, is not able to overcome the syncretic[3] approach, despite attempts related to cross-cultural research (Sarris,

---

2   A helpful tool used in the Biosphere-2 was the paper by W.R. Bion (2013) on identifying competing behavioural modalities in small task groups: (a) task groups guided by conscious acceptance of goals, thinking about reality in relation to time and resources, and managing challenges intelligently; (b) groups unaware of the goals pursued and subordinated entirely to the leader ("group animal").

3   An example of the difficulties involved in breaking down the barrier of syncretic thinking can be found in the experiences of Hans Küng, (1990), who for many years carried out the Weltethos Project, the aim of which had been to find at least some universal values in many leading contemporary cultures, around which a new personality model of value systems could be built. Unfortunately, he did not succeed, although he spoke to all the political and religious leaders of the modern world.

(2007). Contemporary psychology is not yet a global psychology but still a local one, despite the fact that, together with business companies, it tries to master the world and the cosmos effortlessly with comfort understood from the perspective of the vision of positive psychology (Suedfeld, 2001). S.E. Taylor & al. (2000) calls such an approach a "positive illusion" which nevertheless increases the psychological and health resources needed to cope with stress (e.g. illness).[4] The need to re-evaluate contemporary Anglo-Saxon psychology from the perspective of post-modern globalisation of technical civilisation, science and business has been pointed out for twenty years by A. Bandura[5] (2001) and also researchers of resources of coping with extreme stress, which are not unlimited (Palinkas & Browner, 1995). An example of this is the list of sources of stress for Antarctic winter-over participants, recorded at stations by many researchers, which can be described in the shortest terms as follows:

(1) Isolation and/or entrapment (separation from the outside world and the awareness that there is no possibility of external support in a threatening situation) (Weiss & Moser, 1998; Matter, 1991);
(2) Enclosure (habitat with limited privacy and territoriality) (Westin, 1967);
(3) mental fatigue/exhaustion (fatigue associated with prolonged social isolation) (Wilson, 2011);
(4) Low temperatures (very low temperature and its health effects, e.g. hypothermia, frostbite, etc.) (Kowalski, 2002);
(5) Sleep deprivation and or disruption (Pattyn et al., 2017), associated with changes in photo-ecological conditions, among others (Bhattacharyya et al. 2008) and circadian rhythms) (Chen et al., 2016);

---

4 From an empirical study conducted by Shelley E. Taylor et al. (2000), on the implications of cognitive adaptation theory and positive illusions and beliefs related to disease treatment in HIV-infected men, shows, among other things, that even unrealistic but optimistic beliefs about the future protect mental health and the ability to find meaning in the fight against the disease by increasing one's own coping resources.
5 A. Bandura calls for creating a broad vision of a human being and avoiding a reductive approach in psychological research, which is descriptive and normative in nature, and which focuses excessively on "local" cultural problems, an example of which is e.g. personality psychology, the so-called Anglo-Saxon one, which is entangled in European "culturalism" as a relic of a neo-colonial civilisational mission, increasingly incompatible with the vision of the global world. The commencing inter-cultural research, thanks to online tools, are a hope to change many psychological paradigms, more useful for space missions.

(6) Situational hazards (related to the natural environment, e.g. snow blizzards, crevasses, etc., or logistical, e.g. power plant failure, fires, etc.);
(7) Sensory deprivation (known as "sensory starvation," referring to both the lack of stimuli from the Antarctic natural environment (lack of animals) and the social environment (e.g. lack of civilisation and social sources) (Suedfeld, 1979);
(8) Monotony/boredom (repeating the same activities, seeing the same faces) (Sutherland, 2007);
(9) Separation from friends/family (emotional deprivation from friends/family (Taylor & McCormick, 1987) and lack of social support (Uchino, B. N. (2006);
(10) Social tensions (deprivation of many needs gives rise to frustration, which in turn generates aggressive behaviour and conflict).

At the turn of the twentieth and twenty-first centuries, the interests of Antarctic and space researchers, thanks to the development of satellite communications, focused on the problems of communication between isolated habitats and the outside world, in terms of reducing isolation effects and external support in crisis situations. An example is the study by V. Gushin et al. (1997), who, in evaluating an analysis of spacecraft crew members' communication with each other and with the terrestrial mission command centre, drew attention to a sociological phenomenon known as "psychological closure and information filtering," involving a decline in the frequency of communication with the outside world and its subject-matter quality, while maintaining "communicative autonomy." This autonomy manifests itself in certain specific standards, consisting in keeping communication with the command centre to a minimum and increasing egocentrism within the crew. Subsequent studies have confirmed a previously identified sociological phenomenon, which is the adaptation mechanism of a small isolated task group (Kozerenko, Gushin, Sled, 1999).

Both the overview section of the paper and the presentation of the author's own research should be of interest especially to polar and space psychologists who are associated, on a daily basis, with the study of human functioning under conditions of long-term social isolation of small task groups, such as polar explorers and astronauts (Barabasz, 1991). This goal, in the first order, should be served by the theoretical considerations and practical implications contained in the presented work, as well as by an awareness of the new challenges of redefining humans from the perspective of a "space civilisation," since the predictive power of Antarctic research as an analogue of a spacecraft, is considerably limited (Palinkas et al., 2000). The question therefore remains open: Is it possible to predict a universal vision for psychology in the near future? I doubt it, because new civilisations arise in long periods of time,[6] although their heralds include the already functioning universal value of money and a new belief in the unlimited possibilities of new technologies, or the humanist liberalism penetrating people's consciousness, referring to the inalienable right to freedom and the dignity of the human person.

Previous studies of space crews show, among other things, that long-term social isolation adversely affects mood. Iteresting research on the dynamics of five categories of humor according to the Humor Coping Scale (HCS) 46 astronauts staying on the ISS in an international group of active and retired astronauts and cosmonauts[7] presented were presented by J. Brcic, et al. (2018). The following categories of humor were assessed: (1) Affiliative – a meeting for the purpose of interpersonal production; (2) strengthening – expressing a positive attitude to life; (3) aggressive – characterized by a sarcastic attitude to life; (4) self-constructive – expressed by defeatism; (5) rational (problem-oriented) – looking for rational life difficulties. The analysis of the research results looked for demographic responses (professionally active astronauts vs retired people), cultural responses (astronauts vs cosmonauts) and flight phases (beginning, middle and end). Mono other has been found that the use of positive humor more often than negative. In international crews, an aggressive form of humor appeared on the national side of the male, at the expense of affiliate humor. In addition to the fact that the defeatist level of humor, along with the duration of the space flight

---

6 This is emphasised by the psychology of idea, which points to distant sources of knowledge about man in time, which were sometimes used by twentieth century psychology. We can trace this back to the mechanism of frustration that was discovered around 500 BC by Siddhartha Gautama, heir to the throne of a small Himalayan principality, later to be called Buddha, who abandoned the palace and led the life of a homeless ascetic, reflecting on the psychological mechanism of people experiencing states of anxiety, frustration and helplessness for some reason. For six years in seclusion, he meditated on the cause of such mental states and came to the conclusion that the habits of our mind related to the satisfaction of achieving our life goals (e.g. possession, meaning, etc.) are responsible. When a new goal appears, the pleasant experience of achieving the previous goal arouses the desire to pursue the new one and the anxiety connected with the difficulties standing in the way of achieving it (Zieba, 2000). This process is cyclical, and the mechanism of frustration is universal, which was alluded to by the first modern major theory of stress called "Frustration Theory" by J. Dollard et al, (1939).
7 The authors named astronauts part of the NASA crew, and cosmonauts – the members of the Russian crew.

and the affiliate and problem-oriented (coping) humor, are more effective in the final space mission, which may be due to social isolation (Abel, 2002). However, extrapolation of conclusions from previous research carried out in clusions from studies conducted in terrestrial habitats and in the real conditions of spacecraft and stations orbiting the Earth is still incomplete. However, experiences from Antarctic longitudinal studies (e.g. Décamps, Rosnet, 2005) as well as long-term flights on orbital stations (e.g. MIR, ISS) have provided evidence that the longer a space mission lasts, the more important knowledge of the effects of sensory deprivation and social isolation is for understanding the neurometabolic mechanisms that control behaviour in this extreme monotony of life (Douglas, 1991).

From the review of research presented in this paper, we now know for sure that psychosocial adaptation is determined on the one hand by the individual's level of emotional adaptation (Zhao, Xu & Liu, 2001; Yang, Ye, & Tang, 2011), and on the other hand by social adaptation and the degree of task group polarisation, which can be both a source of satisfaction, related to a sense of self-worth within the overall goals of the station, and a cause of interpersonal stress (Zimmer et al, 2013), and sometimes psychiatric disorders, which can also occur in the conditions of space isolation (Kanas & Manzey, 2008). This refers to the new trend of interpreting the previous adaptation symptoms to social isolation, which make up the "winter-over syndrome," by applying a new salutogenetic perspective[8] (Suedfeld, 2005) or eustress (Terelak, 2019), pointing to treating the space mission as a confrontation with the resilience resources of modern man and pointing not only to the psychological and health cost of adaptation in terms of negative symptoms (stress psychology) but also symptoms of anticipated joy

---

8 This is a consequence of the introduction, by L.A. Palinkas (2003), of a new classification of adaptation to Antarctic conditions, which distinguishes four criteria for assessing factors mediating the adaptation processes of Antarctic winter-over participants: (1) Seasonality – changes in mood resulting from annual and daily cycles of the environment and the body's own neurohormonal cycles, as well as changes in the psychological perception of mission phases; (2) Situationality – the current situation at the research station, both environmental and social, with the course of which personality traits, interpersonal needs and coping styles interact; (3) Social context – the cohesion of the group and the way it is managed, as well as the quantity and quality of contacts with the outside world; if an expedition has low internal coherence, its members show more symptoms of depression, anxiety, anger than members of highly coherent expeditions; (4) Salutogenesis – positive experiences of individuals seeking challenges and resulting from contact with the natural environment.

and pride of survival ("Homo invictus" positive psychology)[9] (Suedfeld, 1998, 2001; Ihle, Ritsher & Kanas, 2006; Leon et al., 2011).

In recent years, NASA, through international collaboration, more than in the past has come to appreciate the importance of research into psychosocial and cultural factors for the success and safety of space missions, especially as modern human spaceflight has changed significantly in terms of crew composition, duration and task difficulty (Kring, 2001; Summerhayes, 2008). Although the research literature on space psychology has grown considerably over the past several decades, there are still many open questions that require new research, taking advantage of the International Space Station (ISS), commissioned in 2000, and new projects for the exploration of human habitats on other planets in the Solar System (e.g. the Moon and Mars) that are already underway thanks to rapidly developing space technology. However, there still remains, due to the long journeys towards Mars, the problem of long-term isolation, which should continue to be investigated in the Antarctic analogue environment due to, among other things: the possibility of survival due to the constant dependence on technical systems, the degree of isolation and confinement, and the lack of rapid rescue in case of emergency (Kanas, Sandal, Boyd & Wang, 2009). An example of such a modern space analogue could be the Concordia station, where the European Space Agency (ESA) has conducted research in areas such as telemedicine, the psychophysiological cost of wintering-over and multicultural task force dynamics in social isolation. A similar direction is taken by NASA projects which, in addition to the problems of astronaut selection and training, undertake a search for predictors (including genetic ones) of the subsequent development of psychopathology (e.g. certain types of depression and anxiety disorders, and schizophrenia) and the development and implementation of psychological, pharmaceutical countermeasures to solve these potential problems that arise during a long-term space mission. One example of predicting, preventing and treating behavioural health is the development of interactive computer technology as

---

9   Examples are the established societies and clubs of Antarctic polar explorers (cf. in Poland, the Antarctic Brotherhood, with around 1000 members) and astronautics/space explorers (as of 17 November 2016 it included 552 people from 36 countries), which cultivate happy memories and generate positive myths of courage and heroism (Suedfeld, 1998). However, less frequently studied are disrupted family relationships (work-family interactions) due to the year-long absence of the polar explorer, applications of which are important for long-duration interplanetary flights (Johnson et al., 2012) and psychiatric disorders at various time slots after return from an Antarctic mission (Palinkas et al., 2004).

a tool for psychotherapy and treatment of depression. Another example is the conflict resolution programme led by Dr James Carter of Harvard – Beth Israel Deaconess Medical Center. (Ritsher, 2005). Moreover, despite preliminary research related to cultural, gender and racial aspects of crew composition useful for long-term space missions, there is still very little empirical evidence on the ideal composition of crew progress in this field[10] (Harrison, 2005).

Concluding the discussion so far, we are convinced that the literature data presented (Suedfeld & Weiss, 2000; Suedfeld & Weiss, 2000) as well as our own results obtained under natural conditions of Antarctic isolation, support the accepted thesis that they have application value for space exploration[11] both on the International Space Station (ISS) and again on the Moon, Near Earth Objects (NEOs) and Mars. However, this requires further intensive research, especially longitudinal studies that describe the entire dynamics of the winter-over syndrome more accurately than cross-sectional studies (Palinkas et al., 1998), which, for example, when confronted with the Covid-19-related pandemic, have shown that we still know little about the effects of social isolation on specific values in many cultural regions of our globe. An example of such research is the 180-day experiment conducted by among others M. Qianying et al., (2019) on the psychological effects of long-term residence in an isolated environment with a small group on the dynamics of changes in personal values. This is interesting in that the subjects of the study were members of the Chinese Martian habitat crew (n = 4), thus representing a different cultural region from the European and American ones. Respondents completed the Portrait Values Questionnaire (PVQ) once a month. The results showed that group members during isolation can adapt personal values to the demands of a small task group (cf. also: Blackadder-Weinstein et al., 2019). In the same Chinese experiment, W. Ruilin, (2018) examined the importance of leadership roles in Mars mission success and harmonious interpersonal relationships. The study involved four crew members in 180 days of social isolation. It was found that crew members perceived the leadership role more as a supportive one than controlling one, which had a positive effect on the group climate and reduced aggression. The authors see applied significance in

---

10 There is still too little research on the usefulness of partnerships or marriages in space crew from the perspective of populating other planets or returning after decades or even hundreds of years to Earth from open comet space (Leon, & Sandal, 2003; Leon, 2005).
11 G.M. Sandal's (2000) question seems legitimate: "Coping in Antarctica: is it possible to generalise results across settings?".

selecting not only the members of future space missions, but also in establishing the responsibilities of leaders when working with a group characterised by a high degree of autonomy, which astronauts undoubtedly are.

The evaluation of key experiments as more or less adequate lunar and Martian analogues was undertaken by three authors from NASA: Lauren, B. Landon, Kelley J. Slack & Jamie D. Barrett (2018). They reviewed the litereature on the subject for the past 20 years, focusing primarily on NASA's research activities on group dynamics and the resulting practical recommendations for team members selection and screening,[12] training in effective team performance with the use of countermeasures to reduce risk,[13] and the application of new techniques for on-line measuring and monitoring of the team's performance.[14] They consider the following to be the most valuable experiments, the results and recommendations of which could be useful for a planned interplanetary mission to Mars: HERA (Human Exploration Research Analog); HI-SEAS (Hawai'i Space Exploration Analogue & Simulation); IBMP chamber (Russian Institute for Biomedical Problems); NEEMO (NASA Extreme Environment Mission Operations); Submarine; Antarctica (small station vs. large station). The simulation of analogue spaceflight environments in the experiments listed above included such characterisation attributes as: mission duration; degree of real threat; habitat size; crew composition; crew selection methods; crew multiculturalism; degree of ground mission control; crew size; and the round-trip communication delay with the command centre. In addition, the experiments provided detailed knowledge of such issues important for planned Lunar and Martian missions as: distance from Earth; time of return to Earth (hours, days, months, years); mission duration;

---

12 The applied standard methods of psychological selection and screening of astronauts on the ISS and in ground analogues developed at NASA are presented by, among others, T.J., Williams et al. (2017).
13 Once qualified, astronauts undergo an intensive two-year training programme led by experienced NASA Selection Experts, who take into account not only their high mental level but also their experience and social and leadership skills. Training procedures are discussed in detail by, among others, C.S. Burke et al. (2017).
14 Astronauts, as one of the most frequently monitored and measured occupational groups on Earth, require the optimisation of monitoring techniques and data analysis methodologies from the perspective of crew support during long-duration space exploration missions. These problems are addressed in the development of a methodology to analyse not only lexical but also speech and text analysis of crew diaries from the perspective of emotional stress, social behaviour at both individual and team levels (cf. Driskell et al. 2017).

crew autonomy; view of Earth. Based on the above data, the authors define research directions for Mars analogues as the problem of effective teamwork skills when isolated from real-time support from Mission Control (MC) from Earth due to communication delayed for objective reasons. Consequently, more scientifically rigorous studies of autonomous teams capable of maintaining teamwork skills under mission conditions lasting two to three years with limited MC support are required. The second major research problem continues to be the analysis of factors that influence the physiological and psychological functioning of individuals and teams.

Similar conclusions based on the analysis of the literature on the subject were also reached by S.T. Bell et al. (2015), which concludes, among other things, that unlike space missions to date, a mission to Mars requires the extension of Long-Distance Space Exploration (LDSE) issues in terrestrial habitats (e.g. Antarctic expeditions, submarines, etc.) and space habitats (the ISS, Moon). The authors, in the course of performing a meta-analysis of existing literature data, pessimistically conclude that the data collected to date is largely incomparable mainly for methodological reasons (short periods of isolation, cultural variables, gender) as well as for ethical reasons. The lack of data aggregation and coordination of research projects, which still bear reserved national, and not global identifiers, is not insignificant. Other authors emphasise that the essence of LDSEM analogue studies lies in the small sample sizes of astronauts, which requires a change in the existing statistical paradigm (e.g. Brannick, 2001; LePine et al. 2008; Schmidt and Hunter, 2015). Despite those reservations, the final conclusion of all LDSEM study authors is quite optimistic, since the knowledge gathered so far, especially on the dynamics of group behaviour, is already useful for formulating rules for astronauts selection and screening and provides guidance for the development of space psychology (Brown, & Mitchell, 2019).

The latter problem was developed by D.W. Musson & R.L. Helmreich (2005) based on a review of research on the personality of astronauts, pilots, and Antarctic expedition members for a 15-year period conducted at the University of Texas (UT) as part of the Factors Research Project. Particularly controversial from the perspective of the long-term spaceflight analogue in terms of crew coping with critical situations is the problem of long-term forecasting, conditioned not only by methodological constraints but also by ethical ones arising from the confidentiality of personal data and access to its collection. To ensure the continued development of such long-term databases, a fundamental change in the way psychological research results are managed globally would be required. For the above-mentioned reasons, recommendations for personality profiling of astronauts as an analogue of interplanetary missions are severely limited. The need for this type of methodological approach is mentioned also by E.R. Fiedler & A.A. Harrison (2010), suggesting optimising the selection procedures, training and psychological support given to Mars mission crews, based on measures such as: background checks, psychiatric interviews, psychological testing and behavioural observation, supported by new surveillance technologies. Interesting is the stipulation, that the focus should not be on the diagnosis of the astronauts themselves, but on the psychological support given to families before, during and after the mission. The use of new digital technologies in this respect offers online psychological assistance, with confidentiality and access at any time and without communication delays, especially for the flight to Mars, but also for functioning on a biologically and civilisationally alien planet.

The latter problem of delayed communication between the crew of a space habitat (e.g. the ISS) and a ground-based command centre has been addressed, among others, by N.M. Kintz et al. (2016) in an experiment using the ISS. The authors evaluated the performance of three out of 18 astronauts during a mission lasting 166 days in situations with and without communication delays (50 s one way). They found that, among other things, crew well-being and communication quality were significantly reduced when they performed tasks involving communication delays compared to the control situation, which had an impact on the efficiency of the tasks performed as well as on team management. They see remedies in new technological solutions and training in simulated traffic delay conditions. S.P. Chappell et al., (2016) present more detailed data on this subject on the example of the NEEMO 18–21 experiments (NASA Extreme Environment Mission Operations). In this project, crew members lived in the underwater "Aquarius" habitat for up to two weeks, conducting daily simulated activities outside the EVA (extravehicular activity) submersible capsule on the Aquarius coral reef and communicating via one-way light time (OWLT) communications with the MCC (Mission Control Centre), delayed five or 15 minutes with the science team of the experiment. One should agree with the authors' view that the goal of further research conducted on the timeline of information transfer, should be to minimise the delay of messages from Earth to Mars, which is also confirmed by K.H. Beaton et al., (2019), who also points out, using the BASALT (Biologic Analogue Science Associated with Lava Terrains) project as an example, that in addition to new technologies to minimise communication delays between EVA (extravehicular activity) on Mars and the MCC (Mission Control Centre) crew, communication throughput is important[15] (cf. also: Shepard et al., 1996; Lim et al., 2011).

A fundamental existential question arises at the end of this book: Is the Earth an adequate and at the same time universal model for understanding the possibility of *homo sapiens'* habitation, maintaining life and procreation on other planets? The existing knowledge concerning this problem, although dynamically developing, is still not sufficient to explain this issue within a single scientific field. We have known this well since the Renaissance, when the Polish astronomer Nicolaus Copernicus (1473–1543) published his work entitled *De revolutionibus orbium coelestium,* in which he presented in detail the heliocentric model of the universe. At present, in addition to astronomy, knowledge from various scientific disciplines is needed to flesh out a new vision of this model: biology, chemistry, geology, planetology, physics, but also medicine, psychology and philosophy, all of which are useful in identifying a habitable planetary body. This knowledge should primarily concern ecological factors such as the composition of the atmosphere, solar radiation, magnetism, tectonics, mineral composition, and the availability of liquid water in relation to human habitation. Earth will always serve as a reference system. A. Lorek and A. Koncz (2013) suggest that the initial step of this search is to simulate various potential threats to life in space in terrestrial and space habitats and to collect data from the exploration of technically accessible planets for detecting traces of life in discovered habitable niches. Using Mars as a target planet for exploration in the twenty-first century as an example, the authors describe how the Mars Simulation Facility (MSF) at the German Aerospace Center (DLR) simulates Martian environmental conditions: pressure (1–1,060 hPa), temperature (-70 to +130 °C), humidity, light (150 W xenon lamp emitting in the range 250–2190 nm with calibrated intensity) and gas composition (up to five components). The research is intended to support the development and design of experimental equipment (e.g. sensors) under Martian conditions. This is an example of one of the hundreds of scientific centres that will continue to undertake basic and applied research on all possible factors

---

15 The use of existing analogue research for application in future extraterrestrial operations on the impact of communication delay between the spacecraft crew and the ground command centre is discussed by K.H. Beaton et al. (2019) on the example of the BASALT (Biologic Analogue Science Associated with Lava Terrains) project, which aimed to explore geological data collected by an electric vehicle on the surface of Mars and to develop new technologies to minimise communication delays (of five and 15 minutes) between EVA (extravehicular activity) on Mars and the MCC (Mission Control Center) crew as well as its bandwidth.

of extraterrestrial environments that make habitation difficult, in addition to the medical and psychological knowledge on human adaptation to environments of sensory deprivation and social isolation that is presented in great detail in this monograph. It is a priority task for all human and environmental sciences to create a new model of a "space man" (Heppener, 2020). However, this will not be possible, despite the new technological revolutions, without the contribution of philosophy, which must re-evaluate the many myths that have grown up around "Homo Sapiens," who has made a great step in evolution, a transition "from the world of animals to the world of gods" (Harari, 2019). However, the history of humanity teaches us that what is subject to natural selection is not able to free itself from its biologically determined limits, although the beginning of the twenty-first century raises hopes that these natural earthly limits may be transcended by means of the so-called "intelligent design,"[16] which has undertaken attempts to replace natural selection of the world with an intelligent design carried out in many of the world's biological laboratories in a number of ways, two of which come to the fore: bioengineering and the construction of cyborgs (i.e. organisms composed of organic and non-organic elements, such as in the case of bionic human organs supporting human functioning). There is another group of intelligent designs, namely those based admittedly on a biological model but involving the creation of entirely inorganic beings managed by computer programs subject to spontaneous evolution (Harari, 2019, pp. 485–508). It cannot be excluded that in the near future humanoids, resistant to social isolation and extreme conditions of the physical environment, will be a partner of homo sapiens in the exploration of distant regions of the cosmos. Let us hope, however, that the nineteenth century "Frankenstein prophecy" of a scientist creating a monster out of control does not come true. There is also a fear that future technologies will subordinate humans to humanoids who, in the struggle for survival on a planet ecologically degraded by man, may eliminate *Homo sapiens* by way of further evolution, just as in the past Man eliminated *Homo neanderthalensis* from the face of the Earth in a struggle for territory. This is a catastrophic engineering and technological vision, but fortunately the difference between the biological brain and its digital analogue is definitely remote, which prompts other, more optimistic visions of its future: biological philosophical, ethical, psychological.

---

16 In 2000, in defiance of natural selection, a Brazilian bio-art proponent designed a new work of art called the "fluorescent rabbit," which was crafted by French scientists by implanting a gene taken from a green fluorescent jellyfish into the DNA taken from a white rabbit embryo, creating a glowing rabbit that was named Alba.

One can only hope that the "Frankenstein prophecy" does not come true and that humanoids will be helpful in the exploration of space, but are the hopes placed in artificial intelligence (IS) based on algorithms derived from "big date" and bioengineering enough for this? The first concerns arising from confusing such basic concepts as intelligence (at the science fiction level) and consciousness are signaled by space psychologists, who point out to a new phenomenon called technostress - defined as the stress that people experience due to the use of IS causes.[17] The literature on the three aspects of technostress concerns: information systems, organizational behavior and organizational stress. The latter aspect of the use of IS in organizational behavior in the event of organizational stress M. Tarafdar, C.L. Cooper, and J.-F. Stich is described as three-part ("technostress trifecta"), including techno-eustress (motivating innovation at work), techno-distress disorganizing work, and the principles of designing information systems for strengthening the positive and mitigating the negative effects of technostress. According to the authors, future research on technostress should be interdisciplinary, enriching, on the one hand, new IS technologies, and, on the other, knowledge about the essence of organizational stress.

---

17 A detailed review of the literature on the subject in the field of the current state of the art of technostress accompanying the use of information and communication technologies (ICT) in society was carried out by, among others, T.C. Carabel, et al. (2018).

Post Scriptum

Summarising the whole book from the perspective of the Martian analogue against the backdrop of technological postmodernity, it should be emphasised that, under the influence of the vigorously developing technical tools used by modern astronauts, their image in terms of their professional, health and psychological qualifications, and consequently the object of interest of space psychology, has radically changed. Machines produced in the postmodern period, as instruments of action, not only facilitate action itself, but gradually take over many mental functions (e.g. sensors, computer aided decision-making, protection against threats from the space environment, etc.) originally reserved in the human-environment (P-E) system exclusively for humans. "Intelligent design" has created new support for space exploration, often excluding and taking over the task for it (development of astronomy, cosmonautics) (Wallace, 2004). At the other end of the exploration of the macrocosm on a scale hitherto unimaginable in the development of technical civilisation, the miniaturisation of technical tools to the size of micro- and nanotechnology has been created, which has found application, among other things, in micro- and nanomedicine. The latter field in particular has applications in space medicine, such as remote penetration of the astronaut's body for the following purposes: (a) surgical (more efficient non-invasive surgery, faster and safer for the patient, using a micro-robot controlled from the ground on-line); (b) pharmacotherapeutic (e.g. dosing of drugs in diabetes by "cyberpastilles" circulating in the body placed in chips); (c) diagnostic (e.g. chips placed in the body diagnosing health risks and sending information outside to the diagnostic centre); (d) direct stimulation of specific brain centres (e.g. not only as so far in the treatment of disease, but also in states of depression or neurosis); (e) rehabilitation in the case of damage to limbs using prostheses connected via an interface directly to motor centres in the brain). Predictions made by biologist Eric Drexler of the Massachusetts Institute of Technology in his book "Engines of Creation" (2007), concerning the development of nanomedicine, equipped with miniature "machines" to repair errors in DNA associated with the ageing process, may be possible in the near future, which will be extremely important after humans have physically left the solar system for decades. Certainly, these and other predictions, although, as in the case of stem cell banking and cloning, they raise moral questions (Kopania, 2002), will be feasible in the twenty-first century, which will certainly no longer be called the era of technological post-modernity but (to put it colloquially) the era of "super-modernity." The dreams of space exploration are as old as humanity itself, because man in his cognitive curiosity does not admit the limits of mental cognition (which is discussed among others in epistemology, cognitive science), especially that in the period of development of bioengineering working on the network model of the real brain and the creation of the possibility of self-learning computer networks (called the "artificial brain"), man gained an "intelligent" partner for self-creation and "changing the face of this Earth," but also a support tool for space exploration not only on the level of mental cognition (probes constructed by man have been exploring not only some planets and planetoids in the Solar System, but also in open space for several decades now). So far, this self-creation has been promoted in the concept of a cyber-man (humanoid), who will be saturated with various types of micro- and nanochips: (1) micro-devices enhancing the senses (e.g. long-distance, twilight and night vision); (2) microsensors diagnosing the functioning of all organs; (3) nano-robots in blood vessels analysing blood composition and pressure; (4) micro-translators facilitating communication in multicultural crews; (5) bio-cybernetic implants activating thought processes and memory; (6) stimulators in the brain sensory centres simulating real feelings of, e.g., pleasure, satiety, various smells, etc.; (7) brain interface with the Internet facilitating on-line access to vast resources of computer databases and the possibility of communication with any person at any time; (8) nanochips (2.1 mm x 12 mm) implanted under the skin replacing the real-life doctor (e.g. blood type, medical history, DNA); (9) a cybercosmonaut equipped with chips sewn into various parts of the "smart suit:" (1) digital goggles equipped with GPS, displaying an image of the surroundings, the distance to a designated target and the position of objects in relation to each other; (2) a display sewn into a "smart sleeve" transmitting the latest news from the base, electronic commands, TV images, etc.; (3) devices incorporated into the legs, to strengthen muscle strength, etc.

Perhaps the picture of the cyberman sketched here looks too fantastic, but after all, the descriptions of the underwater adventures of Captain Nemo and his "Nautillus" vessel that I read in my youth, created by Verne's imagination, also bore traces of fantasy, and yet they became reality within the lifetime of a generation. It is not only a paradox that the development of micro- and nanomechanics and electronics is stimulated by the army and space agencies (e.g. NASA, ESO, etc.). Fortunately, to some extent it also serves the average person to improve quality of life and health (e.g. infusion pumps facilitating treatment of diabetes, regeneration of bone decalcification in menopause, a bionic retina restoring sight to the blind or a "bionic ear" – restoring hearing, etc.). Thus, one may venture to say that in the period of technological postmodernity, the machine has become more human-friendly than at any time in the development of earthly civilisation. Anticipating further development of new technologies in the 21st century in relation to the intellectual development of humans, we enter the area of futurology, which is the source of many fears as to whether the psycho-biological model of humans promoted so far in psychology, transformed into a modern psycho-biological-cybernetic model, called the "singularity" model or artificial intelligence, will not eliminate humans as "intellectual capital" and especially from self-development (Locke 2003). It also raises concerns and new futurological questions: What happens when a network model of a real brain is connected by means of an interface to a computer network that acquires the ability to "self-evolve?" Will the "eternal" desire for telepathy, the existence of which has been tried unsuccessfully experimentally for decades, be realised? What will happen when cyborgs are no longer human or organic entities? Will a different kind of being emerge in inorganic evolution, which will be a new challenge for philosophy, anthropology, sociology, psychology?

Such an understanding of the human being solely from a cognitive and technological perspective, which does not take into account the content aspects of the emotional, "feeling" sphere and value system, which have been the subject of research conducted over several decades in natural Antarctic experiments, has exhausted its qualities of universality of the only cognitive paradigm in psychology. The search for a new, more useful and complex paradigm for 21st century psychology continues (Norris, Paton & Ayton,2010). How long will this take? As argued by P. Zimbardo J. Boyd (2009), this will depend on the "paradox of time," which is the resultant or compromise between six perspectives of time perception: past positive, past negative, present hedonistic, present fatalistic, past and future transcendental. Evaluating polar research from a positive and transcendental future perspective, it can be concluded that the medical and psychological knowledge gathered on human adaptability to conditions of information deprivation and long-term social isolation is useful for the organisers and participants of future interplanetary missions both in terms of prevention and management of crisis situations during their duration.

In conclusion, I agree with the opinion of C. Tortello, et al. (2018) formulated from a review of the literature that Antarctica is one of the most reliable analogues for assessing the effects of isolation, confinement, the light-dark cycle and extreme environmental conditions for Homo sapiens. An exemplification of this view can be found in the research of A. Alling et al. (2002), carried out as part of an isolation experiment codenamed Biosphere-2, which concluded, among other things, the following: (1) Crew members should be involved in the design and construction of their life support systems to gain maximum knowledge of these systems; (2) People living in closed Martian habitats should have knowledge of the process of physiological and psychological adaptation to the artificial environment; (3) The new photo-ecological conditions of the space habitat require coordination of artifical lighting, supporting the work-rest cycle; (4) Crews with high cultural diversity and high task commitment can not only be a source of conflicts, but above all things a guarantee of mission success and personal satisfaction. (5) Reducing information deprivation requires comprehensive modelling of a "cybersphere" based on information systems that monitor lifesupport devices in real time) and a "noosphere" based on artificial intelligence that will function in sustainable balance with humans.

E.A. Vessel and S. Russo (2015), from the perspective of long-term space missions (LDSM) to Mars and asteroids, highlighted the adverse living conditions and mental disorders that increase risk during missions. Among the main factors contributing to this risk, the authors include low levels of sensory stimulation and monotony associated with ICE (isolated, confined and extreme) environments, requiring the development of measures to increase the level of sensory stimulation during LDSM. Previous research conducted in natural environments, as described in the space flight literature from the perspective of LDSM analogues such as Antarctic stations, nuclear submarines and terrestrial Martian habitats, may be helpful in this respect. The synthesis of this data, performed by experts, may be useful, on the one hand, to understand the mechanisms underlying the psychic and neuromuscular processes that link the levels of sensory stimulation to the changes observed at the behavioural level in perception, cognition, emotion, mood, and, on the other hand, to develop ways and means to satisfy the need for stimulation and reduce "sensory hunger" at different stages of a long-term space mission. To summarise the state of research to date on the involvement of the human factor in the preparations for the conquest of the cosmos, one can recall the maxim of Friedrich Nietzsche (2013, p. 10): "What doesn't kill me makes me stronger," speaking of his views on the "life's school of war," which can be paraphrased according to Y. N. Harari (2018, p. 321) that the natural obstacles that Homo sapiens must overcome on this path, require the astronaut to be "increasingly strong and better adapted," i.e. close to the adaptive capabilities of the "Homo deus" superman. Is this still science based on the evolutionary laws of nature or is it already magical thinking?[18] Perhaps there is hope in the twenty-first

---

18 By magical thinking in this book, we understand cognitive processes that play an important role in human life, because they refer to the universal narrative about the world, contained, among others, in legends, religions, art, politics, business, etc., which

century smart technology revolution, known as the "technoreligion," based on computers, genetics and nanotechnology. This way of thinking is supported by, among others, G.R. Osinski, et al. (2006) and P. Suedfeld (2010), who believe that although ground-based analogues only in some respect bring us closer to understanding the geological and biological environment of specific environments of other planets, long-term exploration in ground-based marine and terrestrial analogues provide experience from extreme environments about human relationships, strategies of coping with stress, the effects of long-term social isolation and monotony, deprivation of many basic needs, physical discomfort (e.g. climate characteristics) and psychological discomfort (e.g. lack of intimacy). It is certainly a training of self-knowledge about the possibilities and limitations of adaptation to extreme situations, which may be useful for future exploration of Mars. In addition, the experience of dealing with interpersonal issues in real space flights is also important, as pointed out by L.A. Palinkas (2001) on the example of implementation of the Shuttle-Mir Space Programme (SMSP) and other long-duration orbital and ground-based missions in ICE environments.

A separate future problem is the intensification of research into the impact of disruption of Earth's circadian rhythms and sleep patterns and their effect on the physiological performance and mental agility of space explorers. Support in this regard is drawn attention to, inter alia, by M.M. Mallis & C.W. De Roshia (2005), postulating the minimisation of the discrepancy between the astronomical time of the terrestrial day and the artificial day depending on the orbit in which the spacecraft will be travelling and the observed day on the Moon or Mars. Data to date has shown that an excessive discrepancy in Earth vs. space sleep patterns and the work vs. rest cycle, significantly reduces astronaut productivity and is a risk factor to crew health and safety. The directions of support from chronobiology (physiological time) and chronopsychology (psychological time) since 2003 are in line with NASA's task directive in this area, although such research, e.g. on multi-shift work, especially in the aviation industry, had already been conducted much earlier (Kwarecki, et al., 1982; Wegmann, et al., 1983). An example of contemporary proposals is the research of S.M. Kelly et al. (1990) on the performance of the operations personnel of the Space Transportation System (STS) and Mission Operations Directorate (MOD) related to sleep and circadian disturbances of space shuttle crews. The research was conducted in a joint

---

subjectively explains and justifies order in it. In this sense, magical thinking has always existed in human minds, personifying ontological subject-object relations by giving them an existential sense of an undivided reality.

project between NASA Johnson Space Centre and NASA Ames Research, and concerned the development of preliminary operational procedures and measures governing the adherence to sleep-work cycles in space shuttle traffic controllers (N=17) representing three orbital shifts. Strategies were developed to minimise the adverse effects of fatigue due to sleep loss and circadian disruption induced by shift operations, with a particular focus on night shift operations and the acute phase required to transition from day to night shift[19].

Chronobiological research from the perspective of the space mission analogue has also been conducted at a number of Antarctic stations. An interesting study on the endocrinologic correlates of circadian activity of winter-over participants of the British Antarctic Station was conducted by A. Harris, et al. (2010). The diurnal rhythm of cortisol (the hormone responsible for diurnal activity) in saliva was recorded in 55 healthy subjects (49 men, six women), which was collected on 3 consecutive days (just after waking up, 15 and 45 min after waking up, and at 3 and 10 pm) during three wintering-over seasons: upon arrival at the station, in mid-winter, and in the last week before departure. Subjective feelings of health, levels of depression and, only in midwinter, the proportions of negative and positive emotions were also studied in the same three wintering-over phases. It was found, among other things, that changes in external lighting (darkness in winter, midnight sun during arrival and departure) immediately after arrival at the station influenced high blood cortisol levels, which were positively correlated with work performance ratings. In the middle of winter, around 58% of people experienced depressive states, seeing sleep problems and fatigue as the main cause. Similar data on behavioural disorders conditioned by circadian rhythm changes at the Arctowski Station, which are presented in this book (cf. chap. 5.4), is presented among others by J. Arendt (2012) and A. Usui, et al. (2000), confirming – from a paucity of publications – that reports of sleep problems are the most worrying. They also suggest that by simulating the time of day and night with artificial lighting (phototherapy), the occurrence of the so-called true seasonal winter depression (SAD – Seasonal Affective Disorder) can be reduced.

---

19 Various concepts of mental support emerge in the stressful conditions of prolonged isolation and exhausting work during a space mission to Mars. An example of such a concept is the Neuroarchitecture System of Vinci Power NAP, consisting in the use of elements of Earth architecture in the construction of space habitats, promoting well-being and minimizing depressive states. The authors of these concepts refer to both neuropsychological research.

In conclusion, it is worth noting the taxonomy of negative and positive consequences of Antarctic isolation made by M. Zimmer, et al. (2013), presenting findings from literature data on this topic. Among the negative effects of functioning under the conditions of isolation stress are the seasonal patterns of symptoms that make up the "winter-over syndrome:" reduced cognitive ability, mood swings and deterioration of interpersonal relationships. Positive effects included salutogenic effects (Suedfeld, 2005), resulting from successful adaptation to the extreme conditions of wintering-over. C. Tafforin, A. Vinokhodova, and V. Gushin, V. (2021) also refer to the salutogenic approach in the interpretation of the effects of isolation and imprisonment, who suggest that in connection with the space colonization projects (Moon, Mars) revert to simulating a mission involving the long-term social isolation and spatial confinement of the crew. An example of the implementation of such a postulate can be the SIRIUS-19 experiment, consisting in the imprisonment of a six-person crew for 133 days (three women of Russian nationality, one Russian, and two Americans). The experiment used the ethological method, consisting in qualitative and quantitative analyzes of non-verbal and verbal behavior recorded on video at breakfast time (twice a month) and during a group discussion (once a month). There were, inter alia, differences between the respondents in the profiles of frequency of interaction, emotional expression, motor skills and communication, characterized by an average linear upward trend, but without the phenomenon of the so-called third quarter. The authors interpret the obtained results in terms of a "salutogenic adaptation step", which does not exclude other interpretations, such as a too short time of imprisonment, or awareness of a low risk of danger in an experimental situation compared to the real situation, etc., which from the perspective of the mission to Mars has vital importance.

This data is consistent with our own research when we take up the dilemma of interpretation of the results obtained at Arctowski station: Is the "winter-over syndrome" an adaptation or a disease? We oppose the new myth of humanism imposed by the so-called positive psychology (Lopez & Snyder, 2009), called the "dictatorship of happiness" (well-being) in journalistic terms (Kahneman, Diener, & Schwartz, Eds., 1999), which suggests that negative emotions should be treated as symptoms of a disease, while empirical data collected in Antarctic natural experiments, proves that in stressful situations they are largely of an adaptive nature. An example can be the anxiety which promotes awareness of threat, enabling self-reflection on one's own resources of counter-measures and social support, facilitating coping with stressful situations. In turn, positive emotions of hope have a value of motivation to survive in difficult situations (Małkiewicz, Terelak, 2018). This is confirmed by other researchers, such as J. Wood, et al.

(2000), who analysed the intensity of positive and negative problems reported by 104 Australian winter-over participants at four Antarctic stations during two Australian winters. The expedition reports were subjected to content analysis using TextSmart software from SPSS, Inc. and showed that although the list of negative experiences is long, the frequency of negative events is relatively rare. On the other hand, although the list of positive emotions[20] is short, the frequency with which they are reported is much higher than for most problems. Also, P. Suedfeld (2002) pointed out the positive experiences of Antarctic expeditions, which were related to such existential attributes of the situation as e.g.: mysteriousness, efficiency, novelty, excitement with nature, acquaintanceship, improvisations, escape from everyday troubles and formal structures of social groups, camaraderie, independence, fulfilment of goals belonging to an elite group, etc. The author suggests that these attributes can be used in the selection procedures of polar explorers and future astronauts.

The new theoretical view of positive emotions within the framework of positive psychology significantly extends the knowledge of the role of positive emotions in coping with stress, as they expand personal physical, intellectual and social resources, and increase the knowledge of adaptability to extreme situations (cf. Frederickson, 2001: Lazarus, 2006), although on the other hand, it builds new myths based on the alleged almost every human being to live in comfort (whatever that may mean), which runs counter to the notion of extreme situations, typical of outer space.

Exploration of the Martian surface by rovers also convinces us visually, by means of real visualisation, that difficult environmental conditions are to be expected. For this reason, among others, in addition to Antarctic research (glacial deserts), an increasing amount of research is carried out in sand-rock deserts, as adequate analogues of Mars. This is pointed out by B. Rai and J. Kaur (2012) who, based on a human factors study on the Mars Desert Research Station (MDRS) analogue Crew 100B, observed the performance of a multinational crew for 15 days under simulated Mars[21] surface exploration conditions. It was concluded,

---

20 The new theoretical view of positive emotions within the framework of positive psychology significantly extends the knowledge of the role of positive emotions in coping with stress, as they expand personal physical, intellectual and social resources, and increase the knowledge of adaptability to extreme situations (cf. Frederickson, 2001: Lazarus, 2006).

21 In 2017, in the MDRS Mars base analogue located in the San Rafael desert in Utah (USA), the first Polish crew consisting of Krzysztof Jędrzejak, Karolina Zawiejska,

among other things, that previous research on this topic must continue to be conducted from a medical and psychological perspective, taking into account the identification of organisational and environmental characteristics that contribute to coping with the discomfort of being and working in a multinational and heterogeneous Martian crew. This requires continuous monitoring of the crew's health and their consent to loss of intimacy in order to develop optimal standards of support from MCC. The authors dream of creating new global standards for mankind, with specific benefits for all Earth inhabitants. This journalistic phrase can be omitted, as so far no one has succeeded with such an idea, as Hans Küng's Weltethos Project (1990) and the revolt against restrictions on personal freedom during the global COVID-19 Pandemic can attest.

In addition, journalistic raising of hopes for the imminent colonisation of Mars (e.g., science fiction literature, movies, Internet, etc.), without scientific analysis of the biological and psychological barriers that prevent adaptation processes to a space environment extremely different from that of Earth, reduce Homo Sapiens to an abstract level and[22] to a belief in a new technoreligion. It is not without significance, as pointed out by the psychology of idea, that man has always attributed divine attributes to extraterrestrials. However, this does not apply to reaching other planets and staying there briefly in habitats that replicate Earth's living conditions. This idea has guided the continuation of building Martian habitats under various extreme conditions on Earth for more than forty years. Here are some examples: (1) The first ever Martian habitat – BIOS-3 – was built in 1972 in Krasnoyarsk (USSR), using an underground niche with dimensions of 315 m, adapted to accommodate 3 people; (2) The first terrestrial habitat – Martian FMARS (Flashline Mars Arctic Research Station) was built within the frames of a Mars Society project on the Arctic Island of Devon near Canada, in which 6–7 crew members were usually stationed; (3). "Mars One" – an unsuccessful commercial project of a Dutch entrepreneur Bas Lansdorp (2011–2013); (4) Mars-500 (2009–2011) located at the Moscow Biological and Medical Institute, concerning a 500-day isolation of 6 volunteers (Russia, ESA and China);

Natalia Zalewska, Michał Kazaniecki and Jędrzej Górski (Zalewska & Kubiak, 2018), with crew number 176, participated in a 2-week experiment called EXO.17. P. 42–49).

22 Thus, for example, in the 3rd millennium BC, Mars as the Sumerian deity of pestilence and war – Nergal; in the 2nd millennium BC. Mars as the Babylonian god of fire and light; Mars in the Aboriginal culture as the deity Kogolongo, likened to a humanoid Wombu bird, called "the mourner"); the Greeks associated Mars with Ares, the god of war; the Romans made the planet Mars the apotheosis of the god of war; nowadays it is a myth of gender (men from Mars and women from Venus) and technoreligion.

(5) HI-SEAS I-VI (Hawaii Space Exploration Analogue and Simulation) 2013–2020 – a habitat located on the crater of the Mauna Loa volcano on the Big Island of Hawaii, about 8200 feet above the sea level (missions lasting two to four, eight and 12 months); (6) AMARS – Australia Mars Analogue Research Station was established in 1998 by means of an agreement between the U.S. and Australian sections of the Mars Society on Australian soil due to the very hot climate. The research projects conducted within these Martian habitats are of varying scientific value, and some are more akin to survival training than scientific experimentation in that they do not take into account the actual threat to life and the essential attributes of the physical space environment for human biological existence, namely weightlessness or reduced gravity on other planets and repetitive closed systems for the distribution of water, food, terrestrial air composition, and waste. For these reasons, among others, we cannot talk about adaptation to life on Mars without constructing an artificial environment that meets the above-mentioned conditions. Much less can we assume evolutionary processes on other planets, since even on Mars they would have to take several hundred years.

This does not mean that the research conducted so far in terrestrial habitats from the perspective of Martian analogues, and especially the research conducted in Antarctica and on the ISS, has not contributed greatly to a better understanding of the biological and psychological aspects of the adaptation of Homo Sapiens to extreme environmental conditions, as well as the expectation from intelligent technologies to support the designing of the artificial environment necessary for further exploration of Space.

***

In the last word of considerations on the state of Homo Sapiens consciousness on the colonization of space, it is worth emphasizing on the occasion of the 70th anniversary of the first man's flight into space that during this flight Yuri Gagarin was delighted with the view of the Earth and the beauty of the Cosmos, but after returning to Earth he repeatedly said that we are in space only guests. Sixty years have passed since then, and C. Tafforin, S. Michel and G. Galet G. (2019) return to a similar reflection in the proposed evolutionary model of the human species, which is the sum of many scientific disciplines, reflecting the philosophical, naturalistic and technological achievements of Homo Sapiens. This model, called the humanistic ethology, can be illustrated by the Aristotle (epistemological) taxonomy[23] of scientific disciplines describing the multidisciplinary understanding

---

23 The Aristotle taxonomy of scientific disciplines describing the multidisciplinary approach to human ethology concerns such areas as: philosophy – human

of modern man, who, using artificial intelligence, will reach the last stage of integrating the knowledge necessary to implement the next stage of the evolution of the human species towards Homo Deus, characterized by " features of god" enabling the realization of the age-old obsessions of the colonization of space (Tafforin, 1994). In the first place on the way to fulfill these dreams is Mars, the settlement of which will not be so easy, because its surface is cold, deserted, and open to deadly cosmic radiation and solar ultraviolet. If we want to survive there, then only in some underground niches or in specially constructed habitats that artificially recreate the earth's conditions. Artificial intelligence will certainly help us in the construction of space habitats, but for survival in extreme cosmic (Martian) conditions, awareness is responsible as the ability to feel various mental states, such as stress, pain, joy, excitement, love, cognitive curiosity, etc. This perception of different things in the process of solving current problems is the basis for individual differences in terms of aparty's creativity in heuristic thinking, as opposed to artificial intelligence, which solves problems in an algorithmic way (cf. Harari, 2018, p. 101). Although it must be admitted that both artificial intelligence and bioengineering as well as big date can significantly improve the information base in the process of selecting and selecting candidates for future space expeditions, improving the prognostic accuracy of psychological tests. Questions about the time to fulfill these dreams are still open[24]. A small part of the answer, from the perspective of the adaptation mechanisms of the human individual and small task groups to the extreme conditions of space flights and the predicted colonization of other planets, is highlighted in this book, which

---

science – ecology – technology, such as: logic, methodology of sciences, physiology, neurology, anthropology, telemedicine, psychology, sociology, cultural studies, ergonomics, physics, astronomy, computer science, ecology, technology, robotics, artificial intelligence (Tafforin, 2020).

24 One of the important questions facing Homo sapiens is breaking down the barriers of "cultureism:, which, although it grew out of the rejection of the biological concept of racism, remained in the form of "cultural racism", which above all emphasizes the important differences between cultures at the anthropological and sociological levels, psychological and behavioral obstacles in the process of globalization. This undoubtedly poses a threat to the future colonization of planets and the organization of the functioning of small task groups, which requires the intensification of cross-culture research in this regard from the perspective of the settlement of future space habitats, which have been initiated for several decades in Antarctic stations and the ISS. Otherwise, the imagination of Homo sapiens will remain at the level of the films on Star Wars (cf. Harari, 2018, pp. 196–206).

aims to make people aware that Mars as the biblical "promised land" that offers us life in a closed habitat is not identical to the matrix as a product of science fiction, and artificial intelligence and bioengineering will not replace our biological senses, especially the inner "I" responsible for authentic reality24.[25]

Answer to the question "Will we build a Homo sapiens-friendly image of the world beyond the barrier of Martian habitat and when?" remains compelling and open to all human sciences. Beginning with philosophy, which has always been a "friend" of all sciences, I am troubled by an existential question concerning the identity of future cosmic colonizers: Will they be representatives of all mankind, or rather some specific culture that has left a mark on their personalities? The answer to each of these questions will draw a completely different story about the "colonization" of space by Homo sapiens. For example, the aforementioned historian of ideas, YN Harari (2018, pp. 392–393) emphasizes from the introspective point of view that culture was responsible for the development of his personality, which provided "only constructed myths: religious (about gods and heavens), nationalist (about the motherland and its historical mission), romantic (about love and adventures) or capitalist (about economic growth and the fact that I find happiness in buying and consuming). I was sane enough to realize it was probably all fiction. I had no idea how to get to the truth. Is Homo sapiens ready to break away from the terrestrial culture of a regional nature in order to search for the truth about the Universe? If not – then we will transfer the "colonization" of space to the Star Wars level, as a science fiction prophecy, and in the near regions of the planet Earth. Fortunately, further exploration of the open cosmos (outside the solar system) will be possible thanks to mental processes and the rebuilding of the consciousness of the universe from anthropocentric to cosmocentric.

Mars is the first to fulfill these dreams, the colonization of which depends on the construction of a habitat that reflects the earth's conditions "in a nutshell." Questions about the time to

fulfill these dreams are still open. This book draws attention to a small part of the answers from the perspective of the sensual and social man.

---

25 I stand in a position close to the philosophical sensualism of George Berkeley that "It is unjustified to distinguish the mental world from the material world, /.../ Only persons really exist" (qtd. after Flage, Madison, 2011).

References

Aartsen, M., & Jylha, M. (2011). "Onset of loneliness in older adults: results of a 28 year prospective study." *European Journal of Ageing*, 8, 31–38.

Abel, M.H. (2002). "Humor, stress, and coping strategies." *Humor*, 15, 365–381.

Adams, O.S., Chiies, W.D. (1963). "Prolonged human performance as a function oj the work-rest cycle," *Aerospace Medicine*, 34(2), 132–138.

Agadżanian, N.A., Bizin. J.P., Doronin, G.P., Ilyin, E.A., Kuźniecow, A.G., Ezepczuk, N.I. (1965). "O wlijanii na organizm czełowieka dlitielnogo probywanija w zamknutoj kamierie małego objema (About the impact on the human body of being locked for a long time in a camera with a small space)," *Problemy Kosmiczeskoj Biołogii*, 4, 31–43.

Alfano, C.A., Bower, J.L., Cowie, J., Lau, S., Simpson, R.J. (2018). "Long-duration space exploration and emotional health: Recommendations for conceptualizing and evaluating risk." *Acta Astronautica*, 142(2), 289–299. https://doi.org/10.1016/j.actaastro.2017.11.009

Alling, A., Nelson, M., Silverstone, S. & Van Thillo, M. (2002). "Human factor observations of the Biosphere 2, 1991–1993, closed life support human experiment and its application to a long-term manned mission to Mars." *Life Support Biosph Sci.*, 8(2), 71–82.

Almond, Jr., H.H. (1991). Antarctica and Outer Space: Emerging Perspectives and Perceptions (pp. 373–382), In: A.A. Harrison, Y.A. Clearwater & Ch.P. McKay (Eds.): *From Antarctica to Outer Space: Life in Isolation and Confinement*. New York: Springer Verlag Inc.

Altman, L, Haythorn, W.H. (1965). "Interpersonal exchange in isolation." *Sociometry*, 28, 411–426.

Allport, G. W. (1937). *Personality: A psychological interpretation*. New York: Holt.

Altman, I., Smith, R.W., Meyers, R.L., McKenna, F.S., Bryson, S.(1960). Psychological and social adjustment in a simulated shelter. A Research Report, Pittsburg–Pennsylvania, American Institute for Research.

Altman, I., & Taylor, D. A. (1973). *Social penetration: The development of interpersonal relationships*. New York: Holt, Rinehart & Winston.

Andersen, D.T., McKay, C.P., Wharton, Jr.R.A., Rummel, J.D. (1990). "An Antarctic Research as a Model for Planetary Exploration." *Journal of the British Interplanetary Society*, 43, 499–504.

Angiboust R., Saumande P. (1965). "Speleologie expedition in the vitarelles caverns in August 1964 – biological and psychological results," *Revue de Medicine Aeronautique*, 4, 22–24.

Arendt, J. (2012). "Biological rhythms during residence in polar regions." *Chronobiol Int*, 29(4), 379–394. doi: 10.3109/07420528.2012.668997.

Arnhoff, F.N., Leon, H.V., Brownfield, C.A., (1962). "Sensory deprivation: Its effects on human learning," *Science*, 138, 35–43.

Aronson, E. (2009). *Człowiek – istota społeczna* (Man – a social being). Warszawa: Wydawnictwo Naukowe PWN.

Aschoff, J, (1965). "Circadian rhythms in man." *Science*, 11(148), 1427–1432. doi: 10.1126/science.148.3676.1427.

Aschoff, J. (1979). "Circadian rhythms: influences of internal and external factors on the period measured in constant conditions." *Z Tierpsychol*. 49, 225–249.

Atkinson J. W., Feather N. T. (ed.) (1996). *A theory of achievement motivation*, New York: J. Willey.

Atlas of Antarctica (2020). *Wikimedia Coommonons Atlas for the World*.

Bair, J.T., Gallagher, T.J. (1960). "Volunteering for extra-hazardous duty." *Jour. Appl. Psychol.*, 44, 329–331.

Baisden, D.L., Beven, G.E., Campbell, M.R., Charles, J.B., Dervay, J.P., Foster, E.F., Gray, G.W., Hamilton, D.R., Holland, D.A., Jennings, R.T., Johnston, S.I., Jones, J.A., Kerwin, J.P., Locke, J., Polk, J.D., Scarpa, Ph.J., Sipes, W., Stepanek, J., Webb, J.T. (2008). "Human Health and Performance for Long-Duration Spaceflight." *Aviation, Space, and Environmental Medicine*, 79(6), 629–635. DOI: https://doi.org/10.3357/ASEM.2314.2008.

Bales, R.F. (1951). *Interaction process analysis*, Cambridge, Mass.: Addłson-- Wesley.

Bales, R.F. (1970). *Personality and Interpersonal Behavior*. New York: Holt, Rinehart and Winston.

Bandura, A. (2001). "The Changing Face of Psychology at the Dawning of a Globalization Era." *Canadian Psychology*, 42, 1–24.

Bandura, A. (1965). Vicarious processes: A case of no-trial learning (pp. 1–55). In: L. Berkowitz (Ed.), *Advances in experimental social psychology*. Vol. 2. New York: Aademic Press.

Bańka, A. (2002). *Społeczna psychologia środowiskowa* (Social environmental psychology). Warszawa: Wydawnictwo Naukowe Scholar.

Bańka, A. (2018). *Psychologia środowiskowa jakości życia i innowacji społecznych* (Environmental psychology of quality of life and social innovation). Poznań-Katowice: Stowarzyszenie Psychologia i Architektura.

Barabasz, A.F., Gregson, R.A, M. & Mullin, Ch.M. (1984). "Questionable Chronometry: Does Antarctic Isolation Produce Cognitive Slowing?" *New Zealand Journal of Psychology*, 13, 71–73.

Barabasz, A.F. (1991). A Review of Antarctic Behavioral Research (pp. 21–30). In: A.A. Harrison, Y.A. Clearwater & Ch.P. McKay (Eds.): *From Antarctica to Outer Space: Life in Isolation and Confinement*. New York: Springer-Verlag Inc.

Baranov, V., Demin, E., Gushin, V., Belakovsky, M., Sekigushi, C., Inoue, N., Mizuno, K., Vachon, M., Tomi, L. (2001) SFINCSS-99 (simulation of flight on international crew on Space Station): experience and lessons learned. In: Baranov VM (ed.) *Simulation of extended isolation: advances and problems*. Moscow: Firm SLOVO, pp. 524–530.

Barbarito, M., Baldanza, S. & Peri, A. (2001). "Evolution of the coping strategies in an isolated group in an Antarctic base." *Polar Record*, 37(201), 111–120.

Bargagli, R. (2008). "Environmental contamination in Antarctic ecosystems." *Science of the Total Environment*, 400(1), 212–226. https://doi.org/10.1016/j.scitotenv.2008.06.062

Barnett, J.S. & Kring, J.P. (2003). Human performance in extreme environments: A preliminary taxonomy of shared factors. In: *Proceedings of the Human Factors and Ergonomics Society 47th Annual Meeting–2003961*.

Barrick, M. R., & Mount, M. K. (1991). "The big five personality dimensions and job performance: A meta-analysis." *Personnel Psychology*, 44(1), 1–26.

Barkaszi, I., Takács, E., Czigler, I. & Balázs, L. (2016). "Extreme Environment Effects on Cognitive Functions: A Longitudinal Study in High Altitude in Antarctica." *Front Hum Neurosci.*, 10, 331–338.

Basar, E. (2012). "A review of alpha activity in integrative brain function: Fundamental physiology, sensory coding, cognition and pathology." *International Journal of Psychophysiology*, 86(1), 1–24. doi: 10.1016/J.Ijpsycho.2012.07.002

Basner, M., Dinges, D.F., Mollicone, D., Ecker, A., Jones, Ch.W., Hyder, E.C., Di Antonio, A., Savelev, I., Kan, K., Goel, N., Morukov, B.V., Sutton, J.P. (2013). Mars 520-d mission simulation reveals protracted crew hypokinesis and alterations of sleep duration and timing. In: *Proceedings of the National Academy of Sciences of the United States of America(PNASP)*. February 12, 2013. 110 (7) 2635–2640; https://doi.org/10.1073/pnas.1212646110.

Bażela, N., Graczak, P., Sławęcki, P. (2021). Mirosław Hermaszewski – object of pride, object of controversy. The ambiguous history of first Polish cosmonaut under the wings of the Soviet Union. AIAA SciTech Forum, 11–15 & 19–21 January 2021, VIRTUAL EVENT. Crossref DOI link: https://doi.org/10.2514/6.2021-2019.

Beach, H.D., Lucas, R.A. (Eds,) (1960). *Individual and group behavior in a coal mine disaster*, Washington, D.C., Disaster. Study No. 13 National Academy of Sciences.

Bhattacharyya, M., Pal, M. S., Sharma, Y. K., & Majumdar, D. (2008). "Changes in sleep patterns during prolonged stays in Antarctica." *International Journal of Biometeorology*, 52(8), 869–879.

Beaton, K.H., Chappell, S.P., Abercromby, A.F.J., Miller, M.J., Kobs Nawotniak, S.E., Brady, A.L., Stevens, A.H., Payler, S.J., Hughes, S.S. & Lim, D.S.S. (2019). "Using Science-Driven Analogue Research to Investigate Extravehicular Activity Science Operations Concepts and Capabilities for Human Planetary Exploration." *Astrobiology*, 19(3), 300–320. DOI: 10.1089/ast.2018.1861

Beaty, D.W., Ammannito, E., Carrier, B.L., Anand, M. (2018). The Potential Science and Engineering Value of Samples Delivered to Earth by Mars Sample Return. Final Report (white paper). August 14, 2018, at: https://www.researchgate.net/publication/327622468

Bechtel, R.B., Berning, A. (1991). The third-quarter phenomenon: Do people experience discomfort after stress has passed? (pp. 261–266). In: A.A. Harrison, Y.A. Clearwater, C.P. McKay (eds), *From Antarctica to outer space: life in isolation and confinement*, New York: Springer-Verlag.

Behrendt, J.C. (1999). *Innocents on the ice: A memoir of Antarctic exploration, 1957*. Niwot, Colo: University Press of Colorado.

Bell, S.T., Brown, S.G., Abben, D.R. & Outland, N.B. (2015). "Team Composition Issues for Future Space Exploration: A Review and Directions for Future Research." *Aerosp Med Hum Perform*. 86(6), 548–56. doi: 10.3357/AMHP.4195.2015.

Bell, S.T., Brown, S.G. & Mitchell, T. (2019). "What We Know About Team Dynamics for Long-Distance Space Missions: A Systematic Review of Analogue Research." *Front Psychol.*, 15, 10: 811. doi: 10.3389/fpsyg.2019.00811.

Beltramino, J.C.M. (1966). "Mortality in Antarctica since and the and of the nineteenth century." *Antarctic Journ.* 11, 268–271.

Bernd, J., Baarsen, B. (2020). Psychological Monitoring (pp. 421–432). In: A. Chouker (ed.), *Stress Challenges and Immunity in Space: From Mechanisms to Monitoring and Preventive Strategies*. New York City: Springer.

Bessone, L., De Waele, J., & Sauro, F. (2018). "The ESA CAVES Astronaut Training Program: Speleology as an Analogue for Space Missions." *Lunar and Planetary. Science Conference*, 49(2083), 49–50.

Bettelheim, B. (1943). "Individual and mass behavior in extreme situations," *Journ. of Abnorm. and Social Psychol*, 38, 453–457.

Bexton, W.H., Heron, W., Scott, T.H. (1954). "Effects of decreased variation in sensory environment," *Canadian Journal of Psychology*, 8(2), 70–76. https://doi.org/10.1037/

Bhattacharyya, M., Pal, M. S., Sharma, Y. K., & Majumdar, D. (2008). "Changes in sleep patterns during prolonged stays in Antarctica." *International Journal of Biometeorology*, 52(8), 869–879.

Bidłow, E. (1970). *Wojennaja inżeniernaja psichologia* (Engineering military psychology). Moskwa: Wojenizdat.

Biela, A. (1981). *Psychologiczne podstawy wnioskowania przez analogię* (Psychological basis of inference by analogy). Warszawa: PWN.

Bingham. C, Arbogast. B, Guillaume. GC, Lee. JK, Halberg. F. (1982). "Inferential statistical methods for estimating and comparing cosinor parameters." *Chronobiologia*, 9, 397–439.

Binsted, K., Kobrick, R. L., Griofa, M. Ó., Bishop, S., & Lapierre, J. (2010). "Human factors research as part of a Mars exploration analogue mission on Devon Island." *Planetary and Space Science*, 58(7), 994–1006.

Bion, W.R. (2013). *Experiences in Groups: and Other Papers*. London: Routledge.

Bishop, S.L. (2011). From Earth Analogs to Space: Getting There from Here (pp. 47–77). Ed. by D. A. Vakoch, *Psychology of Space Exploration: Contemporary Research in Historical Perspective*. Washington, DC. The NASA History Series. NASA SP-2011-4411

Bishop, S.L., Kobrick, R., Battler, M., Binsted, K. (2010). "FMARS 2007: Stress and coping in an arctic Mars simulation." *Acta Astronautica*, 66(9–10), 1353–1367.

Blackadder-Weinstein, J., Leon, G.R., Norris, R.C., Venables, N.C. & Smith, M. (2019). "Individual Attributes, Values, and Goals of an All-Military Women Antarctic Expedition." *Aerosp. Med. Hum. Perform.* 90, 18–25. doi: 10.3357/AMHP.5248.2019.

Blaekburn, A.B., Shurley, J.T., Natani, K. (1973). Psychological adjustment at a small Antarctic station: An MMPI study (p. 369–383). In: O.G. Edholm and E.K.E. Gunderson (Eds): *Polar human biology*, London: William Heinemann Medical Books, Ltd.

Blake, M.J. (1967). "Relationship between circadian rhythm of body temperature and introversion/extraversion." *Nature*, 215, 896–897.

Bogosłowskij; M.M., Kriwo; W.M. (1975). "Psichołogiczeskije issledo-wanija zimowczikow obserwatora "Mirnyj" w 14 SAE (Sowietskaja Antarkticzeskaja Ekspedicia) (Psychological tests of winter workers at the "Mirnyj" station in 14 SAE (Soviet Antarctic Expedition)," *Informacionnyj Biuletyn*, 91, 91–93.

Boivin, D.B, James, F.O. (2002). "Circadian adaptation to night-shift work by judicious light and darkness exposure." *J Biol Rhythms*. 17, 556–567.

Bombard, A. (1953). *The voyage of the Heretique*, New York: Simon and Schuster.

Boriskin, W.M. (1973). *Żyzn czelowieka w Arktikie i Antarktikie* (Human life in the Arctic and Antarctic). Leningrad: Medicina.

Boriskin, W.M., Slewicz, S.B. (1968). "Czelowiek w Antarktikie (Man in Antarctica)." *Priroda*, 12, 28–37.

Borland, R.C., Rogers, A.S., Nicholson, A.N., Pascoe, P.A., Spencer, M.B. (1986). "Performance overnight in shift workers operating a day-night schedule." *Aviation, Space and Environmental Medicine*, 57 (3), 241–249.

Borucki, Z. (1978). Pomiar indywidualnej preferencji potrzeb (pragnień). Raport z badań: (maszynopis). (Measurement of individual preferences of needs (desires), Research report: typescript). Szczecin: WSP.

Borucki, Z. (1986). *Osobowość a przystosowanie zawodowe marynarza* (Personality and professional adaptation of a seafarer). Wrocław: Zakład Narodowy im. Ossolińskich.

Bouchard, S., Martin, T., Perreault, M. (2009). "Transcultural group performance in extreme environment: Issues, concepts and emerging theory." *Acta Astronautica*, 64(11–12), 304–1313. https://doi.org/10.1016/j.actaastro.2009.01.002

Boyd, J. E., Kanas, N., Gushin, V. I., & Saylor, S. (2007). "Cultural differences in patterns of mood states on board the International Space Station." *Acta Astronautica*, 61, 668–671.

Boyd, J.E., Kanas, N.A., Salnitskiy, V.P., Gushin, V.I., Saylor, S.A., Weiss, D.S. & Marmar, C.R. (2009). "Cultural differences in crewmembers and mission control personnel during two space station programs." *Aviat Space Environ Med*. 80(6), 532–540. doi: 10.3357/asem.2430.2009.

Bradbury, J. (2002). "Utter isolation in a cold climate: The Antarctic challenge." *The Lancet*, 359(9312), 1130–1135.

Brady, J.V. & Anderson, M.A. (1991). Small Groups and Confined Microsocieties (pp. 161–176), In: A.A. Harrison, Y.A. Clearwater & Ch.P. McKay (Eds.): *From Antarctica to Outer Space: Life in Isolation and Confinement*. New York: Springer-Verlag Inc.

Brandes, U. (2001). "A faster algorithm for betweenness centrality." *Journal of Mathematical Sociology*, 25(2), 163–177.

Brand-Persson, A. (1960). The shelter program and shelter occupancy experiments in Sweden, (In) G.W. Baker and J.H. Rohrer (Eds): *Symposium on human problem in the utilization of fall-out shelter*, Disaster Study, 12, Nation. Acad. Sci.

Brannick, M. T. (2001). "Implications of empirical Bayes meta-analysis for test validation." *J. Appl. Psychol.* 86, 468–480. doi: 10.1037/0021-9010.86.3.468

Brcic, J., Suedfeld, P., Johnson, Ph., Huynh, T. & Gushine, V. (2018). "Humor as a coping strategy in space flight." *Acta Astronautica*, 152, 175–178. doi.org/10.1016/j.actaastro.2018.07.039.

Brooks, R.E., Natani, K., Shurley, J.T., Pierce, C.M., Joern, A.T. (1973). An Antarctic sleep and dream laboratory (pp. 322–341). In: O.G. Edholm & E.K.E. Gunderson (Eds): *Polar human biology*, London: William Heinemann Medical Books, Ltd.

Brownfield, C.A. (1964). "Deterioration and facilitation hypothesis in sensory-deprivation research," *Psychol. Bull., (Calif.)*, 61, 4, 304–313.

Brzeziński, T. (red.) (2014). *Historia medycyny* (History of Medicine). Warszawa: PZWL Wydawnictwo Lekarskie.

Buckey, J.C. (2006). *Space physiology*. New York: Oxford University Press.

Budd, G.M. (1966). "Skin temperaturę, łhermal comfort, sweating, clothing and aciivity in Antarctic sleding," *Journ. Physiol.*, 186, 201–215.

Buguet, A., Rivolier, J. & Jouvet, M. (1987). "Human sleep patterns in Antarctica." *Sleep*, 10: 374–382.

Burke, C. S., Shuffler, M., Wiese, C., Hernandez, C., & Flynn, M. (2017). Shared leadership in isolated, confined environments (ICE). Presented at the NASA Human Research Program Investigators' Workshop, Galveston, TX.

Burns, N.M., Gifford, E.C. (1961). Environmental reąuirements of sealed cabins for space and orbital flights – A second study, II: Effects on long-term confinement on performance. Philadelphia, Report No. NAMC--ACEL-414, Naval Air Materiał Center, Air Crew Eguipment Lab.

Burns, N.M., Kimura, D. (1963). Isolation and sensory deprivation, (In:) N.M. Burns, E.M. Chambers and E, Hendler (Eds): *Unusual environments and human behavior* (pp. 167–192). London: Collier-Macmillan. Ltd.

Burns, R. & Sullivan, P. (2000). "Perceptions of Danger, Risk Taking, and Outcomes in a Remote." *Environment and Behavior*, 32(1), 32–71. DOI: 10.1177/00139160021972423

Bussmann, U. & Schweighofer, S. (2014). *Group dynamics: the nature of groups as well as dynamics of informal groups and dysfunctions*. Hamburg (Germany): Anchor Academic Publishing.

Byrd, R.E. (1938). *Alone*, New York: G.P. Putman's Sons.

Cable, D.M., DeRue, D.S. (2002). "The convergent and discriminant validity of subjective fit perceptions." *Journal of Applied Psychology*, 87(5): 875–884.

Cacioppo, J. T., & Patrick, W. (2009). *Loneliness: Human nature and the need for social connection*. New York, NY: W. W. Norton.

Caldwell, B.S. (2005). "Multi-team dynamics and distributed expertise in imission operations." *Aviat Space Environ Med.* 76(6 Supplement): B145–53. PMID.

Campbell, B.G. (1995). *Human Ecology: Story of Our Place in Nature from Prehistory to the Present*. Second Edition. Publisher: Aldine Transaction Pub.

Caplan, R. D. (1983). Person-environment fit: Past, present, and future (pp. 35–78). In C. L. Cooper (Ed.), *Stress research*. New York: Wiley.

Carabel, T.C., Martínez, N.O., García, S.A., Suárez, I.F. (2018). "Tecnoestrés en la Sociedad de la Tecnología y la Comunicación: Revisión Bibliográfica a partir de la Web of Science." *Arch Prev Riesgos Labor*, 21(1), 268–275. DOI: 10.12961/aprl.2018.21.01.4

Carrere, S., Evans, G.W. & Stokols, D. (1991). Winter-over stress: physiological and psychological adaptation to an Antarctic isolated and confined environment (pp. 229–237). In: Harrison, A.A., Clearwater, Y.A., and McKay, C.P. (edit.) *From Antarctica to outer space: life in isolation and confinement*. New York: Springer-Verlag.

Cattell, R. B. (1946). *The description and measurement of personality*. New York: Harcourt, Brace, & World. View

Cernan, E. & Davis, D. (2000). *The Last Man on the Moon: Astronaut Eugene Cernan and America's*. New York: St. Martin's Press. ISBN: 9780312263515.

Chai, Sh.Y. & Flaherty, G.T. (2019). "Lagging Behind – The Emerging Influence of Jet Lag Symptoms on Road Safety." *International Journal of Travel Medicine and Global Health*. 7(2), 39–44. doi 10.15171/ijtmgh.2019.09TMGHI.

Chakraborty, A., Dutta, T., Mondal, S., & Nath, A. (2018). "Application of graph theory in socialmedia." *International Journal of Computer Sciences and Engineering*, 6, 722–729.

Chappell, S.P., Graff, T.G., Beaton, K.H., Abercromby, A.J.F., Halcon, C., Miller, M.J., and Gernhardt, M.L. (2016). NEEMO 18-20: analogue testing for mitigation of communication latency during human space exploration. In: *IEEE Aerospace Conference*, IEEE, Big Sky, MT.

Chen, N., Wu, Q., Xiong, Y., Chen, G., Song, D., & Xu, C. (2016). "Circadian rhythm and sleep during prolonged Antarctic residence at Chinese Zhongshan station." *Wilderness & Environmental Medicine*, 27(4), 458–467.

Cheng, S. K. W., Wong, C. W., Tsang, J., & Wong, K. C. (2004). "Psychological distress and negative appraisals in survivors of severe acute respiratory syndrome (SARS)." *Psychological Medicine*, 34, 1187–1195. https://doi.org/10.1017/S0033291704002272.

Choynowski, M., Kosowski, M., Ostrihańska, Z., Wójcik, D. (1972). *Skala agresji A.H. Buss i A. Durkee* (The aggression scale of A.H. Buss and A. Durkee). Warszawa: Pracownia Psychometryczna PAN.

Ciosek, M. (1993). *Izolacja więzienna. Wybrane aspekty izolacji więziennej w percepcji więźniów i personelu* (Prison isolation. Selected aspects of prison isolation as perceived by prisoners and Staff). Gdańsk: Wyd. Uniwersytetu Gdańskiego.

Ciosek, M. (1977). *Percepcja sytuacji i sprawność psychofizyczna marynarzy w warunkach długotrwałego rejsu morskiego* (Perception of the situation and psychophysical efficiency of seafarers in the conditions of a long sea voyage. Unpublished doctoral dissertation). Warszawa: Uniwersytet Warszawski.

Chodoff, P. (1970). "The German concentration camp as a psychological stress," *Arch. Gen. Psychiat.*, 22, 78–87.

Clarc, B., Graybiel, A. (1957). "The break-off phenomenon: A feeling of separation from the earth, experienced by pilots at high attitudes," *Jour. Aviat. Med.*, 28, 121–126.

Claustrat, B. & Leston, J. (2015). "Melatonin: Physiological effects in humans." *Neurochirurgie*, 61(2–3), 77–84. doi: 10.1016/j.neuchi.2015.03.002. Epub 2015 Apr 20.

Cleveland, S.E., Boy d I., Sheer D., Reitman E.E. (1963). "Effects of fallout shelter confinement on family adjustment," *Arch. Gen. Psychiat.*, 8, 54–62.

Clifton, E.H., Mahnken, C.V., Yanderwalker, J.C., Wallei, E. A. (1979). "Tektite-1, man-in-the-sea project: Marine science program," *Science*, 168(3932), 659–663.

Coburn, K.R. (1967). *A report of the physiological, psychological, and bacteriological aspects of 20 days in fuli pressure suits, 20 days at 27.000 feet on 100% oxygen, and 34 days of confinement.* Washington D.C.: NASA CR-708.

Cockell, C.S., Bush, T., Bryce, C., Direito, S., Fox-Powell, M., Harrison, J.P, Lammer, H., Landenmark, H., Martin-Torres, J., Nicholson, N., Noack, L., O'Malley-James, J., Payler, S.J., Rushby, A., Samuels, T., Schwendner, P., Wadsworth, J., and Zorzano, M.P. (2016). "Habitability: A Review." *Astrobiology*, 16(1), 1–29. DOI: 10.1089/ast.2015.1295.

Collier, B.D., Cox, G.W., Johnson, A.W. and Miller, P.C. (1973). *Dynamic Ecology*. New Jersey: Prentice-Hall.

Colquhoun, W.P. (1971). Circadian variations in mental efficiency (pp. 39–107), (In:) W.P. Colquhoun (ed.): *Biological rhytms and human performance*. London–New York: Academic-Press.

Cong, Z. & Silverstein, M. (2015). "End-of-life co-residence of older parents and their sons in rural China." *Canadian Journal on Aging*, 34, 31–341.

Connors, M.M., Harrison, A.A. & Akins, F.R. (1985). *Living aloft: Human requirements for extended spaceflight*. Washingto, D.C.: NASA Scientific and Technical Information Branch.

Convertino, V.A. (2007). Space, health risk of. In: G. Fink (Ed.): *Encyclopedia of stress*. Second Edition (Vol. 3, pp. 548–554). New York: Academic Press.

Cook, F. A. (1909). *Through the first Antarctic night, 1898–1899: A narrative of the voyage of the" Belgica" among newly discovered lands and over an unknown sea about the South pole*. New York: Doubleday, Page & Company.

Corneliussen, J.G, Leon, G.R, Kjærgaard, A., Fink, B.A, & Venables, N.C. (2017). "Individual Traits, Personal Values, and Conflict Resolution in an Isolated, Confined, Extreme Environment." *Aerosp Med Hum Perform*. 1, 88(6), 535-543. doi: 10.3357/AMHP.4785.2017. PMID: 28539141.

Cowan, T.A., Strickland, D.A. (1965). The legal structure of a confined microsociety (A report of the cases of Penthouse II and III). International working Paper No. 34. Berkeley: Space Sciences Lab. Social Scientes Project, University of Berkeley.

Crocą, L., Rivolier, J., Cazes G. (1973). Selection and psychologieal adjustment of individuals living in smali isolated groups in the French Antarctic Statians (pp. 362–368). In: O.G. Edholm and E.K.E. Gunderson (Eds): *Polar human biology*, London: William Heinemann Medical Books, LTD.

Czeisler, CA, et al. (1999). "Stability, precision, and near-24-hour period of the human circadian pacemaker." *Science* 284(5423), 2177–2181.

Décamps, G. & Rosnet, E. (2005). "A longitudinal assessment of psychological adaptation during a winter-overin Antarctica." *Environment and Behavior*, 37(3), 418–435.

Darley, J., Berscheid, E. (1967). "Increased liking as a result cf the anticipation of personal contact." *Human Relations*, 20, 29–40.

Davidová, L. (2017). *Intragroup conflicts of teams in extreme conditions – analysis and identification of their causes*. Charles University. Unpublished bachelor thesis. https://is.cuni.cz/webapps/zzp/detail/178063

de Gerlache, A., Arctowski, H. & Rakusa-Suszczedwski, S. (2016). *Antarktyczna wyprawa statku "Belgica:" Zbiór relacji* (Antarctic expedition of the ship "Belgica:" A collection of reports). Warszawa: Wyd. Aspra.

De Jong Gierveld, J., Van der Pas, S., & Keating, N. (2015). "Loneliness of Older Immigrant Groups in Canada: Effects of Ethnic-Cultural Background." *Journal of Cross-Cultural Gerontology*, 30, 251–268.

Decamps, G., & Rosnet, E. (2005). "A longitudinal assessment of psychological adaptation during a winter-over in Antarctica." *Environment and Behavior*, 37(3), 418–435. doi: 10.1177/0013916504272561.

de Gerlache, A., Arctowski, H. & Rakusa-Suszczewski, S. (2016). *Antarktyczna wyprawa statku "Belgica": Zbiór relacji* (Antarctic expedition of the ship "Belgica": A collection of reports). Warszawa: Wyd. Aspra.

Dekert, T., Gronowski, M.T., Hrżycki, S. red. (2013). Zachodnie reguły monastyczne (Western monastic rules). In: M. Starowiejski, (Ed. Serii): *Źródła monastyczne 50*. Wydanie 2 (Monastic Sources 50. 2nd edition). Kraków-Tyniec: Wydawnictwo Benedyktyńskie.

De la Torre, G.G., Van Baarsen B., Ferlazzo F., Kanas N., Weiss K., Schneider S., Whiteley I. (2012). "Future perspectives on space psychology: recommendations on psychosocial and neurobehavioural aspects of human spaceflight." *Acta Astronautica*, 81: 587–599.

Del Vecchio, R.J., Goldman A., Philips C.J., Seitz C.P. (1970). Life science investigations during the thlrty-Say Gulf Stream Drift Mission of the Grumman PX-15 (Ben Franklin) submernble. Reprints of Scientific Program Annual Scientific Meeting, St. Luis, April 27–30.

Deriap, N.R., Matusow, A.Ł., Tichomirow L.I. (1978). *Adaptacja i zdorowie czełowieka w Antarktydie* (Adaptation and Human Health in Antarctica) (p. 237–248), W: Antarktyka: Osnownyje itogi izuczenia Antarktyki za 20 let, Moskwa: Nauka.

Derlega, W., Chaikin, A.L. (1975). *Sharing Intimacy: What We Reveal to Others and Why*. New Jersey: Prentice-Hall Inc. Englewood ClitTs. Wprowadzić do tekstu.

Derlega, W., Chaikin, A.L. (2010). "Privacy and Self-Disclosure in Social Relationships." *Journal of Social Issues*, 33(3), 102–115. DOI: 10.1111/j.1540-4560.1977.tb01885.x.

Derlega, V., Grzelak, J.Ł. (red.) (1982). *Cooperation and helping behavior. Theory and research*. New York: Academic Press.

de Vera, J.-P. and Seckbach J. eds. (2013). *Habitability of other planets and satellites: Series: Cellular Origin, Life in Extreme Habitats and Astrobiology*, Vol. 28. Springer.

Dijk D-J, Duffy, J.F., Silva, E.J., Shanahan, T.I., Boivin, D.B. & Czeisler, Ch.A. (2012). "Amplitude reduction and phase shifts of melatonin, cortisol and other circadian rhythms after a gradual advance of sleep and light exposure in humans." *PLoS ONE*, 7(2), e30037. doi: 10.1371/journal.pone.0030037.

Dijk, D.J., Neri, D.F., Wyatt, J.K., Ronda, J.M., Riel, E., Ritz-De Cecco, A., Hughes, R.J., Elliott, A.R., Kim Prisk, G., West, J, B. & Czeisler, Ch.A. (2001). "Sleep, performance, circadian rhythms, and light-dark cycles during two space shuttle flights." *Am J Physiol Regul Integr Comp Physiol* 281(5), R1647–R1664.

Dobruszek, Z. (1968). *Test Kraepelina: Podręcznik (do badań grupowych). Materiały Pracowni Psychometrycznej PAN* (Kraepelin's Test: Textbook (for group studies). Materials of the Psychometric Laboratory of the Polish Academy of Sciences). Warszawa: Wydawca Centarlny Ośrodek Pedagogiczny Szkolnictwa Artystycznego. Sekcja Psychologiczno-Pedagogiczna.

Doli, R.E., Gunderson, E.K.E. (1971). "Group size, occupational status and psychological sympthomatology in an extreme environment," *Journ. Clin. Psychol.*, 27(2), 196–198.

Dollard, J., Doob, L.W., Mikker, N.E., Mower, O.H., Sears, R.R. (1939). *Frustration and aggression*. New Haven: Yale University Press.

Dołęga, J.M. (2005). *Zarys sozologii systemowej* (Outline of systemic sozology). Warszawa: Wyd. UKSW.

Doli, R.E. & Gunderson, E.K.E. (1971). "Group size, occupational status and psychological sympthomatology in an extreme environment." *Journ. Clin. Psychol.*, 27(2), 196–198.

Douglas, W.K. (1991). Psychological and Sociological Aspects of Manned Spaceflight (pp. 81–87), In: A.A. Harrison, Y.A. Clearwater & Ch.P. McKay (Eds.): *From Antarctica to Outer Space: Life in Isolation and Confinement*. New York: Springer-Verlag Inc.

Drexler, K.E. (2007). *Engines of Creation: The Coming Era of Nanotechnology*. New York: Anchor Books.

Driskell, T., Salas, E., Driskell, J., & Iwig, C. (2017, January). *Inter- and intra- crew differences in stress response: A lexical profile*. Presented at the NASA Human Research Program Investigators' Workshop, Galveston, TX.

Earls, J.H. (1969). "Human adjustment to an exotic enmronment." *Arch.Gen. Psychiat*, 20, 117–123.

Edwards, J. R., Cable, D. M., Williamson, I. O., Lambert, L. S., & Shipp, A. J. (2006). "The phenomenology of fit: Linking the person and environment to the subjective experience of person-environment fit." *Journal of Applied Psychology*, 91(4), 802–827.

Edholm, O.G. (1965). "Medical research by the British Antarctic survey," *Polar Rec.*, 12, 575–582.

Edwards, J. R. (1991). Person-job fit: A conceptual integration, literature review, and methodological critique (pp. 283–357). In C. L. Cooper & I. T. Robertson (Eds.), *International review of industrial and organizational psychology*. Vol. 6. John Wiley & Sons.

Edwards, J. R., Cable, D. M., Williamson, I. O., Lambert, L. S., & Shipp, A. J. (2006). "The phenomenology of fit: Linking the person and environment to the subjective experience of person-environment fit." *Journal of Applied Psychology*, 91(4), 802–827.

Ehmann, B., Altbäcker, A, Balázs, L. (2018). "Emotionality in isolated, confined and extreme (ICE) environments: Content analysis of diaries of Antarctic Winteroverers." *Journal of Environmental Psychology*, 60, 112–115.

Ehmann, B., Balazs, L., Fülöp, É., Hargitai, R., Kabai, P., Peley, B., László, J.(2011). "Narrative psychological content analysis as a tool for psychological status monitoring of crews in isolated, confined and extreme settings." *Acta Astronautica*, 68(9), 1560–1566.

Eliasz, A. (1985). Transactional model of tempetrament. In: J. Strelau (Ed.), *Temperamental bases of behawior: Warsaw studies on individual differences* (pp. 41–78). Lisse: Swets & Zeitlinger.

Ellenberger, H.F. (1971). Sehavior under involuntary confinement, (In:) A.W. Esser (ed.) *Behavior and environment. The use of space by animals and man* (pp. 188–203), New York–London: Plenum Press.

Esser, A.W. (ed.) (1971). *Behavior and environment. The use of space by animals and man*, New York–London: Plenum Press.

Eyal, S. & Derendorf, H. (2019). "Medications in Space: In Search of a Pharmacologist's Guide to the Galaxy." *Pharm Res*, 36(10), 148-. doi: 10.1007/s11095-019-2679-3.

Eysenck, H.J. (1990). Biological dimensions of personality. W: Pervin L.A. (ed.): *Handbook of Personality. Theory and Research*. New York: The Guilford Press.

Fan, J., McCandliss, B.D., Sommer, T., Raz, A, & Posner, M.I. (2002). "Testing the Efficiency and Independence of Attentional Networks." *Journal of Cognitive Neuroscience*, 14(3), 340–347. https://doi.org/10.1162/089892902317361886.

Festinger, L. (2007). *Teoria dysonansu poznawczego* (A Theory of Cognitive Dissonance). Wyd. 1. Watszawa: PWN.

Feustl-Beuchl, J. (2003). "25 Years of European Human Spaceflight." *ESA Bulletin*, 11, 6–16.

Fiedler, E. R., & Harrison, A. A. (2010). "Psychosocial Adaptation to a Mars Mission." *Journal of Cosmology*, 12, 3694–3710. Retrieved from http://journalofcosmology.com/Mars100.html.

Filcek, M. (2020). Cutting edge solution for effective and safe missions – Neuroarchitecture System of Vinjci Power NAP – Rrevolution in fast stress reduction, regenerating body and mind which can help you, pilots, astronauts: before, during an after space travels. 71st International Astronautical Congress (IAC) – The CyberSpace Edition, 12–14 October 2020. Copyright ©2020 by the International Astronautical Federation (IAF). IAC-20-D1.1 (56737).

Fisher, W.A, Byrne, D, White, L.A, Kelley K. (1988). "Erotophobia-erotophilia as a dimension of personality." *The Journal of Sex Research*, 25(1), 123–151.

Fiske, D.W. (1967). Effects of monotonous and restricted stimulation, (In:) D-W. Fiske and S. Madd i (Eds): *The functions of varied experience* (pp. 106–144), Homewood: Dorsey Press Inc. 3th ed.

Fiske, D.W., Maddi, S. (Eds) (1967). *The functions of varied experience*, Homewood. Dorsey Press Inc.

Flage, D., Madison, J. (2011). *George Berkeley: Oxford Bibliographies Online Research Guide*. Oxford: Oxford University Press, Inc. ISBN: 9780199810260.

Flinn, D.E., Monroe, J.T., Cramer, E.H., Hagen, D.H. (1961). "Observations in the SAM Two-Man space cabins simulator: IV. Behavioral factors in selection and performance," *Aerospace Medicine*, 32, 7, 610–615.

Flynn, C.F. (2005). "An operational approach to long-duration mission behavioral health and performance factors." *Aviat Space Environ Med*, 76(6) Supplement, B42–51.

Forster, H.A. (1959). *Wysoki biegun. Historia odkrycia Terra Australis* (High pole: The story of the discovery of Terra Australis). Warszawa: Nasza Księgarnia.

Fraser, T.M. (1968). "Confinement and free-volume requirements," *Space Life Sciences*, 1, (2–3), 428–466.

Frederickson, B.F. (2001). "The role of positive emotions in positive psychology: the broaden-and-build theory of positive emotions." *American Psychologist*, 56, 218–226.

Freedman, S.J. Grunebaum, H.U., Greenblatt, M. (1961). Perceptual and cognitive changes in sensory deprivation, (In:) P. Solomon, P.E. Kubzansky, P.H. Leiderman, (Eds): *Sensory deprivation* (p. 58–71). Cambridge, Mass.: Harvard Univ. Press.

Friederici, A.D., Levelt, W.J. (1987). "The cognitive mechanism of spatial orientation." *Aviation, Space and Environmental Medicine*, 58 (9), 164–169.

Fuerst, K., Zubek, J.P. (1968). "Effects of sensory and perceptual deprivation on a battery of open-ended cognitive tasks." *Canad. Journ. Psych.*, 22, 2, 122–130.

Galarza, L., Holland, A.W. (1999). *Selecting Astronauts for Long-Duration Missions: SAE International Document 1999-01-2097*. Presented at the International Conference on Environmental Systems, Denver, CO, July 1999.

Gasiul, H. (2007). *Teorie emocji i motywacji* (Theories of emotions and motivations). Warszawa: Wyd. UKSW.

Gavalas, A. (2011). "Researching team dynamics, personality and behaviour of team members working on Antarctic research stations: A literature review and suggestions for a future study." *OKS Review*, 1(1), 1-12.

Genefke, I., Marcussen, H., Rasmussen, O.V. (2007). Torture. W: G. Fink (ed.): *Encyclopedia of stress*. Second edition (Vol. 3, pp. 749-756). New York: Academic Press.

Gilluly, R.H. (1970). "Tektite: Unique Observations of Men Under Stress," *Science News*, 98, 400-401.

Godorowski, K. (1985). *Psychologia i psychopatologia hitlerowskich obozów koncentracyjnych: Próba analizy postaw i zachowań w warunkach ekstremalnych obciążeń* (Psychology and psychopathology of Nazi concentration camps: An attempt to analyze attitudes and behavior under extreme load conditions). Warszawa: Wyd. ATK.

Goel, N., Bale, T. L., Epperson, C. N., Kornstein, S. G., Leon, G. R., Palinkas, L. A., & Dinges, D. F. (2014). "Effects of Sex and Gender on Adaptation to Space: Behavioral Health." *Journal of Women's Health*, 23(11), 975-986. https://doi.org/10.1089/jwh.2014.4911.

Goemaere. S., Brenning, K., Beyers, W., Vermeulen, A.C.J., Binsted, K., Vansteenkiste, M. (2019). "Do astronauts benefit from autonomy? Investigating perceived Autonomy supportive communication by Mission Support, crew motivation and collaboration during HI-SEAS1." *Acta Astronautica*, 157(4), 9-16.

Goemaere, S., Van Caelenberg, T., Beyers, W., Binsted, K., Vansteenkiste, M. (2019). "Life on mars from a Self-Determination Theory perspective: How astronauts' needs for autonomy, competence and relatedness go hand in hand with crew health and mission success – Results from HI-SEAS IV," *Acta Astronautica*, 159 (7), 273-285.

Goldfried, M.R. (1960). "A psychoanalytic interpretation of sensory deprivation." *Psychological Record*, 10, 211-214.

Goldstein, D.S. (2004). Merging of the Homeostat Theory with the Concept of Allostatic Load (pp. 99-112). W. J. Schulkin (ed.): *Allostasis and the costs of physiological adaptation*. Edinburg: Cambridge University Press.

Grant, I., Eriksen, H.R., Marquis, P., Orre, I, J., Palinkas, L.A., Suedfeld, P., Svensen, E. & Ursin, H. (2007). "Psychological selection of Antarctic personnel: The "SOAP" instrument." *Aviation Space and Environmental Medicine*, 78(8),793-800.

Green, D. E., Walkey, F. H., McCormick, I. A., & Taylor, A. J. (1988). "Development and evaluation of a 21-item version of the Hopkins Symptom Checklist with New Zealand and United States respondents." *Australian Journal of Psychology*, 40(1), 61–70.

Gregson, R.A.M. (1978). "Monitoring cognitive performance in Antarctica." *New Zealand Antarctic Record*, 1(3), 24–32.

Gríofa, M. O., Blue, R. S., Cohen, K.D., O'Keeffe, D.T. (2011). "Free Content Sleep Stability and Cognitive Function in an Arctic Martian Analogue." *Aviation, Space, and Environmental Medicine*, 82(4), 434–444.

Gryziak, G. (2009). "Colonization by mites of glacier-free areas in King George Island, Antarctica." *Pesquisa Agropecuária Brasileira*, 44(8), 891–895.

Grzelak, J.Ł. (1982). Preferences and cognitive processes in social interdependence situations, w: V. Derlega, J.Ł. Grzelak (red.), *Cooperation and helping behavior. Theory and research* (pps. 95–122). New York: Academic Press.

Gunderson, E.K.E. (1963). "Emotional symptoms in extremely isolated groups." *Arch. Gen. Psychiat.*, 9, 362–368.

Gunderson, E.K.E. (1964a). *Performance evaluations of Antarctic volunteers*, San Diego, Calif.: Report No. 64–19, Navy Medical Neuro-psychiatric Research Unit.

Gunderson, E.K.E. (1964b). "Personal and social characteristics of Antarctic volunteers," *Journ. Social Psychol.*, 64, 325–332.

Gunderson, E.K.E. (1965). "Determinants of reliability in personality ratings." *Journ. Clin. Psychol.*, 21, 164–169.

Gunderson, E.K.E. (1966a). *Adaptation to extreme environments: prediction of performance*, San Diego, Calif.: Report No. 68–11, US Navy Medical Neuropsychiatric Research Unit.

Gunderson, E.K.E. (1966b). *Selection for Antarctic service*, San Diego, Calif.: Report No. 65–15, US Navy Medical Neuropsychiatric Research Unit.

Gunderson, E.K.E. (1966c). *Small group structure and performance in extreme environments*, San Diego, Calif.: Report No. 66–3, US Navy Medical Neuropsychiatric Research Unit.

Gunderson, E.K.E. (1968). "Mental health problems in Antarctica." *Arch. Environmental Health*, 17, 558–564.

Gunderson, E.K.E. (1969). "Relationship between expressed personality needs and social backround and military status variables," *Journal of Psychology*, 71, 2, 217–224.

Gunderson, E.K.E. (1973a). "Human Adaptability to Antarctic Conditions." Washington, DC: American Geophysical Union.

Gunderson, E.K.E. (1973b). Individual behavior in confined or isolated groups (p. 145–164). In, J.E. Rasmussen (ed.): *Man in isolation and confinement*. Chicago: Aldine Publ. Co.

Gunderson, E.K.E. (1973c). Psychological studies in Antarctica: A rewiew (pp. 352–361). In: O.G. Edholm and E.K.E. Gunderson (Eds): *Polar human biology*. London: William Heinemann Medical Books, Ltd.

Gunderson, E.K.E. ed. (1974). *Psychological and physiological research in Antarctica*, Washington, D.C.: National Science Foundation.

Gunderson, E.K.E., Mahan, J.L. (1968). "Cultural and psychologicdl differences among occupational groups," *Journal of Psychology*, 62, 287–304.

Gunderson, E.K.E., Nelson, P.D. (1962). *Adjustment criteria in Antarctica*, San Diego, Calif.: Report No. 62-1, US Navy 'Medieal Neuro-psychiatric Research Unit.

Gunderson, E.K.E., Nelson, P.D. (1963). "Adaptation of small groups to extreme environments." *Aerospace Medicine*, 34(12), 1111–1115.

Gunderson, E.K.E., Nelson, P.D. (1964a). *Adaptation of scientist to the Antarctic*, San Diego, Calif.: Report No. 65-5, US Navy Medical Neu-ropsychiatric Unit.

Gunderson, E.K.E, Nelson, P.D. (1964b). "Clinical agreement in assesing for an unknown environment." *Journal of Clinical Psychology*, 20, 290–295.

Gunderson, E.K.E., Nelson, P.D. (1965b). "Measurement of group effectiveness in natural isolated groups," *Journal of Social Psychology*, 66, 241–249.

Gunderson, E.K.E., Nelson, P.D. (1966a). "Criterion measures for extremely isolated groups." *Personnel Psychology*, 19(2), 67–80.

Gunderson, E.K.E., Nelson, P.D. (1963). "Adaptation of small groups to extreme environments," *Aerospace Medicine*, 34(12), 1111–1115.

Gunderson, E.K.E., Nelson, P.D. (1985a). "Biographical predictors of performance in an extreme environment," *Journ. Psychol.*, 61, 59–67.

Gunderson, E.K.E. Ryman, D.H. (1971). "Convergent and discriminant validities of performance evoluations in, extremely isolated groups." *Personnel Psychology*, 24, 715–724.

Gushin, V, Zaprisa, N.S, Kolinitchenko, T.B., Efimov, V.A., Smirnova, T.M., Vinokhodova, A.G., Kana, N. (1997). "Content Analysis of the Crew Communication with External Communicants Under Prolonged Isolation." *Aviation, Space, and Environmental Medicine*, 68,1093–1098.

Gushin, V., Yusupova, A., Pustinnikova, J., Popova, I. (2005). *Crew-ground control communication styles: Preliminary results in psychosicial area*. Presented at the 56th International Astronautical Congress, Fukuoka, Japan, October 17–21, 2005.

Gushin, V.I., Yusupova, A.K., Shved, D.M., Shueva, L.V., Vinokhodova, A.G., & Bubeev, Y.A. (2016). "The evolution of methodological approaches to the psychological analysis of the crew communications with Mission Control Center REACH." *Reviews in Human Space Exploration*, 1, 74–83.

Gussow, Z. (1963). "A preliminary report of "kayak-angst" among the Eskymo of West Greenland: A study, in sensory deprivation," *Int. Journ. Soc. Psychiat*, 9, 18–26.

Halberg, F. (1969). "Chronobiology." *Ann Rev Physiol*. 31, 675–725.

Halberg, F. (1980). "Chronobiology: methodological problems." *Acta Med Rom*. 18, 399–440.

Halberg, F., Tong, Y.L., Johnson, E.A. (1967). Circadian system phase in aspect of temporal morphology: Procedure and illustrative examples (pp. 20–48). In: H. Von Mayersbach (ed.): *The cellular aspect of biorhythms*. Berlin: Verlag Springer.

Halter, J., Terelak, J. (1980). "Nabór kandydatów do Polskich Wypraw Polarnych (Recruitment of candidates for Polish Polar Expeditions)." *Medycyna Lotnicza*, 1980, 69 (4), 42–49.

Hachisuka, H. (1971). "Group psychology of wintering team." *Polar News*, 6, 2, 9–16.

Hammes, J.A., Watson, J.A. (1965). "Behavior patterns of group expe-rimentally confined," *Perceptual and Motor Skills*, 20, 1269–1272.

Hannum, K.M. (2008). "Review of evaluating human resources programs: A 6–phase approach for optimizing performance." *Personnel Psychology*, 61(2), 459–462.

Hare, P.A. (1962). *Handbook of smali group research*, New York: The Free Press of Glencoe.

Harari, Y. N. (2018). *Homo deus: Krótka historia jutra* (Homo deus: A Brief History of Tomorrow) (Translated by M. Romanek). Kraków: Wydawnictwo Literackie.

Harris, Ph. R. (1991). Personnel Deployment Systems: Managing People in Polar and Outer Space Settings (pp. 65–80), In: A.A. Harrison, Y.A. Clearwater & Ch.P. McKay (Eds.): *From Antarctica to Outer Space: Life in Isolation and Confinement*. New York: Springer-Verlag Inc.

Harris, A., Marquis, P., Eriksen, H. R., Grant, I., Corbett, R., Lie, S. A., & Ursin, H. (2010). "508 Diurnal rhythm in British Antarctic personnel." *Rural Remote Health*, 10(2), 1351–1363.

Harrison, A.A., Clearwater, Y.A. eds., (1991). *From Antarctica to Outer Space: Life in Isolation and Confinement*. New York: Springer-Verlag.

Harrison, A.A., Clearwater, Y.A. & McKay, C.P. (1989). "The Human Experience in Antarctica: Applications to Life in Space." *Behavioral Science*, 34(4), 253–271.

Harrison, A.A., Fiedler, E.R. (2011). Introduction: Psychology and the U.S. Space Program (pp. 1–16). Edited by D. A. Vakoch, *Psychology of Space Exploration: Contemporary Research in Historical Perspective*. Washington, DC 2011. The NASA History Series. NASA SP-2011-4411.

Hachisuka, H. (1971). "Group psychologu of wintering team," *Polar News*, 6, 2, 9–16.

Harari, Y.N. (2019). *Sapiens. Od zwierząt dp bogów* (Sapiens. From animals to gods). Translated by: J. Hunia. Warszawa: Wydawnictwo Literackie.

Harrison, A.A. (2005). "Behavioral Health: Integrating Research and Application in Support of Exploration Missions." *Aviation, Space, and Environmental Medicine*, 76(6), sect. II, B3–B12.

Hawkes, C. & Norris, K. (2017). "Time-dependent mood fluctuations in antarctic personnel: A meta-analysis." *Polar Record*, 53(5), 534–549.

Hawkes, C., Norris, K., Ayton, J. & Paton, D (2019). "Mood fluctuation in Antarctic expeditioners: does one size fit all?" *Polar Record*, 55(2), 93–101. doi.org/10.1017/S003224741900024X.

Haythorn, W.W. (1970). Interpersonal stress in isolated groups (pp. 159–176), (In:) J.E. McGrath (ed.): *Social and psychological factors in stress*, New York: Holt, Rińehart and Winston.

Haythorn, W. H. (1973). The mini-world of isolation: Laboratory studies. In J. E. Rasmussen (Ed.), *Man in isolation and confinement* (Chapter 8). Chicago: Aldine.

Hebb, D.O. (1965). Drives and the C.N.S., (In:) H. Fowler (ed.): *Curiosity and exploratory behavior*, New York, 176–190, Macmillan Co.

Heppener, M. (2020) Moon, Mars and Beyond (pp. 709–733). In: Choukèr A. (eds) *Stress Challenges and Immunity in Space*. Second Edition, Springer, Cham.

Hermaszewski, M. (2017). *Ciężar nieważkości: Opowieść pilota-kosmonauty* (The Weight of Weightlessness: The Cosmonaut Pilot's Tale). Kraków:Universitas.

Hermaszewski, M. (2013). "Psychological Aspects of Space Flights." *Pol J Aviat Med Psychol*, 19(3), 5–8.

Heron, W. (1961). Cognitive and physiological effects on perceptual isolation, (In:) P. Solomon (ed.): *Sensory deprivation* (pp. 6–33). Cambridge: Havard Univ. Press.

Heron, W., Doane B.K., Scott T.H. (1956). "Visual disturbances after prolonged perceptual isolation," *Canad. Journ. Psychol.*, 10, 13–18.

Herzfeld, U.Ch. (2004). Topographic Maps from Geostatistical Analysis of Satellite Radar Altimeter (pp. 73–255). In: *Atlas of Antarctica*. Berlin: Heidelberg: Springer-Verlag.

Holt-Lunstad, J., Smith, T. B., & Layton, J. B. (2010). "Social relationships and mortality risk: A metaanalytic review." *PLoS Medicine*, 7(7), e1000316.

Horne, J.A., Östberg, O. (1976). "A self-assessment questionnaire to determine morningness-eveningness in human circadian rhythms." *International Journal of Chronobiology*, 4, 97–110.

Horney, K. (1937). *The neurotic personality of our time*. New York: W. W. Norton & Co.

Hossen, A. (2012). "Social Isolation and Loneliness among Elderly Immigrants: The Case of South Asian Elderly Living in Canada." *Journal of International Social Issues*, 1(1), 1–10.

Hraciarek, M. (2003). *Agresja jako mechanizm obronny* (Aggression as a defense mechanism) (pp. 111–118). W: M. Binczycka-Anholecer (red.), Przemoc i agresja jako zjawiska społeczne (Violence and aggression as social phenomena). Warszawa: Polskie Towarzystwo Higieny Psychicznej.

Hryniewicz, J.T. (2013). "Osobowość neurotyczna nowych czasów (Neurotic personality of the new Times)." *Przegląd Zachodni*, 348(4), 21–48.

Ihle, E.C. Ritsher, J.B. & Kanas, N. (2006). "Positive Psychological Outcomes of Spaceflight: An Empirical Study." *Aviation, Space, and Environmental Medicine*, 77, 93–101.

Ilyin, E.A., Egorow, B.B. (1970). "Psychophysiological aspects of extended eiposure of man to an environment with sensory deprivation." *Aerospace Medicine*, 49(9), 1022–1024.

Ilyin, E.A., Poggenpo, l V.S. (1969). *Information bulletin of the Soviet Antarctic Expedition*, (74, 83–88). Leningrad.

Jack, F., Jack, E. (1988). *Mawson's Antarctic diaries*. Sidney: Allen & Unwin.

Jaksic, C. (2018). *Person-environment fit: Needs and challenges in Antarctica. A Thesis submitted in partial fulfilment of the requirements for the Degree of PhD in Psychology at Lincoln University*. Lincoln (England): University of Lincoln.

Janis, I.L. (1972). *Victims of Groupthink*. Boston: Houghton Mifflin Company.

Jankowski, K. (1970). *Psychophysiological aftereffects of prolonged stay in psychiatrie hospital*, Warsaw: State Sanatorium for Nervous Diseases.

Jankowski, K. (1975). *Od psychiatrii biologicznej do humanistycznej* (From biological to humanistic psychiatry). Warszawa: PIW.

John Paul, F. U., Mandal, M. K., Ramachandran, K., & Panwar, M. R. (2010). "Cognitive performance during long-term residence in a polar environment." *Journal of Environmental Psychology*, 30(1), 129–132.

Johnson, J. C., Boster, J. S., & Palinkas, L. A. (2003). "Social roles and the evolution of networks in extreme and isolated environments." *Journal of Mathematical Sociology*, 27(2–3), 89–121.

Jaremka, L. M., Andridge, R. R., Cristopher, P., Alfano, C. M., Povoski, S. P., Lipari, A. M.,... Kiecolt-Glaser, J. K. (2014). "Pain, depression, and fatigue: Loneliness as a longitudinal risk factor." *Health Psychology*, 33, 948–957.

Jarmolenko A.V. (1961). *Oczerki psichołogii slepo-głuchoniemych*. Leningrad: Izdat. Leningrad Uniwersit.

Jaźwiński-Buczyńska, B. (1978). *Poczucie winy i postawa wobec seksu a wpływ stymulacji erotycznej na natężenie agresji* (The feeling of guilt and attitude towards sex and the impact of erotic stimulation on the intensity of aggression). Warszawa: Uniwersytet Warszawski.

Jezierska, S. (1963). *Obserwacje nad rozwojem głucho-ciemnej Krystyny Hryszkiewicz* (Observations on the development of the deaf-dark Krystyna Hryszkiewicz). Warszawa: PWN.

John Paul, F. U., Mandal, M. K., Ramachandran, K., & Panwar, M. R. (2010). "Cognitive performance during long-term residence in a polar environment." *Journal of Environmental Psychology*, 30(1), 129–132. https://doi.org/10.1016/j.jenvp.2009.09.007.

Jones, D.R., Annes, C.A. (1983). "The Evolution and Present Status of Mental Health Standards for Selection of USAF Candidates for Space Missions." *Aviation, Space, and Environmental Medicine*, 54, 730–734.

Johnson, P.J., Asmaro, D., Suedfeld, P., & Gushin, V. (2012). "Thematic content analysis of work–family interactions: Retired cosmonauts reflections." *Acta Astronautica*, 81(1), 306–317.

Johnson, J.C., Boster, J. & Palinkas, L. A. (2003). "Social roles and the evolution of networks in extreme and isolated environments." *Journal of Mathematical Sociology*, 27(2–3), 89–121. DOI: 10.1080/00222500390213056

Johnson, J.C. & B.R. Finney (1986). "Structural approaches to the study of groups in space: A look at two analogs," *Journal of Social Behavior and Personality*, 1(3), 325–347.

Jones, A., Wilkenson, H.J., Braden. I. (1961). "Information deprivation as a motivational variable." *Journ. Exp. Psychol.*, 62(2), 126–137.

Jurkowski, M., Bobek-Billewicz, B. (2007). "Naturalne czynniki wpływające na sen (Natural Factors Affecting Sleep)" *Przegląd Lekarski*. 64, 572–582.

Kahneman, D, Diener, E. & Schwartz, N. (Eds.) (1999). *Well-being: Foundations of hedonic psychology*. New York, NY: Russell Sage Foundation

Kamiński, M. (1998). *Moje bieguny. Dzienniki z wypraw 1990-1998* (My poles. Journals from the 1990-1998 expeditions). Gdańsk: Ideamedia.

Kamiński, M. (2005). *Together to the poles*. Gdańsk: Wydawnictwo: Kamiński Marek Foundation. ISBN: 978-83-911009-9-8

Kanas, N. A. (2016). "Psychiatric Issues in Space." *Psychopharmacology*, 33(6), 1-3. http://www.psychiatrictimes.com/psychopharmacology/psychiatric-iss uesspace/page/0/2

Kanas, N., & Manzey, D. (2008). *Space psychology and psychiatry*. Dordrecht, NL: Kluwer Academic Press.

Kanas, N., Weiss, DS. (1996). "Marmar CR. Crewmember interactions during a Mir space station simulation." *Aviation, Space, and Environmental Medicine*, 67, 969-975.

Kanas, N., Gushin, V., Yusupova, A, (2008). "Problems and possibilities of astronauts-Ground communication content analysis validity check." *Acta Astronautica*, 63, 822-827.

Kanas, N, Salnitskiy V, Grund EM, Gushin V, Weiss DS, Kozerenko O, Sled A, Marmar CR. (2000). "Interpersonal and Cultural Issues Involving Crews and Ground Personnel During Shuttle / Mir Space Missions." *Aviation, Space, and Environmental Medicine*, 71, 11-16.

Kanas, N., Sandal, G., Boyd, J.E., Gushin, V.I., Manzey, D., North, R., Leon, G.R., Suedfeld, P.Bishop, S., Fiedler, E.R., Inoue, N., Johannes, B., Kealey, D.J., Kraft, N., Matsuzaki, I., Musson, D., Palinkas, L.A., Salnitskiy, V.O., Sipes, W., Stuster, J., Wang, J. (2009). "Psychology and culture durin long-duration space mission." *Acta Astronautica*, 64(7-8), 659-677.

Kąkolewski, K. (1976). *Historie, które napisało życie* (Stories that life wrote). Warszawa: Iskry.

Kennaway, D.J., Van Dorp, C.F. (1991). "Free-running rhythms of melatonin, cortisol, electrolytes, and sleep in humans in Antarctica." *Am J Physiol* 260(6), 1137-1144.

Kelly, A.D., Kanas, N. (1992). "Crewmember communication in space: a survey of astronauts and cosmonauts." *Aviation, Space, and Environmental Medicine*, 63(8), 721-726.

Kelly, S.M., Rosekind, M.R., Dinges, D.F., Miller, D.L., Gillen, K.A., Gregory, K.B., Aguilar, R.D. & Smith, R.M. (1990). "Flight controller alertness and performance during spaceflight shiftwork operations." *Hum Perf Extrem Environ*. 3(1), 100-106. PMID: 12190073.

Kerkhof, G.K. (1987). "Zmiany przebiegu rytmów okołodobowych człowieka w zależności od chronotypu (Changes in the course of human circadian rhythms depending on the chronotype)." *Medycyna Lotnicza*, 95(2), 1-7.

Kielejnikow, I.K. (1971). Sociomietriczeskaja procedura i indiwidualnoie testirowanie małych grup nachodiaszczichsia w ekstriemalnych uslowiach (Sociometric procedure and individual testing of small groups under extreme conditions). Woprosy Kliniczeskoj i Ekspierimientalnoj Psichiatrii, 57. Nowositarsk.

Kintz, N. M., Chou, C. P., Vessey, W. B., Leveton, L. B., & Palinkas, L. A. (2016). "Impact of communication delays to and from the International Space Station on selfreported individual and team behavior and performance: A mixed-methods study." *Acta Astronautica*, 129, 193-200. https://doi.org/10.1016/j.actaastro.2016.09.018.

Kjærgaard, A., Leon, G.R., Venables, N.C & Fink, B.A. (2013). "Personality, personal values and growth in military special unit patrol teams operating in a polar environment." *Mil Psychol.* 25(1), 13-22.

Klein, D.C., Moore, R.Y. & Reppert, S.M. (Eds.) (1991). *Suprachiasmatic Nucleus. The Mind's Clock.* Oxford: Oxford University Press.

Klein, K.E., Wehmann, H.M., Hunt, B.I. (1972). "Desynchronization of body temperature and performance circadian rhythm as a result of outgoig and homegoing transmeridian flights." *Aerospace Medicine*, 43 (2), 119-132.

Klonolf, H., Mcdougall G., Clark C., Kramer P., Horgan, M.A. (1976). "The neuropsychological, psychiatrie and physieal effects of prolonged and severe stress: 30 years later," *Journ. Nerv. Ment Bis.*, 163, 4, 246-252.

Kłossowski, M. (2002). Zespół długu czasowego (Jet Lag): Przyczyny, objawy, profilaktyka (Time debt syndrome (Jet Lag): Causes, symptoms, prevention((pp. 359-364). In: W. Kowalski (red.): *Medycyna lotnicza: Wybrane zagadnienia* (Aviation medicine: Selected issues). Poznań: Wyd. DWL i OP.

Kobayashi, A. (1969). "Human ecology for the polar traverse." *Polar News*, 5(1), 21-25.

Kofta, M. (1977). "Kontrola psychologiczna nad otoczeniem: Ramy pojęciowe teorii (Psychological control over the environment: The conceptual framework of the theory)." *Psychologia Wychowawcza*, 2, 165-169.

Kopania J. (2002). *Etyczny wymiar cielesności* (Ethical dimension of corporeality). Kraków: Aureus.

Korneeva, Y.A., Simonova, N.N. (2018). "Analysis of Psychological Risks in the Professional Activities of Oil Gas Workers in the Far North of the Russian Federation." *Behavioral Sciences.* 8(84), 1-8. https://doi.org/10.3390/bs8090084.

Kortko, D., Pietraszewski, M. (2016). *Kukuczka – opowieść o najsłynniejszym polskim himalaiście* (Kukuczka – a story about the most famous Polish climber). Warszawa: Wyd. Agora.

Kowalski, W. (1982). *Adaptacja człowieka do pobytu w Antarktyce* (Human adaptation to stay in Antarctica). Warszawa: Wyd. WIML.

Kozerenko, O.P., Miasnikow, W.I. (1977). *Mietodiczeskije woprosy issledowanija psichołogiczeskogo aspiekta adaptacji* (Methodical problems of the psychological aspect of adaptation), (W): Tezisy iż X Simpoziuma Socyalisticzeskich Stron Uczastnikow Issledowanii po Kosmiczeskoj Biologii i Miedicynie, Suchumi (ZSRR), 16–26.05.1977, Interkosmos.

Kozerenko, O.P, Gushin, VI, Sled, AD. (1999). "Some problems of group interaction in prolonged space flights." *Human Performance in Extreme Environments*, 4(1), 123–127.

Kozerenko, O. P., Radkowski, G., Terelak, J., Mikshik, O. (1991). The problem of man's psychic adaptation under the space-flight conditions, (In:) K.Boda and V.M.Baranov (Eds): *Current Trends in Cosmic Biology and Medicine, Slovak Academy of Sciences, Ivanka Pri Dunaji (CSRF)*, Vol.II, 91–94.

Kozerenko, O.P., Sled, A.D., and Mirzadzhanov, Yu.A. (2001). "Psychological support of crews, in Orbital'naya stantsiya Mir." *Space Biology and Medicine*, 1, 365–377.

Kozicka, J. (2008). *Problemy architektoniczne bazy na Marsie jako habitatu w ekstremalnych warunkach. Niepunlikowana dysertacja doktorska* (Architectural problems of the base on Mars as a habitat in extreme conditions. Unfinished doctoral dissertation). Gdańsk: Politechnika Gdańska.

Kozłowski, S. (1986). *Granice przystosowania* (Limits of adaptation). Warszawa: Wiedza Powszechna.

Kraft, J.A. (1959). "Measurement of stress and fatigue in flight crews during confinement," *Aerospace Medicine*, 30(6), 424–430.

Krech, D., Crutchfield, R. S., & Ballachey, E. L. (1962). *Individual in society: A textbook of social psychology*. McGraw-Hill.

Kreutz, M. (1949). *Podstawy Psychologii: Studium nad metodami i pojęciami współczesnej psychologii* (Fundamentals of Psychology: A study of the methods and concepts of modern psychology). Warszawa: Czytelnik.

Kring, J. (2001). "Multicultural Factors for International Spaceflight." *Journal of Human Performance in Extreme Environments*, 5(3), 11–32.

Kristof, A.L. (2006). "Person-Organization Fit: An Integrative Review of Its Conceptualizations, Measurement, and Implications." *Personnel Psychology*, 49(1),1–49. DOI: 10.1111/j.1744–6570.1996.tb01790.x

Kristof-Brown, A. L., Zimmerman, R. D., & Johnson, E. C. (2005). "Consequences of individuals' fit at work: A meta-analysis of person-job, person-organization, person-group, and personsupervisor fit." *Personnel Psychology*, 58(2), 281–342.

Kubiczkowa, J., F. Skibniewski, F. (1979). "Przydatność elektrogustometrii w medycynie lotniczej i kosmicznej (Usefulness of electrogustometry in aviation and space medicine)." *Otolaryngol. Pol.*, 33(2), 61–66.

Kubis, J.F. (1972). "Isolation, Confinement, and Group Dynamics in Long Duration Spaceflight." *Acta Astronautica*, 17, 45–72.

Kubzansky, P.E. (1961). The effects of reduced environmental stimuiation, on human behavior: A review (pp. 51–95). (In:) A. Biderman and H. Zimmer (Eds): *The manipulation of human behavior: The case for interrogation*. New York: J. Wiley.

Kurcz, I., Reykowski, J. (red.) (1975). *Studia nad teorią czynności ludzkich* (Studies on the theory of human activities). Warszawa: PWN.

Kuznetsova, P.G., Gushchin, V.I., Vinokhodova, A.G., Chekalina, A.I. & Shved, D.M. (2017). "Interpersonal Interaction under the Conditions of High Autonomy in Interplanetary Mission Simulation (Mars-500 Experiment)." *Human Physiology*, 43(7), 751–756. ISSN 0362-1197.

Küng, H., (1990). *Projekt Weltethos*. München: Piper Verlag GmbH.

Kwarecki. K. (1980). "Problemy biomedyczne pobytu człowieka w strefach polarnych ze szczególnym uwzględnieniem Antarktyki (Biomedical problems of human stay in the polar zones, with particular emphasis on Antarctica)." *Medycyna Lotnicza*, 69, 1–13.

Kwarecki, K. (1985). *Chronobiologia – perspektywy wykorzystania w medycynie, W: Biochemia. Materiały seminaryjne* (Chronobiology – perspectives of use in medicine, In: Biochemistry. Seminar material) (Tom 3). Warszawa: Wyd. WAM.

Kwarecki, K., Terelak, J. (1980). *Medycyna i psychologia kosmiczna* (Medicine and space psychology). Warszawa: Wiedza Powszechna.

Kwarecki, K. Terelak, J. (1981). "Problems of human biology in Polish polar research," *Postępy Astronautyki*, 1981, 14(3), 35–49.

Kwarecki, K., Święcicki, W., Kłossowski, M., Terelak, J., Zużewicz, K. (1982). "Zdolność do pracy fizycznej i umysłowej w warunkach 24-godzinnej bezsenności oraz pracy zmianowej (Ability to work physically and mentally in conditions of 24-hour insomnia and shift work)." *Medycyna Lotnicza*, 74 (1), 7–14.

Kwek, S.K., Chew, W.M., Ong, K.C., Ng, A. W.K., Lee, L. S.U., Kaw, G., & Leow, M. K. (2006). "Quality of life and psychological status in survivors of severe acute respiratory syndrome at 3 months postdischarge." *Journal of Psychosomatic Research*, 60, 513–519. https://doi.org/10.1016/j.jpsychores.2005.08.020.

Landon, L.B., Slack, K.J. & Barrett, J.D. (2018). "Teamwork and Collaboration in Long-Duration Space Missions: Going to Extremes." *American Psychologist*, 73(4), 563–575. 0003–066X/18/$12.00 http://dx.doi.org/10.1037/amp0000260.

Lam, M. H.-B., Wing, Y.-K., Yu, M. W.-M., Leung, C.-M., Ma, R. C. W., Kong, A. P. S., Lam, S. P. (2009). "Mental morbidities and chronic fatigue in severe acute respiratory syndrome survivors." *Archives of Internal Medicine*, 169, 2142–2147. https://doi.org/10.1001/archinternmed.2009.384.

Lazarus, R.S. (2006). *Stress and emotion: A new synthesis*. New York: Springer.

Lebedev, V. (1988). *Diary of a Cosmonaut: 211 Days in Space*. College Station, TX: Phytoresource Research, Inc.

Leiderman P.H., Mendelson J.H., Wexler D., Solomon P. (1958). "Sensory deprivation: Clinical aspects." *A.M.A.Arch. Int. Med.*, 101, 389–396.

Lem S. (1976). *Opowieści o pilocie Pirxie* (Tales of the Pirx pilot). 3rd Edition. Kraków: Wyd. Literackie.

LeMenager, S., Foote, S. (2012). "The sustainable humanities," *PMLA*, 127(3), 572–578.

Leon, G.R. (2005). "Men and women in space." *Aviat Space Environ Med.*, 76, B84-B88.

Leon, G.R., List, N. & Magor, G. (2004). "Personal experiences and team effectiveness during a commemorative trek in the High Arctic." *Environ. Behav.*, 36, 386–401. Doi: 10.1177/0013916503262215.

Leon, G.R. & Sandal, G.M. (2003). "Women and couples in isolated extreme environments: applications to long-duration missions." *Acta Astronaut.*, 53, 259–267.

Leon, G.R., Sandal, G.M., Fink, B. & Ciofani, P. (2011). "Positive experiences and personal growth in a two-man North Pole expedition," *Environ. Behav.* 43(5), 710–731.

Leon, G.R., Venables, N.C. (2015). "Fearless Temperament and Overconfidence in an Unsuccessful Special Forces Polar Expedition." *Aerosp. Med. Hum. Perform.* 86, 567–570. https://doi.org/10.3357/AMHP.4256.2015.

LePine, J. A., Piccolo, R. F., Jackson, C. L., Mathieu, J. E., and Saul, J. R. (2008). "A meta-analysis of teamwork processes: tests of a multidimensional model and relationships with team effectiveness criteria." *Pers. Psychol.* 61, 273–307. doi: 10.1111/j.1744–6570.2008.00114.x.

Le Scanff, C., Larue, J., & Rosnet, E. (1997). "How to measure human adaptation in extreme environments: The case of Antarctic wintering-over." *Aviation, Space and Environmental Medicine*, 68, 1144–1149.

Leuba, C. (1965). Towar some integration of learning theory: The concept of aptimal stimuiation, (In:) H. Fowler (ed.): *Curiosity and exploratory behavior*, New York, 169–175, Macmillan Co.

Levesque, M. (1991). An Experiential Perspective on Conducting Social and Behavioral Research at Antarctic Research Stations (pp. 15–20), In: A.A. Harrison, Y.A. Clearwater & Ch.P. McKay (Eds.): *From Antarctica to Outer Space: Life in Isolation and Confinement*. New York: Springer-Verlag Inc.

Levy, E.Z., Ruf, f G.E., Thaler, V.H. (1959). "Studies in human isotation." *Journal of the American Medical Assotiation*, 169, 3, 236–239.

Lewin, K. (1951). *Field Theory in Social Science*. New York: Harper.

Li, G. (2007). Airline accidents. In: G. Fink (Ed.): *Encyclopedia of stress*, Second Edition (Vol. 1, pp. 114–118). New York: Academic Press.

Lieberman, P., Morey, A., Hochstadt, J., Larson, M., Mather, S. (2005). "Mount Everest: A Space Analogue for Speech Monitoring of Cognitive Deficits and Stress." *Aviation, Space, and Environmental Medicine*, 76(6), 198–207.

Lifton, R, J. (1954). "Home by ship: Reaction patterns of American priso-ners of war repatriated from North Korea." *Amer. Journ. Psychiat.*, 110, 732–739.

Lim, D.S.S., Brady, A., Abercromby, A., Andersen, D., Andersen, M., Arnold, R., Bird, J., Bohm, H., Booth, L., and Cady, S. (2011). "A historical overview of the Pavilion Lake Research Project–analogue science and exploration in an underwater environment." *Geol Soc Am. Spec Pap*, 483, 85–115.

Livingston, H., Livingston, M. (2007). Lockerbie air crash, stress effects of. W: G. Fink (ed.): *Encyclopedia of stress*. Second edition (Vol. 2, pp. 608–611). New York: Academic Press.

Locke, E.A. (Ed.) (2003). *Postmodernism and Management: Pros, Cons, and the Alternative*. New York: JAI Press.

Logan, M. ed. (2017). *The Story of Polar Conquest, the Complete History of Arctic and Antarctic Exploration*. Published by Andesite Press. ISBN 10: 1376229005 / ISBN 13: 9781376229004.

Lopez, S. J., Snyder, C. R. (2009). *The Oxford Handbook of Positive Psychology*. Oxford University Press.

Lorek, A. & Koncz, A. (2013). Simulation and Measurement of Extraterrestrial Conditions for Experiments on Habitability with Respect to Mars (pp. 145–162). Ed. by de Vera, J.-P. and Seckbach J. *Habitability of other planets and satellites*. Springer.

Lorenz, K. (1986). *Regres człowieczeństwa* (Humanity regression). Warszawa: PIW.

Ludvig III, E.J., Happ, D. (1974). "Extraversion and preferred level of sensory stimuiation." *Brit. Journ. Psychol.*, 65(3), 359–365.

Lugg, D.J. (1973). The adaptation of a small group to life in an isolated Antarctic stations (p. 401–409). (In:) O.G. Edholm and E.K.E. Gunderson (Eds): *Polar human biology*, London: William Heinemann Medical Books, Ltd.

Lugg, D.J. (1991). Current International Human Factors Research in Antarctica (pp. 31–42), In: A.A. Harrison, Y.A. Clearwater & Ch.P. McKay (Eds.): *From Antarctica to Outer Space: Life in Isolation and Confinement*. New York: Springer-Verlag Inc.

Lugg, D. J. (2005). "Behavioral health in Antarctica: Implications for long-duration space missions." *Aviation Space and Environmental Medicine*, 76(6), Supplement:B74-B77. PMID: 15943198.

Lung, F. W., Lu, Y. C., Chang, Y. Y., & Shu, B. C. (2009). "Mental symptoms in different health professionals during the SARS attack: A follow-up study." *Psychiatric Quarterly*, 80, 107–116. https://doi.org/10.1007/s11126-009-9095-5.

Maciejczyk, J., Terelak, J. (1978). "Psychological manifestations of "Microwave neurrosis,"" *Polish Psychological Bulletin*, 9 (3), 157–162.

Machowski, J. (1959). *Zdobywcy białego lądu: Historia wypraw i odkryć antarktycznych* (Conquerors of the white mainland: History of expeditions and Antarctic discoveries). Warszawa: PZWS.

Mallis, M.M., DeRoshia, C.W. (2005). "Circadian Rhythms, Sleep, and Performance in Space." *Aviation, Space, and Environmental Medicine*, 76(6), Supplement: B94–107. PMID: 15943202.

Malzberg, B., Lee, E. (1956). *Migration and mental disease*. New York: Social Science Research Co.

Małkiewicz, M.M., Terelak, J.F. (2018). How the Emotion of Hope Shapes the Way of Coping in a Stressful Situation? Proceedings of the 39th STAR Conference: Stress and Anxiety Research Society, Lublin (Poland), 10–13 July, 2018th.

Małkiewicz, M., Terelak, F.F. (2016). "Stress and anxiety in a changing socjety." Report from the 37th Internat5ional Conference of the Stress and Anxiety Research Society (STAR), Zagreb (Croatia), July 6–8, 2016. *Pol J Aviat Med Bioeng Psychol*, 22(1), 55–57.

Mason, M.K. (1942). "Learning to speak after six and one-half years of silence." *Journ. Speech, Dis.*, 7, 295–304.

Mathewson, K. E., Lleras, A., Beck, D. M., Fabiani, M., Ro, T., & Gratton, G. (2011). "Pulsed out of awareness: EEG alpha oscillations represent a pulsed-inhibition of ongoing cortical processing." *Front Psychol*, 2, Article 99. https://doi.org/10.3389/fpsyg.2011.00099

Mathieu, J. E., Kukenberger, M. R., D'Innocenzo, L., & Reilly, G. (2015). "Modeling reciprocal team cohesion-performance relationships, as impacted by shared leadership and members' competence." *Journal of Applied Psychology*, 100(3), 713–734. https://doi.org/10.1037/a0038898.

Matsuda, T. (1964). "Some notes on the social life of 16 persons af the Japanese Antarctic wintering party." *Antarctic Record*, 20, 1755–1767.

Matter, F. (1991). Isolation and Confinement Effects (pp. 199–200). In: A.A. Harrison, Y.A. Clearwater & Ch.P. McKay (Eds.): *From Antarctica to Outer Space: Life in Isolation and Confinement*. New York: Springer-Verlag Inc.

Matysiak, J. (1993). *Głód stymulacji* (Hunger for stimulation). Warszawa: Wyd. UW.

McAdams, D. P. (2006). "The role of narrative in personality psychology today." *Narrative Inquiry*, 16: 11–18. doi: 10.1075/ni.16.1.04mca.

McFadden, T.J., Helmreich, R., Rose, R.M., Fogg, L.F. (1994). "Predicting Astronaut Effectiveness: A Multivariate Approach." *Aviation, Space, and Environmental Medicine*, 65, 904–909.

McGuiire, F., Tolchin S. (1961). "Group adjustment at the South Pole." *Journ. Ment. Science*, 107(450), 954–960.

McNeil, Jean (2016). *Ice Diaries: An Antarctic Memoir*. Toronto (Canada): ECW Press. ISBN13 (EAN): 9781770413184.

McQuaid, K. (2007). "Race, Gender and Space Exploration: A Chapter in the Social History of the Space Age." *Journal of American Studies*, 41(2), 405–434.

Menaker, M. (1961). "The free running period of the bat clock: seasonal variations at low body temperature." *J Cell Comp Physiol*. 57, 81–86.

Mears, J.D., Cleary, P.J. (1980). "Anxiety as a Factor in Underwater Performance." *Ergonomics*, 23(6), 549–557.

Mendelson, J., Foley, J. (1956). "Abnormality of mental functions, affecting patients with poliemyelitis in tank-type respirators." *Trans. Amer. Neurol. Ass.*, 81, 134–138.

Meyers, D.G. (2006). *Social Psychology*. Nowy Jork: McGraw-Hill.

Mika, S. (2011). *Psychologia społeczna* (Social psychology). Warszawa: Wydawnictwo Naukowe Scholar.

Milgram, S. (1963). "Behavioral Study of Obedience." *Journal of Abnormal and Social Psychology*, 67, 371–378.

Miller, J.G. (1991). Applications of Living Systems Theory to Life in Space (pp. 177–197), In: A.A. Harrison, Y.A. Clearwater & Ch.P. McKay

(Eds.): *From Antarctica to Outer Space: Life in Isolation and Confinement*. New York: Springer-Verlag Inc.

Mills, T.M. (1967). *The socioiogy of smali graups*. Englewood Cliffs, New Jersey: Prentice-Hall, Inc.

Moallef, P. (2015). "Psychological impact on SARS survivors: Critical review of the English language literature." *Canadian Psychology*, 56(1), 123–135. http://dx.doi.org/10.1037/a0037973.

Mocellin, J.S.P., Suedfeld, P. (1991). "Voices from the ice: diaries of polar explorers." *Environ Behav.*, 23, 704–722.

Mocellin, J.S.P., Suedfeld, P., Bernaldez, J.P. & Barbarito, M.E. (1991). "Levels of anxiety in polar environments." *J Environ Psychol.*, 11, 265–275.

Mohino M, Kirchner T, Forns M. (2004). "Coping Strategies in Young Male Prisoners." *Journal of Youth and Adolescence*, 33(1), 41–49.

Mok, E., Chung, B. P. M., Chung, J. W. Y., & Wong, T. K. S. (2005). "An exploratory study of nurses suffering from severe acute respiratory syndrome (SARS)." *International Journal of Nursing Practice*, 11, 150–160. https://doi.org/10.1111/j.1440-172X.2005.00520.x.

Moult, C., Norris, K., Paton, D., & Ayton, J. (2015). "Predicting positive and negative change in expeditioners at 2-months and 12-months post Antarctic employment." *The Polar Journal*, 5(1), 128–145.

Mroziewski, M. (2005). *Style kierowania i zarządzania* (Styles of leadership and management). Warszawa: Difin.

Mullin, C.S. (1960). "Some psychological aspects of isolated Antarctic living." *American Journ. Psychiat*, 117, 323–325.

Mullin, C.S., Connery, H.J. (1959). "Psychological study at an Antarctic IGY station." *US Armed Forces Medical Journal*, 10(3), 290–296.

Mundy-Castle, A.C. (1958). "Psychological problems of space flight." *South African Journal Sci. (Cape Town)*, 54, 9, 225–230.

Musson, D. W., Helmreich, R. L. (2005). "Long term personality data collection in support of spaceflight and analogue research." *Aviation, Space and Environmental Medicine*, 78(6), Sect II, B119-B125.

Myasnikov, V.I., Zamaletdinov, I. S. (1996). Psychological States and Group interactions of crew members in flight (pp. 419–432). Edited by C.S.L. Huntoon, V.V. Antipov & A.I. Grigoriev: *Space Biology and Medicine – Humans in Spaceflight*, Volume III, Books 2. New York: NASA Press. https://doi.org/10.2514/5.9781624104671.0419.0432.

Myklebust, H.R. (1960). *The psychology of deafness: Sensory deprivation learning and adjustment*. New York: Grune and Stratton.

Nardini, J.E. (1952). "Survival factors in American prisoners of war of the Japanese." *Amer. Journ. Psychiat*, 109, 241–248.

Nardini, J.E., Herrmann, R.S., Rasmussen, J.E. (1962). "Navy psychiatrie assesment program in the Antarctic." *American Journal of Psychiatry*, 119, 97–105.

Natani, K., Shurley J.T. (1975). "Extrinsic parameters and the self-regulation of sleep in Antarctica." *Biological Psychology Bulletin*, 4, (1), 16–22.

Natani, K., Shurley, J.T., Joern, A.T. (1973). Interpersonal relationship, job satisfaction and subjective feelings of competence: Their influence upon adaptation to Antarctic isolation (pp. 384–400). In: O.G. Edliolm and E.K.E. Gunderson (Eds): *Polar human biology*. London: William Heinemann Medical Books, Ltd.

Nelson, M., Gray, K., & Allen, J. P. (2015). "Group dynamics challenges: Insights from Biosphere-2 experiments." *Life Sciences in Space Research*, 6, 79–86. doi.org/http://dx.doi.org/10.1016/j.lssr.2015.07.003.

Nelson, P.D. (1962). *Leadership in small isolated groups*, San Diego, Calif.: Report No. 62–13, US Navy Medical Neuropsychiatric Research Unit.

Nelson, P.D. (1963). Human adaptation to Antarctic station life (p. 138–145). In: Various authors eds, *Medicine and public health in the Arctic and Antarctic*. Geneva: Public Health Papers, WHO.

Nelson, P.D., Gunderson, E.K.E. (1963a). *Effective individual perforrnance in smali Antarctic stations: A summary of criterion studies*. San Diego, Calif.: Report No. 63–8, US Navy Medical Neuropsychiatric Research Unit.

Nelson, P.D., Gunderson, E.K.E. (1963b). *Personal history correlates of performance among military personnel in smali Antarctic stations*. San Dłego, Calif.: Report No. 63–20, US Navy Medical Neuropsychiatric Research Unit.

Nelson, P.D., Gunderson E.K.E. (1964). "Analysis of adjustment dimensions in smali confined groups." *Bulletin L'Etudes Recherche de Psychologie*, 13(2), 111–126.

Nelson, W, Tong, Y.L, Lee, J.K, Halberg, F. (1979). "Methods for cosinor rhythmometry." *Chronobiologia*. 6, 305–323.

Newcomb, Th.M. (1968). Stabilities underlying changes in Interpersonal attraction (pp. 547–556). In: D. Cartwright & S. Zander, Eds. *Group dynamics*. New York: Harper & Row. Pub.

Newcomb, Th.M., Turner, R.H., Converse, Ph.E. (1966). *Social Psychology: The Study of Human Interaction*. London-New York: Taylor and Francis Group. Psychology, Press.

Ngai, J. C., Ko, F. W., Ng, S. S., To, K.-W., Tong, M., & Hui, D. S. (2010). "The long-term impact of severe acute respiratory syndrome on pulmonary function, exercise capacity and health status." *Respirology*, 15, 543–550. https://doi.org/10.1111/j.1440-1843.2010.01720.

Nietzche, F. (2013). *Zmierzch bożyszcz, czyli jak filozofuje się młotem* (Twilight idols, or how to philosophize with a hammer) (Translation and introduction by: G. Sowiński). Kraków: Wydawnictwo Uniwersytetu Jagiellońskiego.

Nicolas, M., Martinent, G., Suedfeld, P., Gaudino, M. (2019). "Assessing psychological adaptation during polar winter-overs: The isolated and confined environments questionnaire (ICE-Q)." *Journal of Environmental Psychology*, 65, October, 1013–1017. doi.org/10.1016/j.jenvp.2019.101317.

Nicolas, M., Sandal, G.M., Weiss, C. & Yusupova, A. (2013). "Mars-105 study: Time-courses and relationships between coping, defense mechanisms, emotions and depression." *Journal of Environmental Psychology*, 35, 52–58.

Nicolas, M., Suedfeld, P. & Weiss, K., Gaudino, M. (2015). "Affective, Social, and Cognite Outcomes During a 1-Year Wintering in Concordia." *Environment and Behavior*, 48(8), 1073–1091. ff10.1177/0013916515583551ff. ffhal-01634242f.

Norris, K., Paton, D., & Ayton, J. (2010). "Future directions in Antarctic psychology research." *Antarctic Science*, 22(4), 335–342.

Nowikow, M.A., Izosimow, G.W., Gierasimowicz, A.A. (1977). Sriedstwa optimalizacyi gruppowogo wzaimodiejstwija w usłowijach dlitiejnoj izolacyi (Ways of optimizing group coexistence in conditions of long-term isolation). (In:) W.N. Czernłgowskij (ed.): *Problemy Kosmiczeskoj Biologii* (No. 34, pp. 200–215). Moskwa: Nauka.

Nowlis, V. (1965). Research with the Mood Adjectwe Check List, In: S.S. Tomkins and C.E. Izard (Eds): *Affect, cognition and personality*. New York: Tavistock Publ.

Noyes, A.P., Kolb, L.C. (1969). *Nowoczesna psychiatria kliniczna* (Modern clinical psychiatry). Warszawa: PZWL.

Oatley, K., Goodwin, B.C. (1971). The explanation and investigatio'n of biological rhythms (pp. 1–38). (In:) W.P. Colquhoun (ed.): *Biological rhythms and human performance*. London, New York: Academic Press.

O'Donnell, M.L., Creamer, M., Elliot, P., Atkin, C. (2005). "Health costs following motor vehicle accidents: The role of posttraumatic stress disorder." *Journal of Traumatic Stress*, 18(5), 557–561.

Oliver, Donna, C. (1991). Psychological Effects of Isolation and Confinement of a Winter-Over Group at McMurdo Station, Antarctica (pp. 217–227), In: A.A. Harrison, Y.A. Clearwater & Ch.P. McKay (Eds.): *From Antarctica to Outer Space: Life in Isolation and Confinement*. New York: Springer-Verlag Inc.

Ong, A. D., Uchino, B. N., & Wethington, E. (2016). "Loneliness and health in older adults: A mini-review and synthesis." *Gerontology*, 62, 443–449.

Orne, M.T., Scheibe. K.E. (1964). "The contribution of nondeprivation factors in the production of sensory deprivation effects; The psychology of the "panic-button,"" *Journ. Abnorm. Soc. Psychol.*, 68(1), 1, 3–12.

Osgood, Ch.E., Suci, G.J., Tannebaum, P.H. (1967). *The measurement of meaning*, Urbana: Univ. Illin. Press.

Osinski, G.R., Le´veille´, R., Berinstain, A., Lebeuf, M., Barnsey, M. (2006). "Terrestrial analogues to Mars and the Moon: Canada's role." *Geoscience Canada*, 33, 175–188.

Owens, A., Ho, K., Schreiner, S. (2016). "Olivier de Weck: An independent assessment of the technical feasibility of the Mars One mission plan – Updated analysis." *Acta Astronautica*, 120(3–4), 192–228.

Palinkas, L.A. (1986). "Health and Performance of Antarctic Winter-Over Personnel: A Follow-Up Study." *Aviation, Space, and Environmental Medicine*, 57, 954–959.

Palinkas, L. A. (1991). "Effects of physical and social environment on the health and well-being of Antarctic winter-over personnel." *Environment & Behavior*, 23(6), 782–799.

Palinkas, L.A. (2001). "Psychosocial issues in long-term space flight: overview." *Gravit Space Biol Bull.* 14(2), 25–33. PMID: 11865866.

Palinkas, L.A. (2003). "The Psychology of Isolated and Confined Environments: Understanding Human Behavior in Antarctica." *American Psychologist*, 58(3), 353–363.

Palinkas, L.A., & Browner, D. (1995). "Effects of prolonged isolation in extreme environments on stress, coping and depression." *Journal of Applied Social Psychology*, 25, 557–576.

Palinkas, L. A., Cravalho, M., & Browner, D. (1995). "Seasonal variation of depressive symptoms in Antarctica." *Acta Psychiatrica Scandinavica*, 91(6), 423–429.

Palinkas, L.A., Gunderson, E.K.E., Johnson, J.C., Holland, A.W. (2000). "Behavior and performance on long-duration spaceflights: evidence from analogue." *Environments, Aviation, Space, and Environmental Medicine*, 71(1), 29–36.

Palinkas, L. A., Gunderson, E. K., Holland, A. W., Miller, C., & Johnson, J. C. (2000). "Predictors of behavior and performance in extreme environments: The Antarctic space analogue program." *Aviation, Space and Environmental Medicine*, 71, 619–625.

Palinkas, L. A., & Houseal, M. (2000). "Stages of change in mood and behavior during a winter in Antarctica." *Environment and Behavior*, 32(1), 128–141.

Palinkas, L.A., Houseal, M., Rosenthal, N.E. (1996). "Subsyndromal seasonal affective disorder in Antarctica." *J Nerv Ment Dis.*, 184, 530–534.

Palinkas, L.A., Johnson, J.C., Boster, J.S., & Houseal, M. (1998). "Longitudinal studies of behavior and performance during a winter at the South Pole." *Aviation Space and Environmental Medicine*, 69, 73–77.

Palinkas, L.A., Johnson, J.C., Boster, J.S., Rakusa-Suszczewski, S., Klopov, V.P., Fu, X. Q., et al. (2004). "Cross-cultural differences in psychosocial adaptation to isolated and confined environments." *Aviation Space & Environmental Medicine*, 75(11), 973–980.

Palinkas, L.A., Glogower, F.G., Dembert, M., Hansen, K. & Smullen, R. (2004). "Incidence of psychiatric disorders after extended residence in Antarctica." *Int J Circumpolar Health.*, 63, 157–168.

Palinkas, L. A., Johnson, J. C., & Boster, J. S. (2004). "Social support and depressed mood in isolated and confined environments." *Acta Astronautica*, 54(9), 639–647.

Palinkas, L.A., Makinen, T. M., Paakkonen, T., Rintamaki, H., Leppaluoto, J. & Hassi, J. (2005). "Influence of seasonally adjusted exposure to cold and darkness on cognitive performance in circumpolar residents." *Scandinavian Journal of Psychology*, 46, 239–246.

Palinkas, L. A., Reed, H. L., Reedy, K. R., Do, N. V., Case, H. S., & Finney, N. S. (2001). "Circannual pattern of hypothalamic-pituitary-thyroid (HPT) function and mood during extended Antarctic residence." *Psychoneuroendocrinology*, 26(4), 421–431.

Palinkas, L.A., & Suedfeld, P. (2008). "Psychological effects of polar expeditions." *Lancet*, 371(9607), 153–163.

Palmai, G. (1962). "Thermal comfort and acclimatization to cold in a subantarctic environment." *Med. Journ. Austral.*, 1, 9–12.

Palmai, G. (1963a). "Psychological observations on isolated groups in Antarctica." *British Journal of Psychiatry*, 109, 364–370.

Palmai, G. (1963b). Psychological aspects of transient populations in Antarctica (pp. 146–157). In: Various authors (Eds): *Medicine and Public Health in the Arctic and Antarctic*. Geneva: Public Health Papers, WHO).

Patten, S. B., & Williams, J. V. A. (2007). "Chronic obstructive lung diseases and prevalence of mood, anxiety, and substance-use disorders in a large population sample." *Psychosomatics*, 48, 496–501. https://doi.org/10.1176/appi.psy.48.6.496.

Pattyn, N., Mairesse, O., Cortoos, A., Marcoen, N., Neyt, X., & Meeusen, R. (2017). "Sleep during an Antarctic summer expedition: New light on "polar insomnia."" *Journal of Applied Physiology*, 122(4), 788–794.

Paul, J., Mandal, F.U., Ramachandran, M.K., & M.R. Panwar (2010). "Cognitive performance during long-term residence in a polar environment." *Journal of Environmental Psychology*, 30(1), 129–132.

Pääkkönen, T., Leppäluoto, J., Mäkinen, T.M., Rintamäki, H., Ruokonen, A., Hassi, J. & Palinkas, L.A. (2008). "Seasonal levels of melatonin, thyroid hormones, mood, and cognition near the Arctic Circle." *Aviat Space Environ Med.*, 79(7), 695–699. Doi: 10.3357/asem.2148.2008.

Peeters, M.A.G., Van Tuijl, H.F.J.M., Rutte, C.G. & Reymen, I.M.M. (2006). "Personality and team performance: A meta-analysis." *Eur J Pers*. 20(5), 377–396.

Pell, S. J., & Mueller, F. (2016). Homo Ludens: an analysis of play and performance during spaceflight to inspire the cultural sector to design for new modes of space and spatiality. In S. I. Ramirez Jimenez, & N. Mathers (Eds.), *61st International Astronautical Congress (IAC 2016)* (Vol. 14 of 17, pp. 9923–9934).

Persky, H., Zuckerman, M., Basu, G.K., Thornton, D. (1966). "Psychoendocrine effects of perceptual and social isolation." *Arch. Gen. Psychiat.*, 15, 499–515.

Petrov, B.N., Lomov, B.F., Samsonov, N.D. eds. (1979). *Psychological Problems of Spaceflight*. Moscow: Nauka Press.

Pierce, Ch. M. (1991). Theoretical Approaches to Adaptation to Antarctica and Space (pp. 125–133), In: A.A. Harrison, Y.A. Clearwater & Ch.P. McKay (Eds.): *From Antarctica to Outer Space: Life in Isolation and Confinement*. New York: Springer-Verlag Inc.

Pilkiewicz, M. (1973). Techniki socjometryczne. Wprowadzenie do badań (Sociometric techniques. Introduction to research). In: L. Wołoszynowa (ed.): *Materiały do nauczania psychologii* (Materials for teaching psychology), (Seria III, Tom 2). Warszawa: PWN.

Płużek, Z. (1971). *Wartość testu WISKAD-MMPI dla diagnozy różnicowej w zakresie nozologii psychiatrycznej* (The value of the WISKAD-MMPI test for differential diagnosis in the field of psychiatric nosology). Lublin: KUL.

Poláčková Šolcová, I., Šolcová, I., Stuchlíková, I. & Mazehóová, Y. (2016). "The story of 520 days on a simulated flight to Mars." *Acta Astronautica*, 126, 178–189.

Pope, F.E., Rogers, T.A. (1968). "Some psychiatrie aspects of an Aretic survival experiment." *Journ. Nerv. Ment. Dis.*, 146(6), 433–445.

Pyne, S. J. (1986). *The Ice: A Journey to Antarctica*. University of Washington Press Pub. https://www.jstor.org/stable/j.ctvcwnvc8

Qu, T.Z. (2013). *The 29th Antarctic Scientific Expedition: The Team Leader's Diary*. Beijing: Ocean Press.

Qianying, M., Gro, M.S., Ruilin, W., Jianghui, X., Zi, X., Li, H., Yang, L. & Yinghui, L. (2019). "Personal value diversity in confinement and isolation: Pilot study results from the 180-day CELSS integration experiment." *Acta Astronautica*, July, 164. DOI: 10.1016/j.actaastro.2019.07.013.

Rai, B., Kaur, J. (2012). "Human Factor Studies on a Mars Analogue During Crew 100b International Lunar Exploration Working Group EuroMoonMars Crew: Proposed New Approaches for Future Human Space and Interplanetary Missions." *N Am J Med Sci.*, 4(11), 548–57. doi: 10.4103/1947-2714.103313.

Rakusa-Suszczewski, S. (1973). *Antarktyda 1968-1972* (Antarctic 1968–1972). Warszawa: PWN.

Rakusa Suszczewski, S. (1980). "III Wyprawa Antarktyczna na Stację Arctowskiego (Wyspa Króla Jerzego, listopad 1978 – maj 1979) (3rd Antarctic Expedition to Arctowski Station (King George Island, November 1978 – May 1979." *Polish Polar Research*, 1(1), 127–146.

Rakusa-Suszczewski, S. (2012). *Wiatrem przewiany, słońcem spalony* (Blown by the wind, scorched by the sun). Warszawa: Aspra.

Ramskill, E. A. (ed.) (1962). *Studies of Bureau of Yards' and Docks' Protective Shelter*. San Francisco: Report No 5882, Naval Research Laboratory.

Rasmussen, J.E. (1963). "Psychologie discomforts in 1962 Navy protective shelter tests." *Journ. of American Dietetic Assotiation*, 42, 2, 109–116.

Reed, H. L., Reedy, K. R., Palinkas, L. A., Do, N. V., Finney, N. S., Case, H. S., LeMar, H.J., Wright, J. & Thomas, J. (2001). "Impairment in cognitive and exercise performance during prolonged antarctic residence: Effect of thyroxine supplementation the polar triiodothyronine syndrome." *Journal of Clinical Endocrinology and Metabolism*, 86(1), 110–116.

Refinetti, R., Lissen, G.C. & Halberg, F. (2007). "Procedures for numerical analysis of circadian rhythms." *Biol Rhythm Res.*, 38(4), 275–325. doi: 10.1080/09291010600903692.

Reykowski, J. (1975). Osobowość jako centralny system regulacji i integracji czynności (p. 762–825). (Personality as the central system of regulation and integration of activities). In: T. Tomaszewski (ed.): *Psychologia*. Warszawa: PWN.

Reykowski, J. (1973). Postawy a osobowość (p. 89–121) (Attitudes and personality). In: S. Nowak (ed.): *Teorie postaw* (Theories of attitudes). Warszawa: PWN.

Reykowski, J. (1977). "Spontaniczna agresja i spontaniczne czynniki ją hamujące (Spontaneous aggression and spontaneous management that inhibits it)." *Przegląd Psychologiczny*, 2, 204–228.

Ribeiro, A.C., Pfaff, DW. (2007). Stress and CNS arousal: Genomie contribution. In: G. Fink (Ed.): *Encyclopedia of stress*. Second Edition (Vol. 3, pp. 591–596). New York: Academic Press.

Ritsher, J.B. (2005). "Cultural Factors and the International Space Station." *Aviation, Space, and Environmental Medicine*, 76(6), suplement, 135–144.

Rivolier, J., Bachelard C., Cazes G. (1991) Crew Selection for an Antarctic-Based Space Simulator (pp. 291–296). In: Harrison, A.A., Clearwater Y.A., McKay C.P. (eds) *From Antarctica to Outer Space*. New York: Springer.

Rockwell, D.A., Hodgson, M.G., Beljan, J.R., Winget, C.M. (1976). "Psychologie and psychophysiologic response to 105 days of social isolation." *Aviation, Space and Envinormental Medicine*, 47(10), 1087–1093.

Rogers, A.F. (1973). Antarctic climate, clothing, and acclimatization (pp. 263–271). In: O.G. Edholm & E.K.E. Gunderson (Eds): *Polar human biology*. London: William Heinemann Medical Books, Ltd.

Rohrer, J.H. (1961). Interpersonal relationships in isolated smali groups (In:) B.F. Flaherty (ed.): *Psychophysiological aspects of flight*. New York: Columbia Press.

Rose, R.M., Helmreich, R.L., Fogg, L.F., McFadden, &. (1994). "Psychological Predictors of Astronaut Effectiveness." *Aviation, Space, and Environmental Medicine*, 64, 910–925.

Rosnet, E., Jurion, S., Cazes, G., & Bachelard, C. (2004). "Mixed-gender groups: Coping strategies and factors of psychological adaptation in a polar environment." *Aviation Space and Environmental Medicine*, 75(7), C10–C13.

Rosnet, E., Le Scanff, C., & Sagal, M.S. (2000). "How Self-Image and Personality Influence Performance in an Isolated Environment." *Environment and Behavior*, 32(1), 18–31. doi: 10.1177/00139160021972414.

Rosenthal, N.E., Sack, D.A., Gillin, J.C., Lewy, A.J., Goodwin, F.K., Davenport, Y., Mueller, P.S., Newsome, D.A. & Wehr, T.A. (1984). "Seasonal affective disorder. A description of the syndrome and preliminary findings with light therapy." *Arch. Gen. Psychiatry*, 41, 72–80.

Rotenberg, K.J., & Hymel, S. (Eds.) (1999). *Loneliness in childhood and adolescence*. New York, NY: Cambridge University Press.

Ruff, G.E., Levy, E.Z., Thaler, V.H. (1961). Factors influencing the reactions to altered sensory input (p. 72–90). (In:) P. Solomon et al. (Eds): *Sensory deprivation*. Cambridge, Mass.: Harvard Univ. Press.

Ruilin, W., Qianying, M., Jianghui, X., Zi, X. & Yinghui, L. (2018). "Leadership roles and group climate in isolation: A case study of 4-subject 180-day mission." *Acta Astronautica*, 166, 554–559. DOI: 10.1016/j.actaastro.2018.09.017.

Ryan, C. (1995). *The Pre-Astronauts: Manned Ballooning on the Threshold of Space*. Annapolis: MD: Naval Institute Press.

Sadowski, B. (2000). Mechanizmy nerwowe procesów intelektualnych (Nerve mechanisms of intellectual processes) (Vol. 1, pp. 253–277). W: E.

Szczepańska-Sadowska i E. Koźniewska (red.): *Seminaria z fizjologii.* Warszawa: Wyd. Akademii Medycznej.

Salas, E., Tannenbaum, S. I., Kozlowski, S. W. J., Miller, C. A., Mathieu, J. E., & Vessey, W. B. (2015). "Teams in Space Exploration: A New Frontier for the Science of Team Effectiveness." *Current Directions in Psychological Science*, 24(3), 200–207.

Samel, A., Wegmann, H.M. (1989). Circadian rhythm, sleep, and fatigue in aircrews operating on long-dual routes (pp. 404–422). In: R.S. Jensen (Ed.): *Aviation psychology*. Aldershot: Gower Technical.

Sandal, G.M. (2000). "Coping in Antarctica: is it possible to generalize results across settings?" *Aviat Space Environ Med.*, 71, A37–A43.

Sandal, G.M., Bye, H.H., & van de Vijver, F.J.R. (2011). "Personal values and crew compatibility: Results from a 105 days simulated space mission." *Acta Astronautica*, 69(3–4), 141–149. https://doi.org/10.1016/j.actaastro.2011.02.007.

Sandal, G.M., & Bye, H.H. (2015). "Value diversity and crew relationships during a simulated space flight to Mars." *Acta Astronautica*, 114, 164–173. doi.org/10.1016/j.actaastro.2015.05.004.

Sandal, G.M., Leon, G.R. & Palinkas, L. (2006). "Human Challenges in Polar and Space Environments." *Review Environmental Science and Biotechnology*, 5(2–3). doi: 10.1007/ s11157–006–9000–8.

Sandal, G., Vaernes, R. & Ursin, H. (1995). "Interpersonal Relations During Simulated Space Missions." *Aviation, Space, and Environmental Medicine*, 66(7), 17–24.

Sandal, G.M., Endresen, I.M., Vaernes, R., Ursin, H. (1999). "Personality and Coping Strategies During Submarine Missions." *Military Psychology*, 11, 381–404.

Sandal, G.M., Bye, H.H., & van de Vijver, F.J.R. (2011). "Personal values and crew compatibility: Results from a 105 days simulated space mission." *Acta Astronautica*, 69, 141–149. https://doi.org/10.1016/j.actaastro.2011.02.007.

Santy, P.A. (1994). *Choosing the Right Stuff: The Psychological Selection of Astronauts and Cosmonauts*. Westport, CT: Praeger.

Santy, P.A., Kapanka, H., Davis, J.R., Stewart, D.F. (1988). "Analysis of sleep on shuttle mission." *Aviat Space Environ Med.*, 59: 1094–1097.

Sarris, A. (2006). "Personality, culture fit, and job outcomes on Australian Antarctic stations." *Environment and Behavior*, 38, 356–372.

Sarris, A. (2007). "Antarctic culture: 50 years of Antarctic expeditions." *Aviation, Space, and Environmental Medicine*, 78(9), 886–892.

Sarris, A.(2017). "Antarctic station life: The first 15 years of mixed expeditions to the Antarctic." *Acta Astronautica*. 131, 50–54.

Sarris, A. Kirby, N. (2007). "Behavioral Norms and Expectations on Antarctic Stations." *Environment and Behavior*, 39(5), 706–723.

Schaeffer, H.R. (2009). *Psychologia dziecka, tłum. A. Wojciechowski* (Child Psychology, trans. A. Wojciechowski). Warszawa: PWN.

Schachter, S., Ellertson, N., McBride, D., Gregory, D. (1968). An Experimental Study of Cohesiveness and Productivity (pp. 192–198). In: D. Cartwright & A. Zander, Eds. *Group dynamics*. New York: Harper & Row. Pub.

Schein, E.H. (1957). "Reactian patterns to severe chronić stress in American army prisoners of war," *Journ. Soc. Issues*, 13, 21–25.

Schiermeier, Q. (2004). "Antarctic stations: cold comfort." *Nature*, 431(7010), 734–735. doi: 10.1038/431734a.

Schmidt, F.L. & Hunter, J.E. (2015). *Methods of Meta-Analysis: Correcting Error and Bias in Research Findings*, 3rd Edn. Thousand Oaks, CA: Sage. doi: 10.4135/9781483398105.

Schmidt, W.J., Lugg, I., Ayton, A., Phillips, J., Shepanek, T. & Life, M. (2005). "Survival, and behavioral health in small closed communities: 10 years of studying isolated Antarctic groups." *Aviat Space Environ Med.*, 76: B89–B93.

Schmidt, L.L., Wood, J.A. & Lugg, D.J. (2004). "Team climate at Antarctic research stations 1996–2000: leadership matters." *Aviat Space Environ Med.*, 75, 681–687.

Schmidt, L., Wood, J., & Lugg, D. J. (2005). "Gender differences in leader and follower perceptions of social support in Antarctica." *Acta Astronautica*, 56(9–12), 923–931.

Schneider, B., Goldstein, H. W., & Smith, D. B. (1995). "The ASA framework: An update." *Personnel Psychology*, 48, 747–774. doi: 10.1111/j.1744-6570.1995.tb01780.x.

Scott, T.H., Bexton, W.H., Heron, W., Doane, B.K. (1959). "Cognitive effects of perceptual isolation," *Canad. Journ. Psychol.* 13, 200–209.

Sells, S.B. (1966). "A model for the social system for the multiman extended duration space ship." *Aerospace Medicine*, 37, 11, 1130–1135.

Sells, S.B. (1973). The taxonomy of man in enclosed space (pp. 280–303). (In:) J.E. Rasmuissen (ed.): *Man in isolation and confinement*, Chicago, Illinois: Aldine Publ. Co.

Serafetinides, E.A., Shurley, J.T., Brooks, R., Gideon, W.P. (1972). "Electrophysiological changes in humans during perceptual isolation." *Aerospace Medicine*, 43, 4, 432–434.

Sexner, J.L. (1968). "An Experience in Submarine Psychiatry." *American Journal of Psychiatry*, 1, 25–30.

Seymour, G.E., Gunderson, E.K.E. (1971). "Attitudes as predictors of adjustment in extremely isolated groups." *Journ. Clin. Psychol.*, 27(3), 333–338.

Shears, L.M., Gunderson, E.K.E. (1966). "Stable attitude factors in natur al isolated groups." *Journal of Social Psychology*, 70, 199–204.

Sheng, B., Cheng, S.K.W., Lau, K.K., Li, H.L., & Chan, E.L.Y. (2005). "The effects of disease severity, use of corticosteroids and social factors on neuropsychiatric complaints in severe acute respiratory syndrome (SARS) patients in acute and convalescent phases." *European Psychiatry*, 20, 236–242. https://doi.org/10.1016/j.eurpsy.2004.06.023.

Shepanek, M. (2005). "Human behavioral research in space: quandaries for research subjects and researchers." *Aviat Space Environ Med.* 76(6 Supplement), B25–30. PMID: 15943191.

Shepard, A.N., Dinsmore, D.A., Miller, S.L., Cooper, C.B., & Wicklund, R.I. (1996). *Aquarius Undersea Laboratory: The Next Generation.* American Academy of Underwater Sciences, Dauphin Island, AL.

Short, R.R., Oskamp, S. (1965). "Lack of suggestion effects on perceptual isolation (sensory deprivation) phenomena." *Journal of Nervous and Mental Diseases*, 141, 190–194.

Shorter, E. (2005). *Historia psychiatrii – od zakładu dla obłąkanych po erę Prozacu* (History of Psychiatry – From Asylum to the Prozac Era). 1st Edition. Warszawa: Wydawnictwo WSiP.

Shurley, J.T. (1960). "Profound experimental sensory isolation," *Amer. Journ. Psychiat.*, 117, 539–545.

Shurley, J.T. (1962a). Mental imagery in profound experimental sensory isolation (pp. 153–157). (In:) L.J. West (ed.): *Hallucinations*. New York: Grune and Stratton.

Shurley, J.T. (1962b). Problems and methods in experirnental sensory input alternation and invariance (p. 145–160). (In:) T.T. Tourlentes, S.L. Pollach and H.E. Himwich (Eds: *Research approaches to psychiatrie problems*. New York: Grune and Stratton.

Shurley, J.T. (1970). "Man on the South Polar Plateau: Cuest editorial to symposium "Man on the South Polar Plateau."" *Archives of Internal Medicine*, 125, 625–629.

Shurley, J.T. (1973). Antarctica in also a prime natural laboratory for the behavioural science (p. 430–435. In: O.G. Edholm & E.K.B. Guhderson, eds. *Polar human biology*. London: William Heinemann Medical Books, Ltd.

Shurley, J.T., Pierce, Ch.M., Natani, K. et al. Brooks, R.E. (1970). "Sleep and Activity Patterns at South Pole Station. A Preliminary Report." *Arch Gen Psychiatry*, 22(5), 385–389. doi: 10.1001/archpsyc.1970.01740290001001.

Silverberg, R. (ed.), (1965). *Antarctic conquest: The great explorers in their own words.* New York: Bobbs-Merrill.

Silverman, A.J., Cohen, S.I., Bressler, B., Shmavonian, B.M. (1962). Hallucinations in sensory deprivation (p. 125–134). (In:) L.J. West (ed.): *Hallucinations.* New York: Grune and Stratton.

Simonow, I.M. (1975). "Fiziko-geograficzeskaja charaktieristika połostrowa Fields (Juźnyje Shetlandskije Ostrowa) (Physico-geographic characteristics of the Post-Peninsula Fields (South Shetland Islands)." *Antarktika*, 14, 128–148.

Simons, D.G. (1958). "Pilot reactions during "Manhigh II" baloon flight." *Journ. Aviat. Medic.*, 29, 1–14.

Sipes, W.E & Fiedler, E. (2007). *Current Psychological Support for US Astronauts on the International Space Station.* Paper presented at Tools for Psychological Support During Exploration Missions to Mars and Moon, European Space Research and Technology Centre [ESTEC], Noordwijk: Netherlands.

Siwiak-Kobayashi, M. (1974). *Zmiany w stanie klinicznym a modyfikacja niektórych postaw pacjentów leczonych w klinice nerwic* (Changes in the clinical state and the modification of some attitudes of patients treated in the neurosis clinic), Warsaw, unpublished doctoral dissertation, Instytut Psychoneurologiczny.

Skarżyńska, K. (1979). "Ujawnianie siebie innym (Revealing yourself to others)," *Studia Psychologiczne*, XVII, 1–2.

Skorupa, A. (2015). *Wyznaczniki efektywnego funkcjonowania jednostki w grupie w warunkach polarnych* (Determinants of the effective functioning of an individual in a group in polar conditions). Niepublikowana dysertacja doktorska (Unpublished doctoral dissertation). Katowice: University of Silesia.

Skowroński, B.Ł., Talik, E. (2018). "Radzenie sobie ze stresem a jakość życia osób osadzonych w placówkach penitencjarnych (Coping with stress and the sense of quality of life in inmates of correctional facilities)." *Psychiatr. Pol*, 52(3), 525–542. doi.org/10.12740/PP/77901.

Smith, W.M. (1961). *Scientific personnel in Antarctica: Their recruiiment, selection and performance.* Pensacola: Florida, Report No. 60-9, Naval Scholl of Aviation Medicine.

Smith, N., Barrett, E.C., Sandal, G.M. (2018). "Monitoring daily events, coping strategies and emotion during a desert expedition in the Middle East." *Stress and Health*, 34, 534–544. doi: 10.1002/smi.2814.

Smith, W.M., Jones M.B. (1962). "Astronauts, antarctic, scientistf and personal autonomy," *Aerospace Medicine*, 33(2), 162–166.

Smith, S., Haythorn, W. (1972). "Effects of compatibility, crowding, group size, and leadership seniority on stress, anxiety hostility and annoyance in isolated groups," *Journ. Pers. Soc. Psychol.*, 22, 67–69.

Smith, W.M., Jone, s M.B. (1962). "Astronauts, antarctic scientistf and personal autonomy," *Aerospace Medicine*, 33(2), 162–166.

Smith, N., Kinnafick, F., Cooley, S.J., & Sandal, G.M. (2016). "Reported growth following mountaineering expeditions: The role of personality and perceived stress." *Environment and Behavior*, 1–23. DOI: 10.1177/0013916516670447.

Smith, N., Sandal, G. M., Suedfeld, P., Johnson, P., Brcic, J., & Barrett, E. C. (2019). A systematic review of personal values research in isolated, confined and extreme environments. 70th International Astronautical Congress (IAC), Washington D.C., United States, 21–25 October 2019.

Sołżenicyn, A. (2010). *Archipelag GUŁag 1918-1956* (The Gulag Archipelago 1918–1956). Poznań: Dom Wydawniczy REBIS Sp z o.o. ISBN 978-83-7510-343-4.

Spaulding, R.C., Ford, C.V. (1972). "The "Pueblo" incident: Psychological reactions to the stresses of imprisonment and repatriation." *Amer. Journ. Psychiat.* 129(1), 49–58.

Spielberger, C.D. (1966). *Anxiety and behavior*. New York: Academic Press.

Spielberger, C.D., Gorsuch R.L., Lushene, R.E. (1970). *STAI Manual for the State-Trait Anxiety Inventory (Self-evaluation que-stionnaire)*. Palo Alto: Calii, Consulting Psychologists Press, Inc.

Spitzer, M. (1988). *Halluzinationen: Ein Beitrag zur allgemeinen und klinischen Psychopathologie*. Berlin: Springer-Verlag.

Stawowska, L. (1973). *Diagnoza typów osobowości* (Diagnosis of personality types). Katowice: Wyd. Uniwesytetu Śląskiego.

Steel, G.D. (2001). "Polar moods: Third-quarter phenomena in the antarctic." *Environment and Behavior*, 33(1), 126–133.

Steel, G. D. (2005). "Whole lot of parts: Stress in extreme environments." *Aviation Space and Environmental Medicine*, 76(6), B67–B73.

Steel, G.D., Callaway, M., Suedfeld, P. & Palinkas, L.A. (1995). "Human sleep-wake cycles in the high Arctic: effects of unusual photoperiodicity and time disentrainment." *Biol Rhythm Res.*, 26, 582–592.

Strange, R.E., Klein, W.J. (1973). Emotional and social adjustment of recent U S winter-over parties in isolated Antarctic stations (pp. 410–416). In: O.G.

Edholm and E.K.E. Gunderson (Eds): *Polar human biology*. London: William Heinemann Medical Books, Ltd.

Strelau, J, (1995). Temperament and stress: Temperament as moderator of stressors, emotional states, coping, and costs (Vol. 15, pp. 215–254). In: C.D. Spielberger, I.G. Sarason (Eds.). *Stress and emotion: Anxiety, anger, and curiosity*. Washington: Hemisphere.

Strelau, J. (1969). *Temperament i typ układu nerwowego* (Temperament and type of nervous system). Warszawa: Państwowe Wydawnictwo Naukowe.

Strelau, J., Angleitner, A., Newberry, B.H. (1999). *Pavlovian Temperament Survey (PTS): An international handbook*. Göttingen: Hogrefe & Huber Pub.

Strewe, C., Moser, D., Buchheim, J.I., Gunga, H.C., Stahn, A., Crucian, B.E., Fiedel, B., Bauer, H., Gössmann-Lang, P., Thieme, D., Kohlberg, E. Choukèr, A. & Feuerecker, M. (2019). "Sex differences in stress and immune responses during confinement in Antarctica." *Biology of Sex Differences*, 10(20), 1–17. https://doi.org/10.1186/s13293-019-0231-0.

Stroillo, M. (1963). "Ricerche sull' apprezzamento soggettivo del "tempo" durante prove di confinamento di durata limitata al massirno di sei ore," *Rivista di Medicina Aeronautica e Spaziale, (Roma)*, 2, 256–262.

Strus, W., Cieciuch, J. (2017). "Towards a synthesis of personality, temperament, motivation, emotion and mental health models within the Circumplex of Personality Metatraits." *Journal of Research in Personality*, 66, 70–95.

Stuster, J. (1997). "Human and team performance in extreme environments: Antarctic." Proceedings of the Ninth International Atmposium on Aviation Psychology. Columbus, OH, April 27 – May 1, 1997. *The International Journal Of Aviation Psychology*, 2, 942–946.

Stuster, J. (Principal Investigator) (2010). Behavioral Issues Associated with Long Duration Space Expeditions: Review and Analysis of Astronaut Journals Experiment 01-E104. Final Report. NASA Johnson Space Center.Houston: TX 77058. http://ston.jsc.nasa.gov/collection s/TRS/

Suedfeld, P. (2018). "Antarctica and space as psychosocial analogues." *REACH: Reviews in Human Space Exploration*, 9(12), 1–4. doi: 10.1016/j.reach.2018.11.001.

Suedfeld, P. (2001). "Applying positive psychology in the study of extreme environments." *Journal of Human Performance in Extreme Environments*, 6, 21–25. http://dx.doi.org/10.7771/2327-2937.1020.

Suedfeld, P. (1991). Groups in Isolation and Confinement: Environments and Experiences (pp. 135–146). In: Eds. A. A. Harrison, Y. A. Clearwater, and C. P. McKay, *From Antarctica to Outer Space: Life in Isolation and Confinement*. New York: Springer-Verlag.

Suedfeld, P. (2010). "Historical space psychology: Early terrestrial explorations as Mars analogues." *Planetary and Space Science*, 58(4), 639–645. doi.org/10.1016/j.pss.2009.05.010.

Suedfeld, P. (1998). "Homo invictus: The indomitable species." *Canadian Psychology*, 38, 164–173.

Suedfeld, P. (2005). "Invulnerability, coping, salutogenesis, integration: four phases of space psychology." *Aviat Space Environ Med.*, 76: B61-B73.

Suedfeld, P. (2006). "Space Memoirs: Value Hierarchies Before and After Missions–A Pilot Study." *Acta Astronautica*, 58, 583–586.

Suedfeld, P., Shiozaki, L., Archdekin, B., Sandhu, H. & Wood, M. (2017). "The polar exploration diary of Mark Wood: a thematic content analysis." *The Polar Journal*, 7(1), 227–241. https://doi.org/10.1080/2154896X.2017.1333327.

Summerhayes, C.P. (2008). "International collaboration in Antarctica: The international polar years, the international geophysical year, and the scientific committee on Antarctic research." *Polar Record*, 44 (4), 321–334. https://doi.org/10.1017/S0032247408007468.

Suedfeld, P. (2012). Extreme and unusual environments: Challenges and responses (pp. 348–371). In S. D. Clayton (Ed.), *Oxford library of psychology. The Oxford handbook of environmental and conservation psychology*. Oxford University Press. https://doi.org/10.1093/oxfordhb/9780199733026.013.0019

Suedfeld P., Brcic J., Legkaia K. (2009). "Coping with the problems of space flight: reports from astronauts and cosmonauts." *Acta Astronaut*, 65(3), 312–324.

Suedfeld, P., Johnson, Ph.J., Gushin, V., Brcic, J. (2018). "Motivational profiles of retired cosmonauts." *Acta Astronautica*, 146 (5), 202–205.

Suedfeld, P. & Mocellin, J.S. 1989. The stress of polar life: toward a salutogenic examination. Unpublished manuscript presented at the annual meeting of the American Psychological Association. New Orleans, LA.

Suedfeld, P., Shiozaki, L., Archdekin, B., Sandhu, H. & Wood, M. (2017). "The polar exploration diary of Mark Wood: a thematic content analysis," *Polar J*, 7, 227–241. https://doi.org/10.1080/2154896X.2017.1333327.

Suedfeld, P., Steel, G.D. (2000). "The Environmental Psychology of Capsule Habitats." *Annual Review of Psychology*, 51, 227–253.

Suedfeld, P., & Weiss, K. (2000). "Antarctica: Natural laboratory and space analogue for psychological research." *Environment and Behavior*, 32(1), 7–17. https://doi.org/10.1177/00139160021972405.

Sutherland, V.J. (2007). Understimulation/boredom. W: G. Fink (ed.): *Encyclopedia of stress*. Second edition (Vol. 3, pp. 794–796). New York: Academic Press.

Svab, L., Gross, J. (1964). *Bibliography of sensory deprivation and social isolation*, Prague: Psychiatrie Research Instit.

Swartz, J. (1960). "Emotional reactions of patients and medical personnel to respiratory poliomyelitis." *Ment. Hyg.*, 44, 97–102.

Sybil, C., Evans, G.W., Stokols, D. (1991). Winter-Over Stress: Physiological and Psychological Adaptation to an Antarctic Isolated and Confined Environment (pp. 229–237), in: Harrison, A.A., Clearwater, Y.A., McKay, Ch. P. (Eds.): *From Antarctica to Outer Space: Life in Isolation and Confinement*. New York: Springer-Verlag.

Szczepańska-Sadowska, E., Łoń, S., Sadowski, B. (2000). Kontrola zachowania i emocji (Control of behavior and emotions) (Vol. 1, pp. 199–222). W: E. Szczepańska-Sadowska i E. Koźniewska (red.): *Seminaria z fizjologii* (Physiology seminars). Warszawa: Wyd. Akademii Medycznej.

Szocik, K., Abood, S., Shelhamer, M. (2018). "Psychological and biological challenges of the Mars mission viewed through the construct of the evolution of fundamental human needs." *Acta Astronautica*, 152(11), 793–799.

Szymanik, A., Terelak, J.F. (2015). "Sense of Humor and coping stress among young pilots." *Polish Journal of Aviation Medicine, and Psychology*, 21(3), 13–21.

Tafforin, C. (2002). "Ethological observations on a small group of wintering members at dumont d'urville station (terre Adeilie)." *Antarctic Science*, 14(4), 310–318.

Tafforin, C. (2004). "Ethological analysis of a polar team in the French Antarctic station Dumont d'Urville as simulation of space teams for future interplanetary missions." *Acta Astronautica*, 55, 51–60. https://doi.org/10.1016/j.actaastro.2003.12.0

Tafforin C. (1996). "Initial moments of adaptation to microgravity of human orientation behavior, in parabolic flight conditions." *Acta Astronautica*, 38, 963–971.

Tafforin, C. (2020). Human Missions Analysis for Intelligent Missions Improvement (pp. 203–245). In: G. Pezzella, ed: *Mars Exploration – a Step Forward*. IntechOpen, DOI: 10.5772/intechopen.90795.

Tafforin, C. (2009). "Life at the Franco-Italian Concordia Station in Antartica for a voyage to Mars: Ethological study and anthropological perspectives," *Antrocom*, 5(1), 67–722.

Tafforin, C. (1994). "Synthesis of ethological studies on behavioral adaptation of the astronaut to space flight conditions." *Acta Astronautica*. 1994; 32: 131–142.

Tafforin, C., Michel, S., Galet, G. (2019). Ethological approach of the human factors from space missions to space operations (pp. 779–794). In: Pasquier, H.,

Cruzen, C.A., Schlidhuber, M., Lee, Y.H., editors. *Space Operations: Inspiring Humankind's Future*. Vol. 30. Switzerland: Springer.

Tafforin, C., Vinokhodova, A. G., Chekalina, A., & Gushin, V. I. (2015). "Correlation of etho-social and psycho-social data from "Mars-500" interplanetary simulation." *Acta Astronautica*, 111, 19–28. https://doi.org/10.1016/j.actaastro.2015.02.005.

Tanaka, M., Mizumura, K., Sato, J., Kasai, M., Mohri, M., Naraki, N. (1998). "Psychological and physiological changes during isolation and confinement: I. Group dynamics and member interaction." *Environmental Medicine*, 42(1), 4–7.

Tarafdar, M., Cooper, C.L., J-F. Stich (2019). "The technostres trifecta – techchno eustress, techno distress and design: An agenda for research." *Information Systems Journal*, 29(1), 6–42. DOI: 10.1111/isj.12169

Taylor, A.J.W. (1987). *Antarctic Psychology*. Wellington: Science Information Publishing Centre.

Taylor, A.J.W. (1969). "Professional isolates in New Zealand's Antarctic Research Programme," *Ist. Rev. Apph Psychol.*, 18(2), 135–138.

Taylor, A.J.W. (1973). The adaptation of New Zealand research personnel in the Antarctic (pp. 417–429). (In:) O.G. Edholm and E.K.E. Gunderson (Eds): *Polar human biology*, London: William Heinemann Medical Books, Ltd.

Taylor, A.J.W., Brown, M.M. (1994). Quartets in Antarctic Isolation (pp. 223–250). In: Carlson J.G., Seifert A.R., Birbaumer N. (eds) *Clinical Applied Psychophysiology. The Plenum Series in Behavioral Psychophysiology and Medicine*. Boston, MA: Springer.

Taylor, A.J.W., Duncum, K. (1987). "Brief report: Some cognitive effects of wintering-over in the Antarctic." *New Zealand Journal of Psychology*, 16, 93–94.

Taylor, A.J., Frazer, A.G. (1981). *Psychological sequalae of operation overdue following the DC10 aircrash in Antarctica* (vol 27. pp. 1–72). Wellington: Publications in Psychology Victoria Univ. of Wellington (New Zealand).

Taylor, A. J. W., & McCormick, I. A. (1987). "Reactions of family partners of Antarctic expeditioners." *Polar Record*, 23(147), 691–700.

Taylor, D.A., Wheeler, L., Altman, I. (1968). "Stress relations in socialy isolated." *Journal of Personality and Social Psychology*, 9(4), 369–376.

Taylor, S.E., Kemeny, M.E., Reed, G.M., Bower, J.E., Gruenewald, T.L. (2000). "Psychological resources, positive illusions, and health." *Am Psychol*, 55(1), 99–109. doi: 10.1037//0003-066x.55.1.99.

Tennant, C. (2007). Prisoners of war. W: G. Fink (ed.): *Encyclopedia of stress*. Second edition (Vol. 3, pp. 223–226). New York: Academic Press.

Terelak, J.F. (1971). "Refleksje na temat: Wybór zawodu pilota wojskowego jako czynnik motywacyjny o charakterze neurotyczno-kompensacyjnyn (Reflections on: The choice of the profession of a military pilot as a neurotic-compensatory motivational factor)." *Medycyna Lotnicza*, 33, 83–89.

Terelak, J.F. (1974). Reaktywność mierzona indeksem alfa a cechy temperamentalne (pp. 45–70) (Reactivity measured by the alpha index and temperamental traits). In: J. Strelau (red.), *Rola cech temperamentalnych w działaniu* (The role of temperamental traits in action). Wrocław" Ossolineum.

Terelak, J.F. (1976). "Alpha index and personality traits in pilot." *Aviation Space nad Environmental Medicine*, 47 (2), 133–136.

Terelak, J.F. (1978). ""Głód zmysłów" jako problem długotrwałych lotów kosmicznych ("Hunger for the senses" as a problem of long space flights)." *Astronautyka*, 4, 20–22.

Terelak, J.F., Błoszczyński, R. (1978). "Problemy doboru i selekcji psychologicznej kosmonautów (Problems of selection and psychological selection of astronauts)," *Medycyna Lotnicza*, 59, 18–27.

Terelak, J.F., Kobos, Z., Kozerenko, O. (1978). *Kwestionariusz Medyczno-Psychologiczny* (Medical and Psychological Questionnaire). Warszawa-Moskwa: "Interkosmos."

Terelak, J.F., Kwarecki, K., Rakusa-Suszczewski, S. (1978). "Deprywacja sensoryczna i izolacja społeczna jako problemy psychologii kosmicznej (Eksplikacje badań antarktycznych i eksperymentalnych) (Sensory deprivation and social isolation as problems of space psychology (Explanations of Antarctic and experimental research)." *Postępy Astronautyki*, 33(2), 7–31.

Terelak, J.F. (1980). ""Koszty psychologiczne" długotrwałej izolacji w strefach polarnych. (Badania empiryczne na stacji PAN im. H. Arctowskiego) (Psychological costs" of long-term isolation in polar zones. (Empirical research at the H. Arctowski PAN station)," *Medycyna Lotnicza*, 4(69), 16–24.

Terelak, J.F. (1982). *Człowiek w sytuacjach ekstremalnych: Izolacja Antarktyczna* (The Man in Extreme Situations: Antarctic Isolation). Warszawa: Wydawnictwo Ministerstwa Obrony Narodowej).

Terelak, J.F. (1982). *Introspekcje Antarktyczne. Diariusz III Wyprawy Antarktycznej PAN i zimowania – 1979 na Polskiej Stacji Antarktycznej im. Henryka Arctowskiego* (Antarctic Introspection. Diary of the 3rd Antarctic Expedition of the Polish Academy of Sciences and wintering – 1979 at the Polish Antarctic Station Henryk Arctowski). Warszawa: Wydawnictwo Ministerstwa Obrony Narodowej.

Terelak, J.F., Ruta, A. (1982). "Stymulacyjny mechanizm adaptacji człowieka do sytuacji deprywacji sensorycznej (Stimulating mechanism of human adaptation to the situation of sensory deprivation)." *Postępy Astronautyki*, 15 (4), 19–28.

Terelak, J.F. (1985). "Dinamika nieformalnoj struktury malej celewoj gruppy w jestestwiennych uslowiach stressa obszczestwiennoj izolacji (Dynamics of the informal structure of a small task force under the stress of social isolation)." *Kosmiczeskaja Biologia i Awiakosmiczeskaja Medicina*, 19(6), 90–92.

Terelak, J.F., Turlejski, J., Szczechura, J., Rożyński, J., Cieciura, M. (1985). "Dynamics o simple arithmetic task performance under Antarctic isolation," *Polis Psychological Bulletin*, 16 (2), 123–128.

Terelak, J.F., Koter, Z. (1986). "Wpływ jednorazowego podania etanolu na sprawność umysłową i psychomotoryczną (Impact of single administration of ethanol on mental and psychomotor performance)." *Psychiatria Polska*, XX, 1, 33–37.

Terelak, J.F. (1988). *Podstawy psychologii lotniczej* (Fundamentals of aviation psychology). Poznań: DWL.

Terelak, J.F., Maciejczyk, J., (1989). "Some indicators of level of adjustment to extreme conditions of existence in the Arctic and Antarctic," *Polish Psychological Bulletin*, 1989, 20 (3), 26–31.

Terelak, J.F. (2004). "Problemy symulowania "sztucznego środowiska habitatu z perspektywy sozopsychologii kosmicznej (Problems of simulating the 'artificial habitat environment from the perspective of space sozopsychology)." *Studia Ecologiae et Bioethicae*, 2, 575–594.

Terelak, J.F. (2005). *Psychologia organizacji i zarządzania* (Organizational and management psychology). Warszawa: Difin.

Terelak, J.F. & Steckiewicz, M. (2007). "Charakterystyka stresu ekologicznego i różnice indywidualne w radzeniu sobie z nim na przykładzie osób przebywających w więzieniu (Characteristics of ecological stress and individual differences in coping with it on the example of people in prison)." *Studia Ecologiae et Bioethicae*, 5, 23–41.

Terelak, J.F. (2010). *Dwa tysiące stóp pod ziemią* (Two thousand feet underground), Charaktery, 166(11), 72–76.

Terelak, J.F. (2012). "Norms as criterion of pathology in mental health sciences: A methodology perspective." *Health Sciences (Vilnius)*, 22(2), 40–44.

Terelak, J.F. (2013). "The emergence and development of space psychology in Poland: The significance of historical space flight of Mirosław Hermaszewski." *Polish Journal of Aviation Medicine and Psychology*, 19(3), 23–36.

Terelak, J.F. (2016). "Człowiek w Kosmosie: Bariery adaptacyjne z perspektywy astronautycznej (Man in space: Adaptive barriers from an astronautic perspective)." *Studia Philosophiae Christianae*, 52(3), 111–129.

Terelak, J.F. (2017). "Characteristics of the scientific and implementational activities of aviation psychologists and scientific consultancy from the perspective of the 90 years of existence of the Military Institute of Aviation Medicine." *Pol J Aviat Med Bioeng Psychol*, 23(3–4), 74–87 (DOI: 10.13174/pjambp.)

Terelak, J.F. (2018). "Dzieci w jaskini (Children trapped in the cave)." *Charaktery*, 8(259), 13–14.

Terelak, J.F. (2019). *Eustress and distress: Reactivation*. Berlin-New York-Oxford-Warszawa-Wien: Peter Lang.

Terelak, J.F. (2021). *Antarctic Winter-Over Syndrome: Narrative perspective*. Berlin-New York-Oxford-Warszawa-Wien: Peter Lang.

Tin, T, Fleming, Z.L., Hughes, K.A., Ainley, D, Convey, P, Moreno, C, Pfeiffer, S, Scott, J & Snape, I. (2009). "Impacts of local human activities on the Antarctic environment." *Antarctic Science*, 21(1), 3–33. https://doi.org/10.1017/s0954102009001722.

Tinsley, H.E.A. (2000). "The congruence myth: An analysis of the efficacy of the person–environment fit model." *Journal of Vocational Behavior*, 56(2), 147–179. https://doi.org/10.1006/jvbe.1999.1727.

Tisch, C. (2005). *Is it really so bad? A review of positive experiences of personnel wintering over in Antarctica. Literature review*. Christchurch: New Zealand, University of Canterbury.

Thomas, T.L., Garland, F.C., Mole, D., Cohen, B.A., Gudewicz, T.M., Spiro, R.T., & Zahm, S.H. (2003). "Health of US Navy submarine crew during periods of isolation." *Aviation Space and Environmental Medicine*, 74(3), 260–265.

Tkhomirov, I.I. (1973). The main trends of Soviet medical investigation in Antarctica (pp. 41–47). In: O.G. Edholm, E.K.E. Gunderson (Eds.). *Polar human biology*. London: Heinemann.

Tomaszewski, T. (1978). "Środowiskowe i sytuacyjne uwarunkowania rozwoju i działalności człowieka," *Badania Oświatowe*, l, 25–35.

Tomaszewski, T. (1998). "Uwagi na temat regulacji zachowania człowieka." *Studia Psychologiczne*, 37, 63–66.

Torbjörn, A., Göran, K, Mats, G. (2007). "Sleep and sleepiness in relation to stress and displaced work." *Physiology and Behavior*, 92 (1–2), 250–255.

Tortello, C., Barbarito, M., Cuiuli, J.M, Golombek, D., Vigo, D.E. & Plano, S. (2018). "Psychological Adaptation to Extreme Environments: Antarctica as a Space Analogue." *Psychol Behav Sci Int J*. 9(4), 1–4. DOI: 10.19080/PBSIJ.2018.09.555768.

Trossman, B. (1968). "Adolescent children of concentration camp surmvos," *Can. Psychiat. Assoc. Journ.*, 13, 121–123.

Tuckman, B.W., & Jensen, M.A. (1977). "Stages of small-group development revisited." *Group & Organization Studies*, 2(4), 419–427. https://doi.org/10.1177/105960117700200404.

Uchino, B.N. (2006). "Social support and health: A review of physiological processes potentially underlying links to disease outcomes." *Journal of Behavioral Medicine*, 29(4), 377–387.

Ursin, H., Etienne, J. L., and Collet, J. (1990). "An Antarctic crossing as analogue for long-term manned spaceflight." *European Space Agency Bulletin*, 64, 45–49.

Ursin, H., Bergan, T., Collet, J., Endresen, I. M., Lugg, D. J., Maki, P., Matre, R., Molvaer, O., Muller, H. K., Olff, M., Pettersen, R., Sandal, G. M., Vaernes, R., and Warncke, M. (1991). "Psychological studies of individuals in small, isolated groups–in the Antarctic and in space analogues." *Environment and Behavior, Special Polar Psychology Issue*, 23, 766–781.

Usui, A., Obinata, I., Ishizuka, Y., Okado, T., Fukuzawa, H. & Kanba, S. (2000). "Seasonal changes in human sleep-wake rhythm in Antarctica and Japan." *Psychiatry Clin. Neurosci.* 54, 361–362.

Vakoch, D.A. ed. (2011). *Psychology of Space Exploration: Contemporary Research in Historical Perspective*. Washington, DC: The NASA History Series. NASA SP-2011-4411.

Van Baarsen, B., Snijders, T. A. B., Smit, J. H., & Van Duijn, M. A. J. (2001). "Lonely but not alone: Emotional isolation and social isolation as two distinct dimensions of loneliness in older people." *Educational and Psychological Measurement*, 61, 119–135.

Vernon J., Hoffman J. (1956). "Effect of sensory deprivation on learning rate in human." *Science*, 123, 1074–1076.

Vernon J., McGgill, T.E. (1962). Sensory deprivatian and hallucinatians (pp. 146–152). (In:) L.J. West (ed.): *Hallucinations*. New York: London.

Vessel, E.A. & Russo, S. (2015). *Effects of Reduced Sensory Stimulation and Assessment of Countermeasures for Sensory Stimulation Augmentation. A Report for NASA Behavioral Health and Performance Research: Sensory Stimulation Augmentation Tools for Long Duration Spaceflight*. New York, NY: NASA/TM-2015-218576.

Vgontzas, A.N., Pejovic, S., Karataraki, M. (2007). Sleep, sleep disorders, and stress. In: G. Fink (Ed.): *Encyclopedia of stress*. Second Edition (Vol. 3, pp. 506–514). New York: Academic Press.

Vinokhodova, A.G., & Gushin, V.I. (2021). "Anticipated and perceived personal growth and values in two spaceflight simulation studies." *Acta Astronautica*, 179(2), 561–568. https://doi.org/10.1016/j.actaastro.2020.11.029.

Vinokhodova, A.G., & Gushin, V.I. (2014). "Study of values and interpersonal perception in cosmonauts on board of international space station." *Acta Astronautica*, 93(93), 359–365. https://doi.org/10.1016/j.actaastro.2013.07.026.

Voas, R., Zedekar, R. (1963). Astronaut Selection and Training: chap. 10 in *Mercury Project Summary Including the Results of the Fourth Manned Orbital Flight*, May 15 and 16, 1963. Washington, DC: Office of Scientific and Technical Information, NASA.

Volkov, I.P., Matusoy, A.L., Ryabinin, I.F. (1976). A sociopsychological study of typical motives for joining the Somet Antarctic Expedition (Vol. 42, pp. 84–90). In: A.F. Treshnikov (ed.) *Problems of the Arctic and the Antarctic*. New Delhi-Washington, D.C.: Amerind Publ. Co.

Von Holst, E. (1973). *The behavioural physiology of animals and man: The collected papers of Erich von Holst*. Hardcover. University of Miami Press. ISBN-10: 087024261X

Wallace, P. (2004). *The Internet in the Workplace: How New Technology is Transforming Work*. New York: Cambridge University Press.

Ward, L.M. (2003). "Synchronous neural oscillations and cognitive processes." *Trends in Cognitive Sciences*, 7(12), 553–559. doi: 10.1016/J.Tics.2003.10.012.

Wegmann, H.M., Klein, K.E., Conrad, B. & Esser, P. (1983). "A model for prediction of resynchronisation after time-zone flights." *Aviation, Space and Environmental Medicine*, 54(6), 524–527.

Weiss, K., & Moser, G. (1998). "Interpersonal relationships in isolation and confinement: Long-term bed rest in head-down tilt position." *Acta Astronautica*, 43(3–6), 235–248.

Weltman G., Crooks T., Egstrom G.H. (1969). Diver performance measurement for Sealab III, (In:) J.N. Lythgoe and E.A. Drews (Eds): *Underwater assotiation report 1969* (pp. 53–55). Guildforg: Engl., Science and Technology Publ., Ltd.

West, J.B. (2000). "Historical perspectives: Physiology in microgravity." *Journal of Applied Physiology*, 89. 379–84.

Westin, A. F. (1967). *Privacy and freedom*. New York: Atheneum.

Weybrew, B.B. (1961). "The impact of isolation upon personel." *Journ. Occupational Med.*, 3, 6, 290–294.

Weybrew, B.B. (1963). Psychological problems of prolonged marinę submergence (pp. 87–125). (In:) N.M. Burns and R.M. Chambers and E. Hendler (Eds): *Unusual environments and human behavior – Physiological and psychological problems of mań in space*. New York: Macmillan Company.

Weybrew, B.B. (1991). Three Decades of Nuclear Submarine Research: Implications for Space and Antarctic Research (pp. 103–114). In: Harrison, A.A.,

Clearwater, Y.A., McKay, Ch. P. (Eds.): *From Antarctica to Outer Space: Life in Isolation and Confinement.* New York: Springer-Verlag.

Weybrew, B.B., Molish, ELB., Youniss, R.P. (1961). *Prediction of adjustment to the Antarctic.* New London, Conn.: Report No. 350, US Naval Medical Research Laboratory.

Weybrew, B.B., Parker J.W. (1960). "Bibliography of sensory deprivation, isolation, and confinement," *US Araned Forces Med. Journ.*, 11, 903–911.

Wheaton, J.L. (1959). *Fact and fancy in sensory deprivation studies.* Texas: Brooks, Report No. 5–59, School of Aviation Medicine Brooks Air Force Base.

Wheelan, S. A. (1994). *Group processes: A developmental perspective.* Boston: Allyn & Baco.

White, G.J., Taylor, A.J.W., McCormick, I.A. (1983). "A note on the chronometric analysis of cognitive ability: Antarctic effects." *New Zealand Journal of Psychology*, 12, 36–40.

Whitehead, D.L., Syteptoe, A. (2007). Prison. In: G. Fink (Ed.): *Encyclopedia of stress.* Second Edition (Vol. 3, pp. 217–222). New York: Academic Press.

Whitmire, A.M., Leveton, L.B., Barger, L., Jefferson, G.B., Dinges, D.F., Klerman, E., Shea, C. (2009). Risk of Performance Errors due to Sleep loss, Circadian Desynchronization, Fatigue and Work Overload: NASA Evidenced-Based Review (NASA, Washington). Human Research Program Requirements Document HRP-47052, Chapt. 3, pp. 85–116.

Wichman, H.A. (1992). *Human Factors in the Design of Spacecraft.* Stony Brook: State University of New York. Research Foundation.

Wichman, H. (2011). Managing Negative Interactions in Space Crews: The Role of Simulator Research. (103–123). Edited by D. A. Vakoch, *Psychology of Space Exploration: Contemporary Research in Historical Perspective.* Washington, DC: The NASA History Series.

Williams, T. J., Landon, L. B., Vessey, W. B., Schneiderman, J., & Basner, M. (2017). *Developing behavioral health and performance standard measures for the Team, Sleep, B Med Risks.* Presented at NASA Human Research Program Investigators' Workshop, Galveston, TX.

Willkins, W.L. (1966). Group behavior in long-term isolation (p. 278–296). In: M.H. Appley and R. Trumbull (Eds): *Psychological stress: Issues in research.* New York: Appleton-Century-Crofts.

Willkins, W.L, (1973). Isolation research: the methodological contex (Chapter 12). In: J.E. Rasmussen (Ed.). *Man in isolation and confinement.* Chicago: Aldine.

Wilson, O. (1965). Human adaptation to life in Antarctica, (In:) J. Van Meighen, P. Van Oye, J. Schell and W. Junh (Eds) *Monographia biologia: Biogeography and ecology in Antarctica*, Hague: Netherlands.

Wilson, D. (2011). *The third-quarter phenomenon in Antarctic personnel*. Christchurch: University of Canterbury.

Wittmann, M., Dinich, J., Merrow, M. & Roenneberg T (2006). "Social jetlag: Misalignment of biological and social time." *Chronobiol Int*, 23(1–2), 497–509.

Wood, J., Hysong, S. J., Lugg, D. J., Harm, D. L. (2000). "Is it really so bad? A comparison of positive and negative experiences in Antarctic winter stations." *Environment and Behavior*, 32, 84–110. doi: 10.1177/00139160021972441.

Wood, J., Schmidt, L., Lugg, D., Ayton, J., Phillips, T., Shepanek, M. (2005). "Life, Survival and Behavioral Health in Small Closed Communities: 10 Years of Studying Small Antarctic Groups." *Aviation, Space, and Environmental Medicine*, 76(6), Supplement), B89–93. PMID: 15943201.

Wróblewski, R.J. (2017). *Dzienniki antarktyczne. Dwie zimy w krainie lodu, śniegu i wiatru. Wyd. 2* (Antarctic memoirs. Two winters in the land of ice, snow and wind (2nd ed.). Wrocław: Oficyna Wyd. Atut. ISBN: 978-83-7977-284-1.

Wu, Z., & Penning, M. J. (2015). "Immigration and loneliness in later life." *Ageing & Society*, 35, 64–95.

Wu, R., Ma, Q., Xiong, J., Xu, Z., Li. I. (2020). "Leadership roles and group climate in isolation: A case study of 4-subject 180-day mission." *Acta Astronautica*, 166, 554–559.

Wu, R., Wang, W. (2015). "Psychosocial interaction during a 105-day isolated mission in Lunar Palace 1," *Acta Astronautica*, 113, 1–7.

Von Mayersbach, H., ed. (1967). *The Cellular Aspects of Biorhythms*. Berlin: Springer.

Von Wulfften-PaIthe. P.M. (1968). "Time sence in isolation." *Psychiat. Neurol. Neurochir.*, 71, 221–241.

Xing, W., Hejblum, G., Leung, G. M., & Valleron, A.-J. (2010). "Anatomy of the epidemiological literature on the 2003 SARS outbreaks in Hong Kong and Toronto: A time-stratified review." *PLoS ONE*, 7, Article e1000272.

Yan, G., Wu, S., Wang, T., Zhang, X., & Saklofske, D. H. (2012). "Cognitive effects of long-term residence in the Antarctic environment." *Advances in Polar Science*, 23(3), 170–175.

Yang, K. (2019). *"Loneliness:" The Social Problem*. New York: Routledge Pub.

Yang, G., Ye, Q. & Tang, C. (2011). "Adaptation and coping strategies in Chinese Antarctic expeditioners winter-over life." *Advances in Polar Science*, 22, 111–117. DOI: 10.3724/SP.J.1085.2011.00111.

Yehuda, R., Wong, CM. (2007). Acute stress disorder and posttraymatic stress disorder. In: G. Fink (Ed.): *Encyclopedia of stress*. Second Edition (Vol. 1, pp. 1–7). New York: Academic Press.

Yoneyama S, Hashimoto S, Honma K. (1999). "Seasonal changes of human circadian rhythms in Antarctica." *Am. J. Physiol.* 277, Supplement: R1091–R1097.

Yusupova, K., Shveda, D.M., Gushchin, V.I., Supolkina, N.S. & Chekalkin A.I. (2019). "Preliminary Results of "Content" Space Experiment." *Human Physiology*, 45(7), 710–717. DOI: 10.1134/S0362119719070181.

Zalewska, N. & Kubiak, K. (2018). "Mars Desert Research Station, czyli Mars na Ziemi: Relacja z pobytu w symulowanej bazie marsjańskiej w Utah (USA) pierwszej polskiej załogi EXO (Mars Desert Research Station, or Mars on Earth: An account of the stay in a simulated Martian base in Utah (USA) of the first Polish EXO crew)." *Fizyka w Szkole z Astronomią*, 355 (2), 42–49.

Zamble, E. & Porporino, F.J. (1988). *Coping, behavior, and adaptation in prison inmates*. New York: Springer-Verlag Inc.

Zeger, S.L., Irizarry, R., Peng, R.D. (2006). "On time series analysis of public health and biomedical data." *Ann Rev Publ Health*, 27: 57–79.

Zeitzer, J.M., Dijk, D.J., Kronauer, R., Brown, E. & Czeisler, C. (2000). "Sensitivity of the human circadian pacemaker to nocturnal light: Melatonin phase resetting and suppression." *J. Physiol.*, 1(526), 695–702. doi: 10.1111/j.1469-7793.2000.00695.x.

Zhao, Y., Xu, T.H & Liu, F.Y. (2001). "The study of personality of research members in the Antarctic region." *Heath Psychology Journal*, 4, 309–310. http://www.cnki.net/kcms/doi/10.13342/j.cnki.cjhp.2001.04.042.html

Zięba, M.S. (2000). Budda (pp. 699–703). W: A. Maryniarczyk, M.J. Gondek, T. Zawojska & N. Nawracała-Urban (Red.). *Powszechna Historia Filozofii* (The Universal History of Philosophy), Vol. 1. Lublin: Wyd. Polskie Towarzystwo Tomasza z Akwinu (Ed. Polish Society of Thomas Aquinas).

Zimbardo, P.G., Boyd J. (2009). *The Paradox of Time: The New Psychology of Time That Will Change Your Life*. New York: Free Press.

Zimmer, M., Cabral, J. C. C. R., Borges, F. C., Coco, K. G., & Hameister, B. R. (2013). "Psychological changes arising from an Antarctic stay: Systematic overview." *Estudos de Psicologia Campinas*, 30(3), 415–423. Doi.org/10.1590/S0103-166X2013000300011.

Ziskind, E., Jones H., Filante W., Goldberg J. (1960). "Observations on mental symptoms in eye patched patients: Hypnagogic symptoms in sensory deprivation." *Amer. Journ. Psychiat.*, 116, 893–900.

Zubek, J.P. (1964). "Effects of prolonged sensory and perceptual deprivation." *British Medical Bulletin*, 20, l, 38–42.

Zubek, J.P. (1969). *Sensory Deprivation: Fifteen Years of Research.* New York: Appleton-Century Crofts.

Zubek, J.P. (1970). Behavioural and EEG changes during and after 14 days oj perceptual deprivation and confinement (pp. 165–171). In: A.W. Pressey & J.P. Zubek (Eds): *Readings in general psychology, Canadian Contributions.* Toronto: McClelland and Stewart.

Zubek, J.P., Aftanas, M., Hasek, J., Sanson, W., Schludermann, E., Wilgosh, L., Winocur, G. (1962). "Intellectual and perceptual changes during prolonged perceptual deprivation: Low illumination and noise level," *Perceptual and Motor Skills,* 15(1), 171–198.

Zubek, J.P., Sanson, W., Prysiazniuk, A. (1960). "Intellectual changes during prolonged perceptual isolation (Darkness and silence)." *Canad. Journ. Psychpl.,* 14, 4, 233–242.

Zubek, J.P., Welch, G., Saunders, M.G. (1963). "Electroencephalographic changes during and after 14 days of perceptual deprivctian." *Science,* 139, 355, 490–492.

Zuckerman, M., Albright, R. J., Marks, C. S., & Miller, G. L. (1962). "Stress and hallucinatory effects of perceptual isolation and confinement." *Psychological Monographs: General and Applied,* 76(30), 1–15. https://doi.org/10.1037/h0093851.

Zuckerman, M., Persky, H., Link, K.E., Basu, G.K. (1968). "Experimental and subject factors determining responses to sensory deprivation, social isolation, and confinement." *Journ. Abnorm. Psychol.,* 73, 183–194.

Zuckerman, M., Persky, H., Miller, L., Levine, B. (1970). "Sensory deprivation versus sensory variation," *Journ. Abnorm. Psychol.,* 76(1), 76–82.

Zużewicz, K., Rejment-Kisiel, E., Kwarecki, K., Meller, H. (1978). "Zasady analizy matematyczno-statystycznej rytmów okołodobowych (Principles of mathematical and statistical analysis of circadian rhythms)." *Medycyna Lotnicza,* 1, 17–24.

Annex: Some Abbreviations Used in the Text

BASALT – Biologic Analogue Science Associated with Lava Terrains
CAPCOM – capsule communicator
ConOps – concepts of operations
DRATS = Desert Research And Technology Studies
EAMD – Exploration Analogue and Mission Development
EV – extravehicular
EVA – extravehicular activity
EVIB – EV informatics backpack
FST – field support team
GAT – ground assimilation time
ICE – Isolated, Confined, Extreme
ISS – International Space Station
IV – intravehicular
JSC – Johnson Space Center
MCC – Mission Control Center
MIP – mobile instrument platform
MMT – mission management team
MSC – Mission Support Center
NEEMO – NASA Extreme Environment Mission Operations
PLRP – Pavilion Lake Research Project
RATS – Research And Technology Studies
SA – situational awareness
SCICOM – science communicator
SIMCOORD – simulation coordinator

Index of Names

A
Adams, O.S. 87
Agajanian, N.A. 89
Alling, A. 333
Allport, G.W. 216
Altbäcker, A. 70-72, 145
Altman, I. 98, 148, 250, 251
Amundsen, R. 14, 115
Arnhoff, F.N. 79
Arctowski, H. 13, 17, 22, 68, 69, 73, 115, 117, 148, 149, 151, 154, 155, 161-165, 167, 168, 170-172, 174, 182, 184-186, 191, 198, 202, 203, 219, 221-224, 227, 228, 231, 235, 237, 239, 252, 256, 264-269, 271, 272, 275, 276, 292, 295, 296, 298, 299, 301, 302, 306, 308, 336, 337
Arendt, J. 292, 303, 336
Aronson, E. 149, 218, 279
Atkinson, J.W. 46
Ayton, J. 233, 313, 333

B
Bachelard, C. 119, 122
Baisden, D.L. 114
Balázs, L. 70-72, 145
Baldanza, S. 314
Bales, R.F. 88, 95, 96, 260, 261
Bandura, A. 319
Barabasz, A.F. 248, 249, 320
Barbarito, M. 314
Barnett, J.S. 189
Barrett, E.C. 71
Barrett, J.D. 325
Basar, E. 37
Basner, M. 92, 274, 303
Bazel, N. 111

Beaton, K.H. 327, 328
Behrendt, J.C. 70
Bell, S.T. 326
Berkeley, G. 342
Berscheid, E. 147
Bessone, L. 61
Bettelheim, B. 58
Bexton, W. 21, 51, 56, 74, 81
Biela, A. 46, 71
Bingham, C. 293, 294
Bion, W.R. 318
Bishop, S.L. 96, 102, 132, 314
Blackadder-Weinstein, J. 273, 324
Blackburn, A.B. 202, 310
Blake, M.J. 305
Bombard, A. 70
Boriskin, W.M. 131, 275, 292, 297
Borucki, Z. 219, 220
Boster, J.S. 238, 266, 267, 272, 273, 276
Bouchard, S. 95
Boyd, J.E. 232, 274, 323, 333
Brand-Persson, A. 97
Brcic, J. 92, 215, 321
Brooks, R.E. 192
Brown, M.M. 115, 326
Browner, D. 199, 319
Brownfield, C.H. 76, 79, 83
Burke, C.S. 325
Burns, N.M. 78, 88
Buss, A.H. 211
Bye, H.H. 94, 259, 318
Byrd, R.E. 17, 117, 126, 143

C
Cable, D.M. 137
Caldwell, B.S. 309

Caplan, R.D. 251
Carabel, T.C. 330
Carter, J. 324
Cattell, R.B. 216
Cazes, G. 119, 121, 122, 306, 307
Chaikin, A.L. 257, 262, 268
Chappell, S.P. 327
Chiies, W.D. 87
Ciosek, M. 61, 67, 158, 238, 240
Coburn, K.R. 88
Colquhoun, W.P. 291, 305
Connery, H.J. 116, 117, 124, 129, 223, 251
Converse, Ph.E. 250, 279
Cook, F.A. 68
Copernicus, M. 328
Corneliussen, J.G. 225
Cowan, T.A. 98
Cravalho, M. 199
Crocą, L. 121, 306, 307
Crooks, T. 100

D
Dahlen, M. 264
Darley, J. 147
Davidová, L. 257, 273
De Roshia, C.W. 335
Del Vecchio, R.J. 100
De Palma, S.M. 130
Derendorf, H. 209
Deriap, N.R. 139, 150
Derlega, V. 257, 261, 262, 268
De Waele, J. 61
Dobrowolski, A.B. 115, 154, 161
Doli, R.E. 130, 310
Dollard, J. 321
Duncum, K. 248
Durkee, A. 211

E
Earls, J.H. 101, 240
Edholm, O.G. 117

Edwards, J.R. 136-138
Egorov, B.B. 126, 151, 195, 246
Egstrom, G.H. 100
Ehmann, B. 70-72, 145
Eliasz, A. 44, 191, 216
Elliott, D. 130
Esser, A.W. 58
Eyal, S. 209
Eysenck, H.J. 40

F
Fan, J. 249
Festinger, L. 150, 240
Fiedler, E.R. 89, 96, 327
Fiske, D.W. 34, 40, 49, 76, 78, 191
Flinn, D.E. 86, 259, 260
Flynn, C.F. 308
Foley, J. 64
Ford, C.V. 59, 60
Frazier, A.G. 63
Freedman, S.J. 81, 196
Fuerst, K. 82, 243

G
Galet, G. 340
Gardner, P.J. 66
Gavalas, A. 197, 316
Genefke, I. 60
Gierasimowicz, A.A. 89
Gifford, E.C. 88
Godorowski, K. 59
Goel, N. 273
Górski, J. 339
Gorsuch, R.L. 207, 234
Görana, K. 302
Gössmann-Lang, D. 309
Gracak, P. 111
Graybiel, A. 55
Gregson, R.A.M. 248
Greenblatt, M. 81, 196
Gross, J. 57, 70, 77
Grunebaum, H.H. 81, 196

Grzelak, J. 261, 279
Guilford, J.P. 40, 82
Gunderson, E.K.E. 68, 118-122, 125, 126, 130, 131, 140, 145, 150, 152, 158, 175, 177, 197, 199, 201, 220, 227, 229, 235, 239, 240, 250, 254, 275, 306, 308, 310
Gushin, V. 91, 113, 273, 285, 317, 320, 337
Gussow, Z. 70

H
Hachisuka, H. 121, 221, 223, 308
Halberg, F. 285, 286, 293
Happ, D. 49
Harari, Y.N. 329, 334, 341, 342
Hare, P.A. 276
Harris, A. 173, 293, 336
Harrison, A.A. 88, 89, 135, 144, 306, 324, 327
Hebb, D. 21, 34, 35, 48, 51, 64, 74, 192, 195, 249
Helmholtz, H. 76
Helmreich, R.L. 326
Heppener, M. 24, 103, 329
Hermaszewski, M. 106, 109, 110, 111, 155
Heron, W. 21, 51, 56, 74, 78, 81, 83
Hoffman, J. 51, 75
Horne, J.A. 305
Houseal, M. 70, 72, 233, 313
Hunt, B.I. 291, 303

I
Ilyin, E.A. 119, 126, 148, 151, 195, 240, 246

J
Jack, E. 68
Jack, F. 68
Jaksic, C. 137, 138, 141, 189, 197
Jankowski, K. 65, 66

Jarmolenko, A.V. 57
Jażwiński-Buczyńska, B. 221
Jezierska, S. 48, 57
Joern, A.T. 127, 152, 239, 250, 284
John Paul, F.U. 248
Johnson, J.C. 238, 266, 267, 272, 273, 276, 323,
Jones, M.B. 117

K
Kamiński, M. 71
Kanas, N. 90, 91, 95, 112, 209, 322, 323
Kaur, J. 338
Kazaniecki, M. 339
Kąkolewski, K. 55
Kelly, A.D. 112
Kelly, S.M. 335
Kimura, D. 78
Kintz, N.M. 327
Kirchner, T. 60
Kirby, N. 213, 274
Kjærgaard, A. 226
Klein, W.J. 125, 130, 148, 151, 158, 250, 268, 275
Klimuk, P. 155
Klonoff, H. 59
Kobos, Z. 107, 108
Kofta, M. 220
Koncz, A. 328
Kowalski, W. 143, 300, 301, 302, 319
Kozerenko, O. 89, 91, 107, 108, 110, 112, 320
Kraepelin, E. 243-248
Kreutz, M. 53, 54
Kring, J.P. 189, 323
Kristof, A.L. 136, 137
Kubzansky, P.E. 76
Kuznetsov, O.N. 237
Küng, H. 318
Kwarecki, K. 23, 52, 89, 102, 115, 122, 154, 156, 286, 315, 335

L
Landon, L.B. 325
Lebedev, V.L. 113, 237
Lee, E. 56
Lem, S. 13, 54
Leon, G.R. 72, 116, 226, 323, 324
Leuba, C. 34
Levy, E.Z. 79, 146, 147
Lewin, K. 30, 129
Lissen, G.C. 293
Lomov, B.T. 25
Lorek, A. 328
Lorenz, K. 54, 282
Lugg, D.J. 122, 239, 270, 284, 306, 309
Lung, F.W. 67
Lushene, R.E. 207, 234

M
Maddi, S. 34, 40, 49, 191
Mahan, J.L. 121, 177, 197, 250, 310
Mallis, M.M. 144, 159, 335
Malzberg, B. 56
Małkiewicz, M. 337
Marcussen, H. 60
Martin, M. 70
Mason, M.K. 57
Mathieu, J.E. 317
Mats, G. 32, 138, 159, 302, 305
Matusow, A.Ł. 139, 142, 150
Matysiak, J. 40, 48
McAdams, D.P. 54
McCormick, I.A. 233, 248, 320
McGuire, F. 117
McNeil, J. 21, 70, 130, 139, 146, 185, 186, 247, 270, 275
Michel, S. 340
Milgram, S. 21
Mills, T.M. 170, 311
Moallef, P. 66
Mocellin, J.S.P. 70, 238, 314
Moczydłowski, L. 68
Mohino, M. 60

Molish, E.L.B. 121
Mueller, F. 109
Mullin, C.S. 116, 117, 124, 129, 145, 161, 170, 195, 215, 223-225, 227, 235, 248, 251, 297, 314
Musson, D.W. 326
Myasnikov, V.I. 90

N
Nelson, P.D. 120, 129, 141, 145, 152, 158, 220, 239, 250, 252, 254, 275, 306
Newcomb, T.M. 153, 158, 250, 254, 279
Nicolas, M. 91, 206, 306, 312
Norris, K. 233, 313, 333

O
Orne, M.T. 50
Osgood, Ch.E. 241
Osiński, G.R. 335
Oskamp, S. 80

Östberg, O. 305

P
Padalka, G. 114
Palinkas, L.A. 17, 70, 72, 73, 86, 116, 119, 158, 161, 199, 226, 233, 238, 248, 250, 266, 267, 272, 273, 276, 291, 311, 313, 319, 320, 322, 323, 324, 335
Palmai, G. 117, 121, 126, 130, 142, 145, 147, 161, 223, 240, 243, 268, 297
Paton, K.D. 233, 313, 333
Patten, S.B. 67
Paul, J. 243
Peeters, M.A.G. 226
Pell, S.F. 109
Peri, A. 314
Perreault, M. 95

Persky, H. 84
Petrov, B.N. 25
Pfaff, D.W. 39
Pilkiewicz, M. 254
Płużek, Z. 199
Poláčková Šolcová, I. 93
Polakov, V. 114, 135
Pope, F.E. 122
Porporino, E.J. 61
Prysiazniuk, A. 82

Q
Qianying, M. 324
Qu, T.Z. 70

R
Rai, B. 338
Rakusa-Suszczewski, S. 70, 102, 115, 117, 139, 154, 161, 308
Ramskill, E.A. 97
Rasmussen, O.V. 60, 97, 197
Reuning, H. 247
Reykowski, J. 34, 46, 58, 212, 224
Ribeiro, A.C. 39
Ritter, Ch. 70
Rivolier, J. 119, 121, 122, 306, 307
Rockwel, D.A. 85, 236
Rogers, T.A. 122, 144
Rohrer, J.H. 117, 124, 128
Rosenthal, N.E. 70, 72, 313
Ruff, G.E. 79
Ruilin, W. 324
Russo, S. 334
Ruta, A. 41
Ryabinin, I.T. 123, 150, 177
Ryman, D.H. 120, 128

S
Saint-Exupery, A. 70
Salas, E. 210
Sandal, G.M. 71, 72, 88, 94, 100, 102, 116, 259, 318, 323, 324

Sanson, W. 82
Sarris, A. 137, 197, 213, 274, 318
Saunders, M.G. 79, 83
Sauro, F. 61
Schachter, S. 152
Schaffer, H.R. 58, 275
Scheibe. K.E. 50
Schein, E.G. 59
Schmidt, L.L. 270
Scott, T.H. 51, 56, 74, 78, 81
Seckbach, J. 114
Sells, S.B. 52, 77, 146, 156
Serafetinides, E.A. 84
Shepanek, M. 309
Short, S.S. 80
Shurley, J.T. 51, 54, 75, 76, 79, 126-128, 139, 143, 148, 151, 152, 159, 195, 201, 239, 250, 284, 292, 297, 301, 304, 310, 316
Silverberg, R. 69
Silverman, A.J. 79
Simonov, I.M. 162
Simons, D.G. 55
Siwiak-Kobayashi, M. 197
Skorupa, A. 207
Skowronski, B.Ł. 58, 60
Slack, K.J. 325
Slewicz, S.B. 131, 275
Smith, W.N. 90, 117, 121
Spaulding, R.C. 59, 60
Speil, J. 171, 172
Spielberger, C.D. 86, 207, 234, 236
Spitzer, M. 80
Steckiewicz, M. 60
Steel, G.D. 73, 100, 161, 292
Stich, J.F. 110, 330
Strange, R.E. 125, 130, 148, 151, 158, 250, 268, 275
Strelau, J. 34, 40, 44, 49, 50, 181, 192, 215, 234, 249
Strickland, D.A. 98
Stuster, J. 112, 127

Subramanian, Ł. 67, 264
Suci, G.J. 241
Suedfeld, P. 48, 70-72, 92, 100, 116, 161, 276, 283, 314, 315, 317, 319, 320, 322, 323, 324, 335, 337, 338
Svab, J. 57, 70, 77
Swartz, J. 64

T
Tafforin, C. 92, 105, 249, 282, 337, 340, 341
Talik, E. 58, 60
Tanaka, M. 87
Tannebaum, P.H.
Tarafdar, M. 110, 330
Taylor, A.J.W. 23, 86, 115, 119, 121, 123, 127, 151, 197, 218, 221, 233, 235, 248
Taylor, J.A. 40
Taylor, S.E. 319
Terelak, J.F. 11-14, 17, 19, 23, 32, 33, 37, 38, 40, 42, 48, 52, 56, 60, 62, 70, 71, 89, 90, 102, 103, 107, 108 110, 112, 115, 116, 122, 129, 145, 148, 149, 154-156, 162, 163, 171, 181, 186, 191, 195, 196, 201, 205, 207, 209, 211-213, 215, 223, 224, 230, 231, 237, 240, 257, 258, 261, 268, 270, 271, 272, 278, 282, 308-310, 312, 315, 322, 337
Thaler, V.H. 79, 146
Thomas, T.L. 102
Thurstone, L.L. 40
Tikhomirov, L.I. 139, 142, 150
Tisch, C. 197
Tolchin, S. 117
Tomaszewski, T. 29-31, 34, 35, 46, 138, 139, 157, 190, 218, 220, 312
Torbjörna, A. 302
Tortello, C. 333
Turner, R.H. 250, 279

U
Ursin, H. 88, 116
Usui, A. 336

V
Vaernes, R. 88
Van De Vijver, F.J.R. 94
Vernon, J. 51, 75, 79, 88
Vessel, E.A. 334
Vinokhodova, A. 273, 317, 337
Volkov, I.P. 123, 150, 177
Von Holst, E. 282

W
Wehmann, H.M. 291, 303
Weiss, D.S. 90, 319, 324
Welch, G. 79, 83
Weltman, G. 100
Weybrew, B.B. 100-102, 121
Wheelan, S. 250
Wheeler, L. 148, 250
White, L. 221
Wichman, H. 24, 96
Williams, J.V.A. 67
Wilson, O. 118
Wood, J.A. 270
Wróblewski, R.J. 70
Wu, R. 95

X
Xing, W. 66

Y
Yan, G. 248
Yang, K. 52, 322
Youniss, R.P. 121
Yusupova, A. 91

Z
Zalewska, N. 339
Zamaletdinov, I.S. 90
Zamble, E. 61

Zawadzki, B. 15
Zawiejska, K. 338
Zimbardo, P. 333
Zimmer, H. 152
Zimmerman, W.S. 40

Ziskind, E. 63, 80
Zubek, J.P. 78, 79, 81-83, 85, 243
Zuckerman, M. 39, 49, 84, 211
Zużewicz, K. 293

Studies in Social Sciences, Philosophy and History of Ideas

Edited by Andrzej Rychard

- Vol. 1 Józef Niżnik: Twentieth Century Wars in European Memory. 2013.
- Vol. 2 Szymon Wróbel: Deferring the Self. 2013.
- Vol. 3 Cain Elliott: Fire Backstage. Philip Rieff and the Monastery of Culture. 2013.
- Vol. 4 Seweryn Blandzi: Platon und das Problem der Letztbegründung der Metaphysik. Eine historische Einführung. 2014.
- Vol. 5 Maria Gołębiewska / Andrzej Leder/Paul Zawadzki (éds.): L'homme démocratique. Perspectives de recherche. 2014.
- Vol. 6 Zeynep Talay-Turner: Philosophy, Literature, and the Dissolution of the Subject. Nietzsche, Musil, Atay. 2014.
- Vol. 7 Saidbek Goziev: Mahalla – Traditional Institution in Tajikistan and Civil Society in the West. 2015.
- Vol. 8 Andrzej Rychard / Gabriel Motzkin (eds.): The Legacy of Polish Solidarity. Social Activism, Regime Collapse, and the Building of a New Society. 2015.
- Vol. 9 Wojciech Klimczyk / Agata Świerzowska (eds.): Music and Genocide. 2015.
- Vol. 10 Paweł B. Sztabiński / Henryk Domański / Franciszek Sztabiński (eds.): Hopes and Anxieties in Europe. Six Waves of the European Social Survey. 2015.
- Vol. 11 Gavin Rae: Privatising Capital. The Commodification of Poland´s Welfare State. 2015.
- Vol. 12 Adriana Mica / Jan Winczorek / Rafał Wiśniewski (eds.): Sociologies of Formality and Informality. 2015.
- Vol. 13 Henryk Domański: The Polish Middle Class. Translated by Patrycja Poniatowska. 2015.
- Vol. 14 Henryk Domański: Prestige. Translated by Patrycja Poniatowska. 2015.
- Vol. 15 Cezary Wodziński: Heidegger and the Problem of Evil. Translated into English by Agata Bielik-Robson and Patrick Trompiz. 2016.
- Vol. 16 Maria Gołębiewska (ed.): Cultural Normativity. Between Philosophical Apriority and Social Practices. 2017.
- Vol. 17 Anita Williams: Psychology and Formalisation. Phenomenology, Ethnomethodology and Statistics. 2017.
- Vol. 18 Mikołaj Pawlak: Tying Micro and Macro. 2018.
- Vol. 19 Franciszek Sztabiński / Henryk Domański / Paweł B. Sztabiński (eds.): New Uncertainties and Anxieties in Europe. Seven Waves of the European Social Survey. 2018.
- Vol. 20 Adriana Mica / Katarzyna M. Wyrzykowska / Rafał Wiśniewski / Iwona Zielińska (eds.): Sociology of the Invisible Hand. 2018.

Studies in Philosophy, Culture and Contemporary Society

Edited by Bogusław Paź

- Vol. 21 Jan Felicjan Terelak: Psychology of the Operator of Technical Devices. 2019.
- Vol. 22 Dorota Maria Leszczyna: Del idealismo al realismo crítico. La política como realización en José Ortega y Gasset. 2019.

Vol. 23 Zbigniew Drozdowicz: La république des savants. Sans révérence. Traduit du polonais par Catherine Popczyk. 2019.

Vol. 24 Andrzej Waśkiewicz: The Idea of Political Representation and Its Paradoxes. Translated from Polish by Agnieszka Waśkiewicz and Marilyn Burton. 2019.

Vol. 25 Ilona Błocian / Dmitry Prokudin (eds.): Imagination – Art, Science and Social World. 2019.

Vol. 26 Zbigniew Drozdowicz: Faces of the Enlightenment. Philosophical sketches. 2019.

Vol. 27 From Medicine to Sociology. Health and Illness in Magdalena Sokołowska`s Research Conceptions 2020.

Vol 28 Roman Witold Ingarden: Die Mitschriften von den Vorlesungen Martin Heideggers über die Phänomenologische Interpretation von Kants Kritik der reinen Vernunft (Wintersemester 1927/28). Aus dem Manuskript abgeschrieben und das Vorwort verfasst haben: Radosław Kuliniak und Mariusz Pandura. 2020.

Vol 29 Krzysztof Wielecki / Klaudia Śledzińska (eds.): The Relational Theory of Society. Archerian Studies vol. 2. 2020.

Vol 30 Zenon Gajdzica / Robin McWilliam / Miloň Potměšil / Guo Ling: Inclusive Education of Learners with Disability – The Theory versus Reality. 2020.

Vol 31 Jan Felicjan Terelak: Antarctic Winter-Over Syndrome. Narrative Perspective. 2021.

Vol 32 Nuria Sánchez Madrid / Julia Muñoz Velasco / José Luis Villacañas Berlanga (eds.): El ethos del republicanismo cosmopolita. Perspectivas euroamericanas sobre Kant. 2021.

Vol 33 Zbigniew Drozdowicz: Academic Culture. Traditions and the Present Days. 2021.

Vol 34 Jan Felicjan Terelak: Antarctic Isolation as a Mars Habitat Analogue. A Psychological Perspective. 2021.

www.peterlang.com

www.ingramcontent.com/pod-product-compliance
Ingram Content Group UK Ltd.
Pitfield, Milton Keynes, MK11 3LW, UK
UKHW021830210426
5322IPUK00004B/112